POWER ENGINEERING
Advances and Challenges

Part A: Thermal, Hydro and Nuclear Power

Editors

Viorel Badescu

Candida Oancea Institute
Polytechnic University of Bucharest
Bucharest, Romania

George Cristian Lazaroiu

Department of Power Systems
University Politehnica of Bucharest
Bucharest, Romania

Linda Barelli

Department of Engineering
University of Perugia
Perugia, Italy

CRC Press
Taylor & Francis Group
Boca Raton London New York

CRC Press is an imprint of the
Taylor & Francis Group, an **informa** business

A SCIENCE PUBLISHERS BOOK

Cover credit: Ch 4-Fig. 17—Author (*Maurizio Luigi Cumo* and *Renato Gatto*)

MATLAB® and Simulink® are trademarks of The MathWorks, Inc. and are used with permission. The MathWorks does not warrant the accuracy of the text or exercises in this book. This book's use or discussion of MATLAB® and Simulink® software or related products does not constitute endorsement or sponsorship by The MathWorks of a particular pedagogical approach or particular use of the MATLAB® and Simulink® software.

CRC Press
Taylor & Francis Group
6000 Broken Sound Parkway NW, Suite 300
Boca Raton, FL 33487-2742

First issued in paperback 2020

© 2018 by Taylor & Francis Group, LLC
CRC Press is an imprint of Taylor & Francis Group, an Informa business

No claim to original U.S. Government works

ISBN-13: 978-1-138-70585-2 (hbk)
ISBN-13: 978-0-367-78112-5 (pbk)

Visit the Taylor & Francis Web site at
http://www.taylorandfrancis.com

and the CRC Press Web site at
http://www.crcpress.com

The most important discoveries will provide answers to questions that we do not yet know how to ask and will concern objects we have not yet imagined.

John N. Bahcall, Astrophysicist (1935–2005)

Foreword

It is indeed a pleasure for me to write a Foreword to this book. As a seasoned (some say "old"!) theoretician and practitioner in Energy Systems, I feel, as others do, the need for a continuous renewal of the tools we can avail ourselves of in this field to keep ourselves updated on the continuous flow of information about the advancements, both theoretical and operational, in the general process of energy conversion: indeed, this is a very active area, not only for the immense economic implications that new developments have on our everyday life, but also for the importance that practical implementations have on the well-being of our human species and on a more balanced—I am tempted to say "intimate and respectful"—relationship with our environment, be it mother Earth today or the future worlds we are going to explore in the not so distant future. In my over 40 years of academic tenure, I have had the unique opportunity to meet scholars who have literally shaped our knowledge in the field, and was fortunate enough to learn something from all of them. The same goes for the present Editors: I met professor Badescu in the 90s when he was studying solar energy conversion, a topic in which most of his research focused over the years to come. Some ten years later, I met professor Lazaroiu, at the time he was completing his Ph.D. at the Bucuresti Polytechnic in final energy uses and optimization techniques. And professor Barelli was a graduate student as well when I first met her at the 2002 ASME-IMECE conference. Since then, of course, I have studied most of their publications, and could verify first-hand how their interests were on the one side expanding, and on the other side converging towards a possible synthesis of their diverse interests and backgrounds. The last decades saw a growing awareness of climate change phenomena that resulted in modifications in the social attitudes, among which were an increasing interest in the preservation of the natural cycles of the environment and a string concern about the depletion of fossil energy resources. Consequently, the power production sector was forced to revise its attitude towards nuclear energy, about the degree of penetration of electricity generated by renewable sources and about the importance of distributed power generation. Academic scholars and practicing engineers are still in search for a solution to some very relevant and poignant problems related to the "value choices" in this topics. New research directions emerged, and the declared objective of this book is to present some of them in an analytical, logical and rigorous fashion.

And here we come to the specific reason for which I am glad to introduce the readers to this book. "Power Engineering: Advances and Challenges" is a multi-Author compendium of the most relevant and recent techniques in all of the topics of today's Energy Conversion Systems. It discusses in great depth and with broad coverage all of the current issues, from fossil fuels use (including the most recent developments in environmental topics), to nuclear power, to the exploitation of geothermal sources, to renewable sources proper. But it also contains valuable information about the challenges that each single technology is confronted with: in this sense, it is not only a "textbook" for graduate students, but a sort of handbook for both theoreticians and practitioners in the field. Even the most advanced topics (stationary Fuel Cells and hybrid systems, biopower technologies, energy plantations value-added options, solid biomass-hydrogen conversion, tidal power) are treated in detail and with much dispatch. A series of chapters deals with electrical machines and sub-systems, including domotics, smart cities and demand response. A specific chapter is devoted to the analysis of the market operation with electricity generated by renewable sources, a topic of major and immediate relevance to the densely populated and energy-hungry Europe.

I am not able to offer detailed comments on every single chapter, because the broadness of the coverage exceeds by far my own qualifications, but I call the attention of the perspective readers to what is offered here: an omni-comprehensive, multi-disciplinary approach that is the essence of modern research. In today's very complex and interconnected world, each single scientific advancement can only be produced by research teams consisting of several domain experts with a variety of specializations, and of course the task of the coordinators is that of unifying the results, of maximizing the synergetic effects, and of directing readers to a better comprehension of the intertwining of the individual lines of research. This is a task that the present Editors have definitely achieved, with an amount of effort and dedication that is very clear to specialists in the field.

I would like to conclude this foreword on a personal note: in 1972, I graduated with a M.Eng. from the University of Roma with a thesis on a numerical analysis of a depressurization accident in the cold leg of a BWR reactor, and my operative advisor was professor Maurizio Cumo, one of the major experts in nuclear power who earned worldwide respect and fame in the nuclear field: he happens to be the author of chapters 3 and 4 in this book. Soon after graduation, I started my working and teaching peregrinations in Germany and in the US before returning to my alma mater, but throughout the years I have managed to maintain a close friendship with him, to whom goes my highest respect as a scientist and as a person: I am glad to close, so to say, the circle by dedicating this foreword to him, 45 years after his exhortation to leave for the US to enrol in a Ph.D. program there.

Enrico Sciubba, Ph.D.
Professor of Turbomachinery and Energy Systems
Dept. of Mechanical and Aerospace Engineering
University of Roma Sapienza

Preface

Power engineering is a subfield of energy engineering and electrical engineering. It deals with the generation, transmission, distribution and utilization of electric power and the electrical devices connected to such systems including generators, motors and transformers. This perception is associated with the generation of power in large hydraulic, thermal and nuclear plants and distributed consumption. In the last few decades mankind has faced the climate change phenomena and seen changes in social attitudes including interest in environment protection, and the depletion of classical energy resources. These have had bearings on the power production sector, and resulted in extensive changes and the need to adapt.

Future energy systems must take advantage of the changes and advances in technologies like improvements in natural gas combined cycles and clean coal technologies, carbon dioxide capture and storage, advancements in nuclear reactors and hydropower, renewable energy engineering, power to gas conversion and fuel cells, energy crops, new energy vectors biomass hydrogen, thermal energy storage, new storage systems diffusion, modern substations, high voltage engineering equipment and compatibility, HVDC transmission with FACTS, advanced optimization in a liberalized market environment, active grids and smart grids, power system resilience, power quality and cost of supply, plugin electric vehicles, smart metering, control and communication technologies, new key actors as prosumers, smart cities. These advances will enhance the security of power systems, safety in operation, protection of the environment, high energy efficiency, reliability and sustainability.

The book is a source of information for specialists involved in power engineering related activities and a good starting point for young researchers. The content is structured along logical lines of progressive thought. It presents the current developments and active technological advances in the main aspects of energy engineering, both in thermal and electrical engineering, with contributions from highly qualified experts in each field of study.

The principal audience consists of researchers, engineers, educators involved with the curriculum and research strategies in the field of power engineering. The book is useful for industry managers and developers interested in joining national or international power engineering development programs. Finally, the book can be used for undergraduate, postgraduate and doctoral teaching in faculties of engineering sciences.

Viorel Badescu
George Cristian Lazaroiu
Linda Barelli

Acknowledgments

A critical part of writing any book is the review process, and the authors and editors are very much obliged to the following researchers who patiently helped them read through subsequent chapters and who made valuable suggestions: Viktor Bolgov (Tallinn University of Technology), Yacine Chakhchoukh (University of Idaho, USA), Minh Quan Duong (The University of Da Nang, Danang City, Vietnam), Roberto Faranda (Politecnico di Milano, Italy), Mahmud Fotuhi-Firuzabad (Sharif University of Technology, Tehran), Nicolae Golovanov (University Politehnica of Bucharest, Romania), Francesco Grimaccia (Politecnico di Milano, Italy), Youguang Guo (University of Technology, Sydney, Australia), Bogdan Ionescu (University "Politehnica" of Bucharest, Romania), Marty Page (Southernco, USA), Ivo Palu (Tallinn University of Technology), Agis M. Papadopoulos (Aristotle University, Thessaloniki, Greece), Ion Petre (Abo Akademi University, Turku, Finland), Gustavo A. Ramos (University of the Andes, Columbia), Mariacristina Roscia (Universita di Bergamo, Italy), Vladislav Samoylenko (Ural Federal University, Russia), Giorgos Stavrakakis (Technical University of Crete, Greece), Gorazd Stumberger (University of Maribor, Slovenia), Lee Taylor (Southernco, USA), Fernando Lessa Tofoli (Universidade Federal de Juiz de Fora, Brazil), Geraldo Leite Torres (Universidade Federal de Pernambuco, Brazil), Michael von Spakovsky (Virginia Tech, USA), Dario Zaninelli (Politecnico di Milano, Italy), and Wenxiang Zhao (Jiangsu University, China).

Substantial help and guidance has been received from Vijay Primlani, from the Science Publishers team, to whom our thanks are kindly addressed.

The editors, furthermore, owe a debt of gratitude to all authors. Collaborating with these stimulating colleagues has been a privilege and a very satisfying experience.

Viorel Badescu
George Cristian Lazaroiu
Linda Barelli

Contents

III. Storage of Thermal Energy

List of Contributors

D. Apostolou
Soft Energy Applications and Environmental Protection Laboratory, Piraeus University of Applied Sciences, Athens, 12201, Greece
E-mail: j.apostolou@puas.gr

Kostantin G. Aravossis
National Technical University of Athens, 9, Iroon Polytechniou str, 15780 Zografou, Greece
E-mail: arvis@mail.ntua.gr

Lukasz Bartela
Silesian University of Technology, Akademicka 2A, 44-100 Gliwice, Poland
E-mail: Lukasz.Bartela@polsl.pl

Mikel Belsué Echevarria
Tecnalia R&I, Mikeletegi Pasealekua, 2, E-20009 Donostia-San Sebastián—Guipúzcoa, Spain
E-mail: mikel.belsue@tecnalia.com

David Bullejos Martín
University of Córdoba, Spain, Campus de Rabanales, 14071 Córdoba, Spain
E-mail: bullejos@uco.es

David Chiaramonti
University of Florence, RE-CORD and Department of Industrial Engineering, Viale Morgagni 40, 50134 Firenze, Italy.
E-mail: david.chiaramonti@unifi.it

Viviana Cigolotti
ENEA, Italian National Agency for New Technologies, Energy and Sustainable Economic Development, Piazzale Enrico Fermi 1, 80055 Portici (Napoli), Italy
E-mail: viviana.cigolotti@enea.it

Maurizio Luigi Cumo
Sapienza University of Rome, Via Eudossiana, 18 - 00184 Roma, Italy
E-mail: maurizio.cumo@uniroma1.it

Franco Donatini
University of Pisa, Largo Lucio Lazzarino, 56122 Pisa, Italy
E-mail: franco.donatini@unipi.it

Renato Gatto
Sapienza University of Rome, Via Eudossiana, 18 - 00184 Roma, Italy
E-mail: renato.gatto@uniroma1.it

Timothy J. Held
Echogen Power Systems, 365 Water Street, Akron, OH 44308, United States
of America
E-mail: theld@echogen.com

Vasilis C. Kapsalis
National Technical University of Athens, 9, Iroon Polytechniou str, 15780
Zografou, Greece
E-mail: bkapsal@mail.ntua.gr

Kosmas A. Kavadias
Soft Energy Applications and Environmental Protection Laboratory,
Piraeus University of Applied Sciences, Athens, 12201, Greece
E-mail: kkav@puas.gr

Gheorghe Lazaroiu
University Politehnica of Bucharest, Spl. Independentei 313, Bucharest
060042, Romania
E-mail: glazaroiu@yahoo.com

Jorge M. Llamas Aragonés
University of Córdoba, Spain, Campus de Rabanales, 14071   Córdoba,
Spain
E-mail: p52llarj@uco.es; jllaragones@yahoo.es

Lucian Mihaescu
University Politehnica of Bucharest, Spl. Independentei 313, Bucharest
060042, Romania
E-mail: lmihaescu@caz.mecen.pub.ro

Jarosław Milewski
Warsaw University of Technology, 21/25 Nowowiejska Street, 00-665
Warsaw, Poland
E-mail: milewski@itc.pw.edu.pl

Mariagiovanna Minutillo
University of Naples Parthenope, Centro Direzionale di Napoli, Isola C4,
80143 Napoli, Italy
E-mail: mariagiovanna.minutillo@uniparthenope.it

Angelo Moreno
ENEA, Italian National Agency for New Technologies, Energy and Sustainable Economic Development, Via Anguillarese 301, 00123 Roma, Italy
E-mail: angelo.moreno@enea.it

Gabriel Paul Negreanu
University Politehnica of Bucharest, Spl. Independentei 313, Bucharest 060042, Romania
E-mail: gabriel.negreanu@upb.ro

Susana Pérez Gil
Tecnalia R&I, Mikeletegi Pasealekua, 2, E-20009 Donostia-San Sebastián—Guipúzcoa, Spain
E-mail: susana.perez@tecnalia.com

Alessandra Perna
University of Cassino and Southern Lazio, Viale dell'Università, 03043 Cassino (FR), Italy
E-mail: perna@unicas.it

Ionel Pisa
University Politehnica of Bucharest, Spl. Independentei 313, Bucharest 060042, Romania
E-mail: ipisa@caz.mecen.pub.ro

Matteo Prussi
University of Florence, RE-CORD and Department of Industrial Engineering, Viale Morgagni 40, 50134 Firenze, Italy
E-mail: matteo.prussi@unifi.it

Andrea Maria Rizzo
University of Florence, RE-CORD and Department of Industrial Engineering, Viale Morgagni 40, 50134 Firenze, Italy
E-mail: andreamaria.Rizzo@unifi.it

Goran Strbac
Imperial College London, EEE Building, South Kensington Campus, SW7 2AZ, United Kingdom
E-mail: g.strbac@imperial.ac.uk

Fei Teng
Imperial College London, EEE Building, South Kensington Campus, SW7 2AZ, United Kingdom
E-mail: fei.teng09@imperial.ac.uk

I

Progress in Thermal, Hydro and Nuclear Classical Technologies

CHAPTER 1

Understanding of the Flexibility from Combined Cycle Gas Turbine Plant

*Fei Teng** and *Goran Strbac*

1. Introduction

The target to decarbonise the electricity system is expected to be largely achieved by integrating high penetration of renewable energy sources (RES). Due to the variability, uncertainty and limited inertia capability of RES, the need for ancillary services will be significantly increased. These services are traditionally delivered through part-loaded online plants with reduced efficiency, and/or through standing plants with higher operating costs. This not only leads to an increase in real-time balancing costs but also may eventfully limit the ability of the system to absorb RES, particularly when high output of RES coincides with low demand. As shown in Fig. 1, alternative flexible technologies, including energy storage (ES), Demand Side Response (DSR), network technologies and flexible generation, have been proposed and investigated to mitigate these challenges (Strbac et al. 2012a).

Previous study (Strbac et al. 2012b) suggests that ES may have an important role to play in facilitating the transition to the future low-carbon power system. ES can deliver cost savings across the whole electricity

Imperial College London, EEE Building, South Kensington Campus, SW7 2AZ, United Kingdom.
Email: g.strbac@imperial.ac.uk
* Corresponding author: fei.teng09@imperial.ac.uk

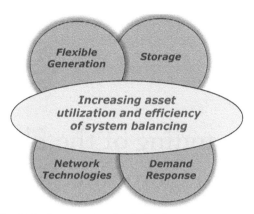

Fig. 1: Flexible technologies to support system balancing.

system due to the ability to offset the need for generation, transmission and distribution investment while at the same time contribute to operating cost savings through reducing RES curtailment and delivery of reserve and frequency regulation services.

DSR can be provided by a number of potentially flexible loads, such as flexible industrial and commercial (I&C) loads, flexible heat pump or HVAC systems, electric vehicles following smart charging strategies, smart domestic appliances, etc. DSR can support the integration of low-carbon generation by providing both energy arbitrage (load shifting or peak load reduction) and ancillary services (Strbac 2008).

A variety of network solutions have been considered to address the challenges of integrating RES. Conventional measures include the reinforcement of transmission and distribution grids in order to enable the connection of the increasing amount of wind and PV generation. There are also a number of advanced network technologies (e.g., FACTs) that facilitate a smarter and more efficient management of transmission and distribution networks.

Flexible generation is another key technology to support the integration of high penetration of RES. The operating flexibility of a thermal generator is determined by its operational limitations (e.g., minimum stable generation), the ability to provide ancillary services, ramping rates, minimum up/down time, and the reduced efficiency when running part-loaded. The manufacturers have identified the potential technology innovations to increase the flexibility of thermal plants (Probert 2011, Stevens et al. 2012).

Due to the electrification of transport/heating sectors and the upcoming retirement of coal plants under European Industrial Emissions Directive in Europe, significant amount of investment is required to build new thermal plants to cope with the potential capacity shortage. Although flexible versions of thermal generators have already been developed at a moderately higher cost than its less flexible alternatives, the limited needs of flexibility under the present market does not justify the adoption of this type of generation. However, this may change in the future with further expansion of intermittent RES. A clear understanding on the role and value of flexible thermal plants in the future low-carbon electricity system is required in order to inform the correct investment decisions.

Some previous studies have been carried to investigate the flexibility of thermal plants (Denholm and Hand 2011) demonstrate that high penetration of base-load thermal plants can cause significant RES curtailment. The authors in Eamonn et al. (2012) develop a flexibility index for the system consisted of thermal plants and the result indicates that the need for flexible plants increases along with the higher penetration of RES. Furthermore, Ma et al. (2013) propose a Unit Construction and Commitment model to simultaneously optimise the investment and operation of power plants. The case study demonstrates that the increased penetration of wind generation leads to a higher capacity of flexible plants to be built, although they are associated with higher investment and operation costs. Rautkivi and Kruisdilk (2013) analyse the value of Smart Power Generation in the future systems of UK and California and conclude that flexible plants can potentially reduce the balancing cost by up to 19%.

However, the previous studies are mainly based on the traditional deterministic scheduling methods, which rely on pre-selected requirements on the ancillary services to main the system security. Recent development of stochastic optimisation in the electricity sector (Meibom et al. 2011, Papavasiliou et al. 2011) may fundamentally change the way to schedule and operate the system, which in turn changes the value and the need for flexible power plants. Moreover, the increasing requirements of frequency regulation due to the declining system inertia have not yet been considered when assessing the value of flexible plants.

In this context, this chapter focuses on assessing the technical potentials and commercial benefits of the increasing flexibility of the combined cycle gas turbine (CCGT). An advanced multi-stage stochastic scheduling framework (Teng et al. In Press) is applied to analyse the benefits of enhancing ramp rate, minimum stable generation (MSG), frequency response capability, commitment time, idle state capability and part-load efficiency. A wide range of sensitivity studies are carried out across two representative systems.

The rest of this chapter is organised as following: Section 2 introduces the flexibility features that are assessed in this chapter. Section 3 describes the modelling framework and system assumptions. The main results are presented and discussed in Section 4, while Section 5 concludes this chapter.

2. Key Flexibility Features of Natural Gas Combined Cycle Plants

This section presents a detailed discussion on the key flexibility features of a CCGT plant and the technical potentials to enhance these features. This section focuses on ramp rate, MSG, frequency response capability, commitment time, idle state capability and part-load efficiency.

2.1 Ramp Rate

Ramp rate describes the speed at which the thermal plant can change its output between the minimum and maximum load levels. It is normally expressed as percentage of the maximum capacity per minute. The rate ramp for a typical CCGT plant is around 4%/min (Pierre et al. 2011). The higher variability of net demand driven by the increasing RES requires more ramping capability. Moreover, higher ramp rate also means more contribution to the spinning reserve provision. The main limitation to achieve higher ramp rate for a CCGT plant is the steam cycle, as a gas turbine can ramp very fast. Some modifications are suggested in NETL (2012) to enhance the ramping capability of CCGTs. It is also worth noting that the ramp rate constraint on a CCGT plant is hardly binding if the system operation is simulated in hourly resolution. Therefore, more granular time resolution is required in order to fully understand the value of higher ramp rate.

2.2 Minimum Stable Generation (MSG)

MSG is defined as the lowest level of output that a thermal plant can continuously operate at. Due to MSG constraint, provision of the synchronized ancillary services is inevitably accompanied by the delivery of electricity production. This may lead to RES curtailment during low demand and high RES production periods. Lower MSG may not only increase the amount of ancillary services provided by a CCGT plant, but also reduces the associated electricity production. Currently, MSG of a CCGT plant is limited at around 50% (Pierre et al. 2011). Some incremental improvements can reduce it to 40% or below by 2020 (Brouwer et al. 2015). The latest version of KA26 CCGT plant is expected to be capable to continuously operate at around 20% of the nominal power (Mark et al. 2011).

2.3 Frequency Response Capability

Frequency response capability of a thermal plant is defined by two parameters, maximum response and response slope (Doherty et al. 2005).

Due to the physical constraints, such as governor speed, there is a limitation on the maximum response that each plant can provide. At the same time, spinning headroom is required to deliver the frequency response. Response slope is used to define the ratio of frequency response availability over the required spinning headroom. Due to the limited inertia capability of RES, frequency response requirements are expected to significantly increase in the future low carbon systems (Teng and Strbac (under review)), leading to a higher demand on the frequency response contribution from thermal plants. For a CCGT plant, frequency control is normally provided by the gas turbine, while the steam turbine is operated with fully opened control valves. As discussed in Henkel et al. (2007), the steam turbine can also be enabled to participate in frequency control in order to enhance the frequency response capability of a CCGT plant. A technology called Turn Up is discussed in Michalke and Schmuck (2012) to enable the frequency response provision even at full-load operating condition. Moreover, a modified controller is introduced in Carmona et al. (2010) to improve primary frequency regulation from a CCGT plant. Figure 2 presents the typical and enhanced frequency response characteristics of a CCGT plant (Erinmez et al. 1999).

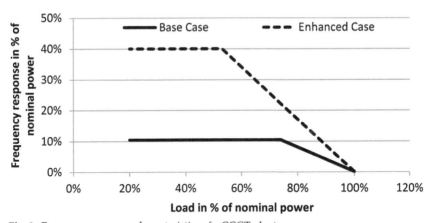

Fig. 2: Frequency response characteristics of a CCGT plant.

2.4 Commitment Time

Commitment time describes the time that a thermal plant takes from being turned on to reach MSG. The commitment time of a thermal plant highly depends on how much time has elapsed since its last shutdown. Different commitment times are used for different unit conditions, normally distinguished among hot start-up, warm start-up and cold start-up.

Similar as in Brouwer et al. (2015), hot start-up time is used in the system simulation. Due to the commitment time constraint, thermal plant needs to be turned on before the real time operation. In the future low carbon system, the variability of RES increases the number of start-ups of thermal plants and the uncertainty of RES increases the challenge of making optimal start-up decision long time ahead of real-time operation. The start-up decision of a thermal plant with shorter commitment time can be made nearer to real-time operation, which significantly reduces the uncertainty faced by the system operator. Potential improvements have been suggested in Probert (2011) and NETL (2012) to shorten the commitment time of a CCGT plant.

2.5 *Idle State Capability*

Idle state is a hold point in the plant start-up procedure. The HV breaker is closed and the auxiliary power for the plant is taken from the grid. The steam turbine is at standstill and the clutch is open. The shaft speed is controlled by the gas turbine. A plant under idle state has not yet been synchronized with the grid but can complete the synchronization with very short notice. This capability allows a thermal plant to provide operating reserve services without any energy delivered. This feature is very beneficial during low demand and high RES conditions to avoid the RES curtailment.

2.6 *Part-load Efficiency*

Part-load efficiency is defined as the ratio of the efficiency at part-load operation over the efficiency at full-load operation. As the increased requirements on ramping, operating reserve and frequency response in the future low carbon system, part-load operation of a thermal plant will become more common in order to provide these services. Gas turbine based technologies show relatively poor part-load performance due to a lower turbine inlet temperature (Brouwer et al. 2015). Significant amount of work have been conducted to improve the part load efficiency of a CCGT plant. The authors in Variny and Mierka (2009) conclude that, with limited modifications and capital expenses, more than 2% of gas consumption can be reduced during part-load operation. An innovative solution is introduced in Barelli and Ottaviano (2015) to increase the operational flexibility as well as the part load efficiency. A typical and an improved part-load efficiency curve of a CCGT plant are shown in Fig. 3.

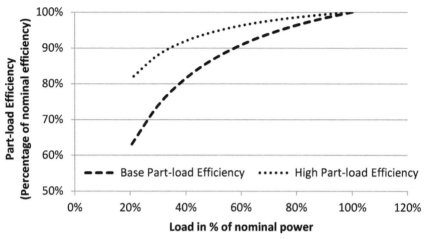

Fig. 3: A typical part-load efficiency curve of a CCGT plant.

3. Modelling Framework and System Assumptions

3.1 *Modelling Framework*

The value of a CCGT plant with enhanced flexibility is assessed by applying the least-cost annual generation system scheduling approach, capable of considering both the delivery of electricity and the provision of ancillary services. Generation scheduling determines the commitment and dispatch decisions of generators to minimize the system operation costs, with respect to dynamic operating constraints, e.g., commitment time for a thermal unit. The stochastic scheduling simulation tool is designed to provide optimized generation operation with explicit consideration of the uncertainty associated with RES, demand and generator outage. RES realizations, RES forecasts and generator outages are synthesized from the relevant models and fed into the scheduling tool. A scenario tree is built by quantile-based method to represent the possible outcomes of the stochastic variables (e.g., available RES output). This model minimizes the expected operation cost over the scenario tree by simultaneously scheduling electricity production, standing/spinning reserve service and inertia-dependent frequency response service. The simulations of system operation are carried out over a year time horizon in order to capture the variations in demand and RES generation.

The advantage of this model is its capability to optimally schedule the provision of different types of ancillary services to achieve the minimum operation cost. Compared with deterministic approaches that rely on the pre-defined requirements, the provision of ancillary services is endogenously optimised within the model. As the increased need on these services is the main driver to enhance the flexibility, their optimal scheduling is critical to fully understand the benefits of flexible generation. Furthermore, this model explicitly considers the frequency response requirements in the system, taking into account the levels of system inertia. Given that RES is expected to gradually replace conventional generation, the system inertia provided by rotating synchronous machines will decline. This not only increases the required amount of frequency regulation to maintain the frequency within the statutory limits, but also leads to more volatility in the needs of frequency response across different hours. It is therefore important to consider this effect when quantifying the benefits of flexible generation. Figure 4 provides a schematic illustration of the components of the stochastic scheduling tool.

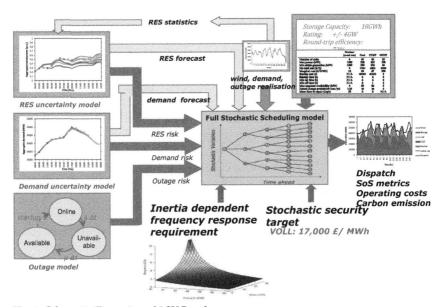

Fig. 4: Schematic illustration of ASUC tool.

3.2 System Assumptions

The value of CCGTs with enhanced flexibility is quantified in two representative systems. The generation mixes are summarised in Table 1. Peak demand in both systems is assumed to be 50 GW with annual energy consumption of 293 TWh. 80% of hydro plants are equipped with 10 h

Table 1: Generation Mix of Flexible System and Inflexible System.

	Nuclear	CCS	GAS	COAL	OCGT	Hydro
Flexible System (GW)	0	7.2	16.8	12	7.2	16.8
Inflexible System (GW)	33.6	5.7	6.3	2.4	2.7	9

Table 2: Main Economic Assumptions.

	CO_2	COAL	GAS	Nuclear
Price	74.2 €/T	3.23 €/GJ	8.85 €/GJ	0.256 €/GJ

reservoir, while another 20% are run-of-river. Fuel price and carbon cost are chosen to match the predictions in years 2020–2030 of the International Energy Agency for the 450 scenario (IEA 2013). Unless otherwise specified, the forecast error of wind generation is assumed to be 10% of installed capacity in 4-hour ahead.

The stochastic scheduling tool is first applied to the base-case systems without any improved flexibility features. The results in Fig. 5 show that the operation cost in the flexible system is very high while a relatively small amount of wind generation is curtailed. On the other hand, the operation cost in the nuclear-dominated inflexible system is much lower; however a significant amount of wind generation is curtailed. Furthermore, the emission rate in the flexible system reduces from 310 g/kWh to 90 g/kWh when the wind penetration level increases from 0 to 60%; while that in the inflexible system keeps at around 45 g/kWh regardless of wind penetration levels. The two base case systems with distinguished performances are chosen in order to fully unveil the key drivers for the value of CCGT plants with enhanced flexibility.

4. Valuing the Enhanced Flexibility of CCGT Plant

Main characteristics of flexibility under investigation are defined in Table 3. Unless otherwise specified, 5 GW of CCGT plants are assumed to be equipped with enhanced flexibility. The value of flexibility is calculated as the operation cost saving driven by the enhanced flexibility. More specifically, after solving the scheduling problem for a certain defined baseline, one or more technical parameters of CCGT plants are improved and the system scheduling is performed again. The relative operating cost saving is referred as the value of improving such a technical parameter, or, more simply the value of flexibility in that specific context. The reader should be aware that such a value does not necessary represent a source of extra revenue for the owners of CCGT plants.

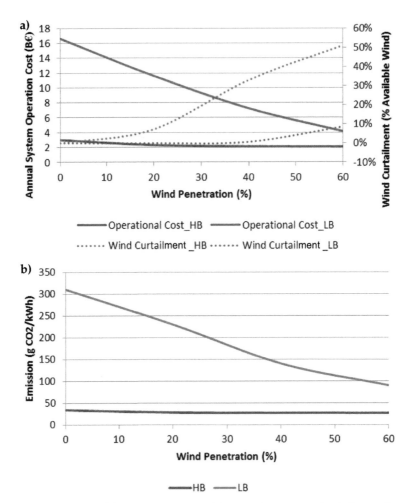

Fig. 5: Performance of base case systems: (a) operation cost and wind curtailment; (b) emission rate.

Table 3: Flexibility features to be investigated.

	Base case	Enhanced Flexibility
Ramp Rate (RR)	32%/10 mins	50%/10 mins
Minimum Stable Generation (MSG)	50%	20%
Max Response Capability (Response)	17%	40%
Commitment Time (CT)	4 hours	2 hours
Idle State (Idle)	No	With
Part-load Efficiency (PE)	Low	High

4.1 Value of CCGTs with Different Enhanced Flexibility

The value of the enhancement on each flexibility feature in the flexible system is shown in Fig. 6. It is clear that the increased penetration of wind generation drives up the value of enhanced flexibility. This is due to the fact that high penetration of wind generation increases the need for ancillary service provision, and hence the CCGT plants with enhanced flexibility become more desirable. However, the value of high response capability is constantly low in this system. This is due to the fact that there are a significant amount of flexible hydro plants to efficiently fulfil the frequency response requirement, there is no need for CCGT plants to enhance their frequency

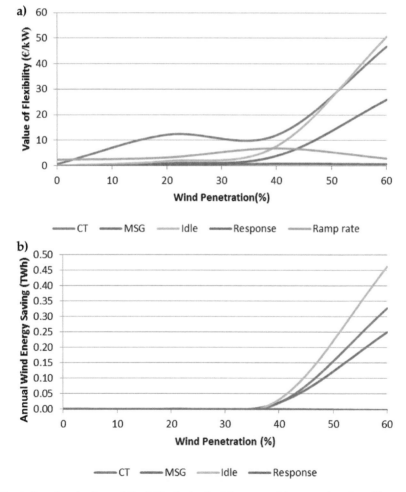

Fig. 6: Benefits of enhanced flexibility in the flexible system: (a) economic value; (b) annual wind energy saving.

response capability. The simulation results in Fig. 6(b) also suggest that the presence of enhanced flexibility from CCGT plants significantly reduces the wind curtailment after the wind penetration reaches 40%.

The benefit of enhanced flexibility in the inflexible system is presented in Fig. 7. Lower MSG and higher frequency response capability show

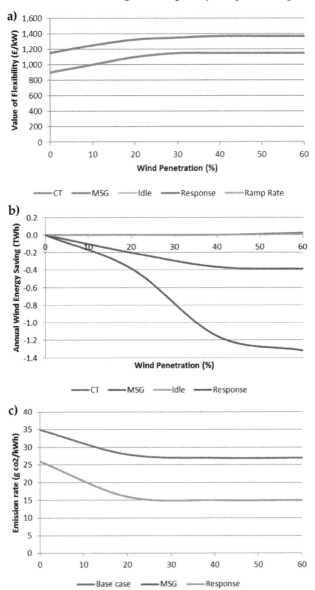

Fig. 7: Benefits of enhanced flexibility in the inflexible system: (a) economic value; (b) annual wind energy saving; (c) system emission rate.

constantly high value regardless of wind penetration levels, while the value of other flexibility features is low. The simulation results suggest that due to the lack of frequency response capability for base-load plants (i.e., nuclear), some CCGT plants are scheduled to run at MSG only to fulfil the frequency response requirements. This not only reduces the operational efficiency of CCGT plants, but also causes the curtailment of low carbon generation. Therefore, higher frequency response capability and lower MSG are extremely valuable in supporting the provision of frequency response. However, Fig. 7(b) suggests that these two enhanced flexibility features actually lead to increased RES curtailment. This is driven by the fact that the provision of frequency response is shifted from OCGT plants to flexible CCGT plants, which leads to more energy production to provide the same amount of frequency response. Although the wind curtailment increases, the overall system emission rate (Fig. 7(c)) significantly reduces with the enhanced flexibility. In this inflexible system, the value of reserve-related features (through commitment time, idle state and ramp rate) is low as de-loaded nuclear plants and curtailed wind generation can provide sufficient operating reserve without incurring high cost.

4.2 Market Volume for Flexible Plants

In order to understand the total market volume, the value of flexibility is calculated when different amounts of CCGT plants are equipped with enhanced flexibility features. As shown in Fig. 8, the marginal value of flexible CCGT plants in the flexible system reduces significantly after about 6% of the total thermal plants have been with the enhanced flexibility features.

The results in the inflexible system are shown in Fig. 9. It is interesting to see that although the high response capability is extremely valuable, the marginal value declines rapidly with increased penetration level and reaches zero after 5% of the total plant capacity. As the volume of frequency response market is relatively limited, once there are enough flexible plants to cost-effectively provide frequency response, the marginal value of additional CCGT plants with enhanced flexibility becomes zero. However, as the penetration of RES increases, the declining system inertia may increase the amount of required frequency response and hence enlarge the market volume for response-related flexibility features.

4.3 Levels of Flexibility of the Plants

Another important question is how flexible the plants need to be. It is clear that as the flexibility of the plant increases, its benefits as well as costs would both increase. To fully capture the value of flexible plants,

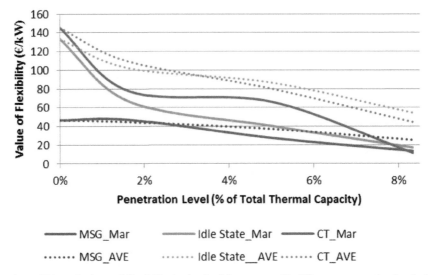

Fig. 8: Value of enhanced flexibility in the flexible system with different penetration level of flexible CCGT plants.

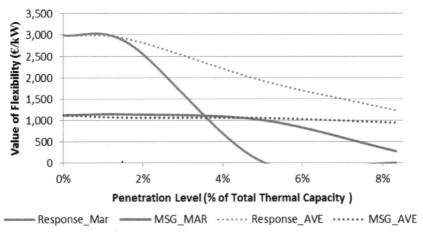

Fig. 9: Value of enhanced flexibility in the inflexible system with different penetration level of flexible plants.

it is critical to balance the benefits from the enhanced flexibility and the associated costs to improve it. Therefore, the improvements of some specific flexibility features are varied and the value is quantified. This study focuses on the selected flexibility features that have been shown to be extremely valuable in the inflexible system. For both the lower MSG and higher frequency response capability, the results in Fig. 10 and Fig. 11 suggest that the value increase almost linearly in the range of

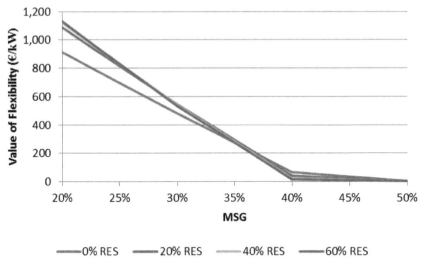

Fig. 10: Value of CCGTs with different levels of MSG in the inflexible system.

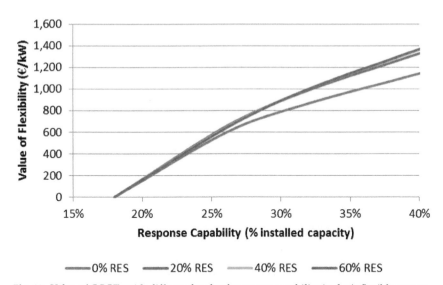

Fig. 11: Value of CCGTs with different levels of response capability in the inflexible system.

interest. However, increasing MSG from 50% to 40% would not make the CCGT plant competitive with other technologies in providing frequency response, and hence shows zero value for this improvement. Given the annualised investment cost associated with different levels of enhanced flexibility, this results can be used as a reference to determine the optimal level of flexibility for a CCGT plant.

4.4 Solar versus Wind Integration

Wind and solar Photovoltaic (PV) are very different in terms of time distribution. PV produces mainly 8–12 hours a day, depending on seasonal and specific weather conditions. Wind power typically produces with no interruptions over a much longer period of time, but low wind periods can last for several days.

From this point of view, PV production is easier to predict, particularly for hourly variations. However, for large shares of penetration, solar is generally more difficult to integrate, compared to wind. This is illustrated with an example in Fig. 12, where the penetration is scaled-up to 50% of the overall energy produced, in the case of wind only (top), and of a mix of 40% PV and 60% wind (bottom). Load demand and wind and PV production data (before being scaled-up) are taken from the German TSO area of Amprion. As clearly shown, PV exceeds the load demand almost every day, while wind production is most of the time below the load demand. As clearly shown, even if the base load (represented as a

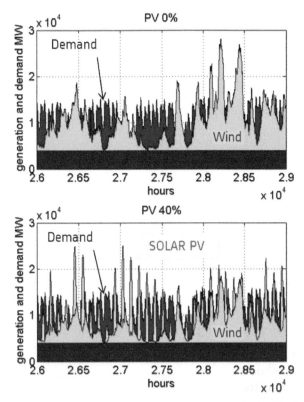

Fig. 12: Example of 50% RES penetration, of which 40% PV and 60% Wind (bottom) and wind only (top).

blue bar) is reduced to zero, solar production would still exceed the load demand, thus no flexibility or variation of fleet composition would be able to eliminate solar curtailment (energy storage, exports, and demand-side-management are not in the scope of this analysis).

The different nature of solar and wind has a major impact on thermal power plants operation if PV or wind is the dominating RES. In particular, thermal plants cycles are expected to be more severe in the case of PV. This is shown in Fig. 13, where annual start-ups for different RES penetrations are displayed, in the case of the flexible system. The increased number of start/stops in the case of PV is due to the fact that solar energy appears and disappears daily, whereas wind has cycles of intermittency more widely distributed. One should also note that start/stops do not increase monotony with wind penetration. For some CCGT, the number of starts/stops decreases when wind penetration is higher than 20–30%. This can be explained by an increased parking time of such power plants. This is clearly not the case for PV, as power plants have to provide power to the grid when after sunset, no matter the capacity of PV installed. However, the enhanced flexibility in the system with PV as dominating RES shows the similar value as that in the system with wind as dominating RES.

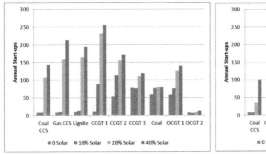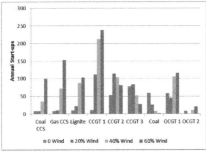

Fig. 13: Example of annual start-ups in the case of solar only (left) and wind only (right), for different level of RES penetration.

4.5 *Impact of Scheduling Methods on the Value of Enhanced Flexibility*

The value of flexibility is primarily driven by the need for ancillary services induced by the integration of RES. Different scheduling methods lead to different allocations of ancillary services among different sources. Recent works Sturt and Strbac (2012), Tuohy et al. (2009) show that stochastic scheduling method results in lower operation cost and lower renewable curtailment than traditional deterministic method, especially with high penetration of RES. Although deterministic approach is still the dominating method in the present power systems, stochastic scheduling is likely to

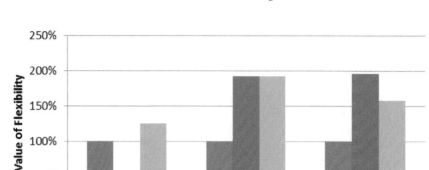

Fig. 14: Relative value of flexibility in flexible system by using different scheduling methods.

be implemented more widely as increasing penetration of intermittent RES. Different scheduling methods show significant impact on the value of storage in Strbac et al. (2012b). Therefore, this section investigates the impact of different scheduling methods and time resolutions on the value of CCGT plants with enhanced flexibility.

The deterministic and stochastic scheduling methods are used to quantify the values of enhanced flexibility in the systems with 60% of wind penetration. The deterministic method here refers to the case that reserve and frequency response requirements are calculated dynamically but only based on a single scenario as current operation practice.

As shown in Fig. 14, in the deterministic scheduling case, the value of MSG and Idle-state shows almost twice of that in the stochastic scheduling case while the value of commitment time reduces to almost zero. This is due to the fact that the deterministic method tends to rely more on spinning reserve, which would increase the values of spinning-reserve related flexibility features (e.g., MSG and idle-state) and decrease the value of standing-reserve related flexibility features (e.g., commitment time). Modelling of 10-min operation in stochastic framework increases the need of reserve and ramps to compensate intra-hour variability and uncertainty of wind generation, and therefore leads to increased value of flexibility. To summarize, the need and the value of CCGT plants with enhanced flexibility show significant differences in the same system by using different scheduling methods.

On the contrary, the value of flexibility in the inflexible system is not highly influenced either by the scheduling method or time resolution. The

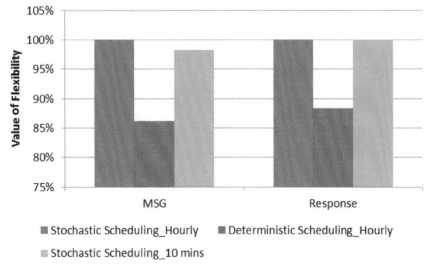

Fig. 15: Relative value of flexibility in inflexible system by using different scheduling methods.

reason is that the value of flexibility in this system is primarily driven by the need of frequency response, which is not highly related to the scheduling method or time resolution. However, lower MSG and high frequency response capability shows slightly lower value in the deterministic scheduling (as shown in Fig. 15) as the deterministic scheduling method tends to keep more generators online and hence reduces the challenge to provide sufficient amount of frequency response.

4.6 Impact of Inertia Reduction on the Value of Enhanced Flexibility

Another issue associated with integration of RES is the declining system inertia, which leads to increased requirements on the provision of frequency response. The impact of this effect on the value of enhanced flexibility is investigated in this section. As shown in Fig. 16, the value of CCGT plants with enhanced flexibility significantly increases in the flexible system, when inertia reduction effect is taken into consideration. It is surprising to see that the reserve-related flexibility features also increases, although the inertia reduction is expected to only increase the demand for frequency response. This is related to the fact that the requirements of frequency response depend on the system inertia, which is in turn driven by the amount of synchronised conventional plants. Different realisations of RES production can significantly change the schedule of conventional plants, resulting in different levels of system inertia. Shorter commitment time

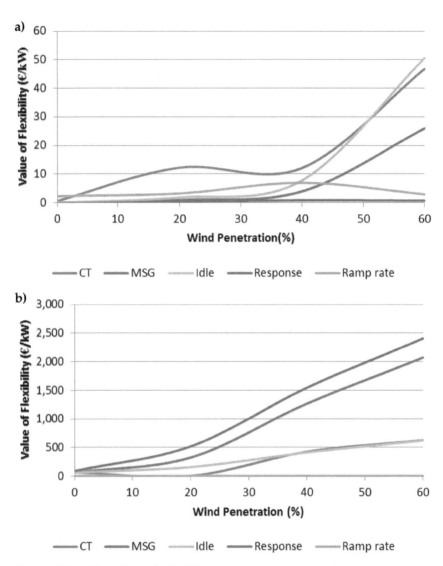

Fig. 16: Value of flexibility in the flexible system: Constant Response Requirement (top) VS Inertia Dependent Response Requirement (bottom).

and idle state capability can be used to reduce the cost associated with this type of uncertainty, increasing value of CCGT plants with those enhanced flexibility features.

The results in Fig. 17 suggest that due to the increased frequency response requirements driven by the inertia reduction, the value of frequency response related flexibility features increase in the inflexible

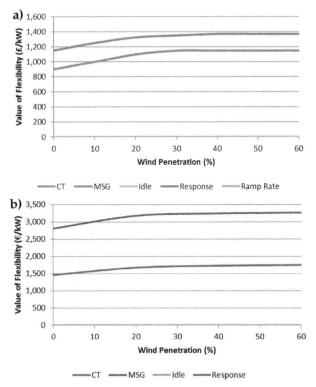

Fig. 17: Value of flexibility in the inflexible system: Constant Response Requirement (top) VS Inertia Dependent Response Requirement (bottom).

system. In particular, the value of high response capability increases by almost 3 times, while the value of MSG increases by around 1.5 times.

4.7 Impact of Renewable Energy Support Scheme

Recently, various RES support schemes have been proposed and implemented all over the world. Carbon tax is one of the most widely implemented schemes. In the current power system, the nuclear and coal-fired plants serve as base load due to the low operation cost, while CCGT plants are used to supply peaking demand. The introduction of carbon tax may change this situation due to the high emission rate of coal-fired plants. Therefore, two different carbon taxes, 73 €/tonne and 20 €/tonne, are introduced in this section to investigate their impact on the value of flexibility.

The results in Fig. 18 and Fig. 19 demonstrate that the increased carbon price leads to higher value of flexibility from CCGT plants, while reduces that associated with coal-fired plants. This is due to the fact that

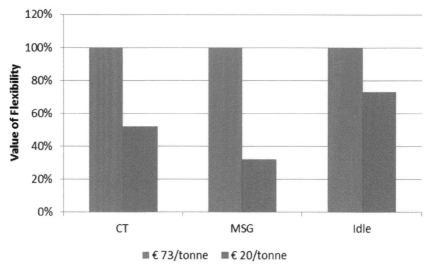

Fig. 18: Impact of carbon price on the value of flexible CCGTs in the flexible system.

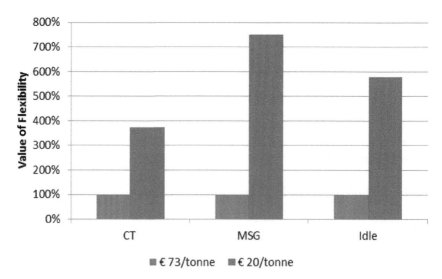

Fig. 19: Impact of carbon price on the value of flexible coal plants in a flexible system.

high carbon price increases the operation cost of coal-fired plants above that of CCGT plants. Hence, this significantly reduces the operation hours of coal-fired plants, even if they are equipped with enhanced flexibility. We can therefore conclude that the increased carbon price shifts the value of flexibility from coal-fired plants to CCGT plants.

As discussed above, the main benefit of flexible CCGT plants in the inflexible system is to replace OCGT plants in providing frequency

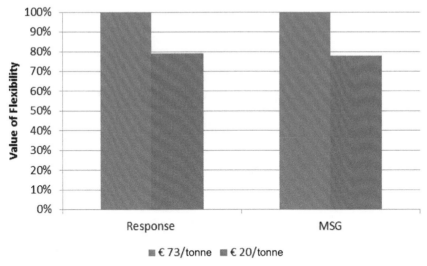

Fig. 20: Impact of carbon tax on the value of flexibility of CCGTs in inflexible system.

response. At the same time, the emission rate of OCGT plant is much higher than that of CCGT plant. The higher carbon price increases the cost difference between OCGT plant and CCGT plant, which is expected to increase the value of flexible CCGT plants. This is demonstrated by the results in Fig. 20.

4.8 Market Regard on Flexibility

The above analysis demonstrates the potential benefits of CCGT plants with enhanced flexibility for the system operation. However, it is not clear that whether these benefits can be fully or partially captured by the owners of the plants that provide the additional flexibility. A typical example is reported in Fig. 21 for the case of 1 GW CCGT plants with reduced MSG from 50% to 20%. After the MSG is reduced, these flexible CCGT plants will be operated more often at reduced load to provide flexibility, thus it will produce less energy (in the figure the reduction at 40% wind penetration is about ~ 500 GWh, equivalent to ~ 5% abs reduction of the capacity factor). Under the present market arrangements, where the revenues are pre-dominantly driven by the energy sold, this clearly represents a disadvantage for the flexible CCGTs. Currently there are different fora, working groups and initiatives with the main aim of suggesting possible ways of modifying the market regulations to recognize the benefits of flexibility for the system and properly reward the providers of flexibility. It should be noticed that a capacity market based on capacity only would not provide any reward to flexibility.

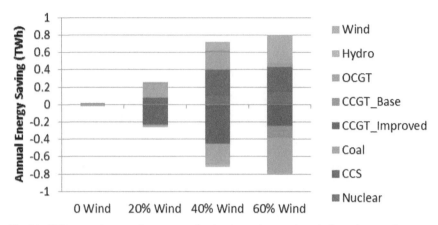

Fig. 21: Differences in annual energy production (negative numbers indicate less production, compared to the same scenario, but with CCGT without improved MSG).

5. Concluding Remark

This chapter investigates the technical potentials and commercial value of enhanced flexibility of CCGT plants in the future low carbon system. The key flexibility features that are investigated in the chapter are: ramp rate, MSG, frequency response capability, commitment time, idle state capability and part-load efficiency. It has been shown that value of flexibility increases with penetration of RES; however, different systems require different types of flexibility. In the coal and gas dominated system, the value of reserve-related flexibility features (shorter commitment time, idle state capability and so on) is higher, while in the nuclear dominated system, frequency response related flexibility features (higher response capability and lower MSG) are more desirable. The analysis also suggests that different system scheduling methods may significantly change the value of different flexibility features. In the flexible system, the deterministic schedule increases the value of lower MSG and Idle state capability, while decreases the value of shorter commitment time. The reduced system inertia driven by the integration of RES significantly increase the value of enhanced flexibility in both systems. Higher carbon price leads to a swifts the value of flexibility from coal-fired plants to CCGT plants.

Europe is going through profound electricity market disturbances, and the recent impressive thermal asset write-offs incurred by most utility companies are a clear signal that "energy only" markets do not function properly with a high level of renewables penetration. Electricity market price formation should include both energy and dependable capacity components. Resolving this question is the most urgent priority. Utilities should be compensated in real time for providing the underlying thermal base power required to ensure a more flexible and reliable source of energy. The so-called capacity market is one of several potential solutions to this issue, but its design should also take into account the remuneration of flexibility.

Acknowledgements

Part of this work is based on results of a cooperation between Alstom Power (now GE Power) and Imperial College. The authors would like to thank Roberto Bove for his scientific contribution and Alstom Power for their financial contribution.

References

Barelli, L. and A. Ottaviano. 2015. Supercharged gas turbine combined cycle: An improvement in plant flexibility and efficiency. Energy 81: 615–626.

Brouwer, A.S., M.v.d. Broek, A. Seebregts and A. Faaij. 2015. Operational flexibility and economics of power plants in future low-carbon power systems. Applied Energy 156: 107–128.

Carmona, S., S. Rios, H. Peña, R. Raineri and G. Nakic. 2012. Combined cycle unit controllers modification for improved primary frequency regulation. IEEE Trans. on Power Syst. 25(3): 1648–1654.

Denholm, P. and M. Hand. 2011. Grid flexibility and storage required to achieve very high penetration of variable renewable electricity. Energy Policy 39(3): 1817–1830.

Doherty, R., G. Lalor and M. O'Malley. 2005. Frequency control in competitive electricity market dispatch. IEEE Trans. Power Syst. 20(3): 1588–1596.

Eamonn, L., D. Flynn and M. O'Malley. 2012. Evaluation of power system flexibility. IEEE Trans. Power Systems 27(2): 922–931.

Erinmez, I.A., D.O. Bickers, G.F. Wood and W.W. Hung. 1999. NGC experience with frequency control in England and Wales—provision of frequency response by generators. In IEEE PES Winter Meeting, New York.

Henkel, N., E. Schmid and E. Gobrecht. 2007. Operating flexibility enhancement of combined cycle power plants. In POWER-GEN Europe.

IEA. World Energy Outlook 2013. 2013. IEA.

Ma, J., V. Silva, R. Belhomme, D.S. Kirschen and L.F. Ochoa. 2012. Evaluating and planning flexibility in sustainable power systems. IEEE Trans. Sustain. Energy 4(1): 200–209.

Meibom, P., R. Barth, B. Hasche, H. Brand, C. Weber and M. O'Malley. 2011. Stochastic optimization model to study the operational impacts of high wind penetrations in Ireland. IEEE Transactions on Power Systems 26(3): 1367–1379.

Michalke, P. and T. Schmuck. 2012. Powerful products for the enhanced flexibility of gas turbines. In POWER-GEN Europe, Cologne.

NETL. 2012. Impact of load following on power plant cost and performance: literature review and industry. National Energy Technology Laboratory.

Papavasiliou, A., S.S. Oren and R.P. O'Neill. 2011. Reserve requirements for wind power integration: A scenario-based stochastic programming framework. IEEE Trans. Power Syst. 26(4): 2197–2011.

Pierre, I., F. Bauer, R. Blasko, N. Dahlback, M. Dumpelmann, K. Kainurinne, S. Luedge, P. Opdenacker, I.P. Chamorro, D. Romano, F. Schoonacker and G. Weisrock. 2011. Flexible generation: Backing up renewables. Eurelectric.

Probert, T. 2011. Fast starts and flexibility—let the gas turbine battle commence. Power Engineering International.

Rautkivi, M. and M. Kruisdijk. 2013. Future market design for reliable electricity systems in Europe. In PowerGen Europe.

Mark Stevens, Frank Hummel, Ralf Jakoby, Volker Eppler and Chr. Ruchti. 2011. Increased operational flexibility from the latest GT26 (2011) upgrade. In PowerGen Europe, Cologne, Germany, 2012.

Stevens, M., F. Hummel, R. Jakoby, V. Eppler, C. Ruchti and A. Power. 2012. Increased operational flexibility from the latest GT26 (2011) upgrade. In PowerGen Europe, Cologne, Germany.

Strbac, G. 2008. Demand side management: benefits and challenges. Energy Policy 36(2): 4419–26.

Strbac, G., M. Aunedi, D. Pudjianto, P. Djapic, S. Gammons and R. Druce. 2012a. Understanding the balancing challenge. Report for the UK Department of Energy and Climate Change.

Strbac, G., M. Aunedi, D. Pudjianto, P. Djapic, F. Teng, A. Sturt, D. Jackravut, R. Sansom, V. Yufit and N. Brandon. 2012b. Strategic Assessment of the Role and Value of Energy Storage Systems in the UK Low Carbon Energy Future. Carbon Trust, London.

Sturt, A. and G. Strbac. 2012. Efficient stochastic scheduling for simulation of wind-integrated power systems. IEEE Trans. Power Syst. 27(3): 323–334.

Teng, F. and G. Strbac. Assessment of the role and value of frequency response support from wind plants. IEEE Trans. on Sustain. Energy, Under review.

Teng, F., V. Trovato and G. Strbac. Stochastic scheduling with inertia-dependent fast frequency response requirements. IEEE Trans. Power Syst., vol. In Press.

Tuohy, A., P. Meibom, E. Denny and M. O'Malley. 2009. Unit commitment for system with significant wind penetration. IEEE Trans. Power Syst. 24(2): 592–601.

Variny, M. and O. Mierka. 2009. Improvement of part load efficiency of a combined cycle power plant provisioning ancillary services. Applied Energy 86: 888–894.

CHAPTER 2

Directions for Improving the Flexibility of Coal-Fired Units in an Era of Increasing Potential of Renewable Energy Sources

Łukasz Bartela[1,*] and *Jarosław Milewski*[2]

For many years, the use of renewable energy sources (Milewski et al. 2013, 2014, Kupecki et al. 2013) has been promoted and supported around the globe. In 2015 worldwide, the total installed power in renewable sources was 1,849 GW, with hydropower plants accounting for more than one half. In recent years, wind and solar energy have outperformed the rest in terms of growth. In 2015 installed wind power stood at 63 GW, which represented at year-end 14.5% of the total installed power in the wind farms. The growth rate was even more spectacular for solar energy in that year, with the headline figure hitting 50 GW, which constituted 22.0% of the total installed power in solar power plants (Eurostat). Wind and solar energy have one great advantage over hydropower energy, namely availability. They can be readily harnessed at points that are far more dispersed around the globe; hence access to these energy resources is significantly more

[1] Silesian University of Technology, Akademicka 2A, 44-100 Gliwice, Poland.
[2] Warsaw University of Technology, 21/25 Nowowiejska Street, 00-665 Warsaw, Poland.
 Email: milewski@itc.pw.edu.pl
* Corresponding author: Lukasz.Bartela@polsl.pl

common. Unfortunately, both solar power plants and wind turbines are characterized by strongly diversified potential over time, which adversely impacts the functioning of energy systems and jeopardizes the continuity of energy supply (Szablowski et al. 2011). Put simply, it may be cloudy and windless during an extended cold snap when energy demand is high. This may be compounded by the poor structuring of generation sources and lack of energy storage (Bertsch et al. 2016). The problem may concern first and foremost systems in which power generation sources with bad regulatory properties are predominant, for example, conventional coal-fired units (Papaefthymiou and Dragoon 2016, Eurelectric 2011). In countries where energy development is dominated by renewable sources, appropriate support mechanisms are needed to maintain a high degree of energy security. Financial support is required, for example, to stimulate the development of energy storage systems (Walker et al. 2016, Kotowicz et al. 2017, Bartela et al. 2016, Després et al. 2016).

In Poland, it was mainly the wind power sector that was growing dynamically until 2015, when a change of government resulted in an abrupt volte-face and an attempted redirection to offshore wind power. 2015 saw an increase in installed wind power of 1.3 GW, marking total installed power of 5.1 GW (a 34% increase year-on-year) (Eurostat). In the arena of world statistics, in 2015 Poland ranked seventh in terms of national growth in the field of the wind energy. Worryingly for the development of wind energy in Poland are the problems identified in the coal-fired generating sector, which experienced an increase in the number of forced shutdowns of power units, caused indirectly by priority network access being given to energy produced from renewable sources (Gawlik et al. 2015). The stop-start situation contributes to the increase in the cost of electricity production and carries with it the risk of outages. The need to ensure the country's energy security forces the system operator to limit the number of concessions issued for hooking up wind farms to the power grid. In Fig. 1, the characteristics of electricity demand in Poland in June 2014 are marked in red (Polskie Sieci Elektroenergetyczne). Here, large differences in demand can clearly be seen   between peak and offpeak demand periods (overnight and weekend). In order to adapt the potential of production to the demand characteristics, Centrally Dispatched Generating Units (CDGU) operate in the electricity system. The power of these units, which in Poland are mainly coal units, is marked in Fig. 1 by an unbroken black line. The power supplied to the system by wind farms was marked by an unbroken green line. Any increase in installed wind power will adversely affect the characteristics of the CDGU's operations, primarily contributing to the increase in working time at low load and

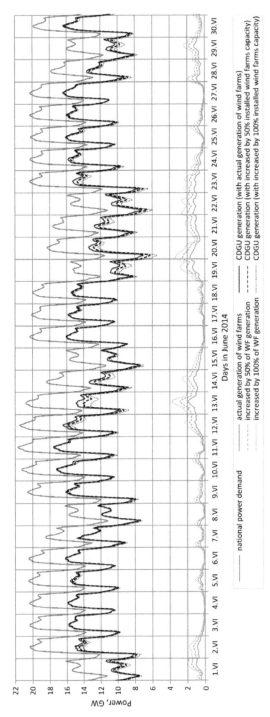

Fig. 1: Characteristics of the National Power System (NPS) for 2014 (Polskie Sieci Elektronergetyczne).

more frequent shutdowns of base load regulatory units. In Fig. 1 the dashed lines depict the effect of what would happen if wind power production increased by 50%, and the dotted lines represent the hypothetical situation in which the power of wind farms is 100% higher.

In Poland, electricity is mainly produced in coal-fired generating units. In 2015, coal accounted for 83.7% of electricity generated. In 2016, this share fell to 81.5%. While the Polish power sector is investing in modern units with supercritical parameters, coal-fired generation is mostly used from class 200 MW units, built in the 1970s and 1980s. The increase in installed power in renewable sources, with intermittent production potential, drove a change in current requirements for thermal power plants (Brouwer et al. 2015, Huber et al. 2014). The ability of a power unit to switch from base to regulatory work is particularly required. The weak regulatory properties of the 200 MW coal-fired power units now contribute to the need for frequent shutdowns, which are observed mainly during weekend valleys of electricity demand. In the weekend valleys, the situation will become more serious yet, as the potential of wind energy generation expands. In the planned program to upgrade the coal-fired power units, much emphasis is being put on increasing their operational flexibility. All of this is to be delivered through making design changes in key elements of the units or by adding intermediate dust trays and heat accumulators. Energy storage systems may also find a place among the solutions beneficial to the regulation abilities of the system, which compensate for the adverse effects of intermittent renewable energy sources.

An important aspect for improving the regulatory capacity of the system is the replacement of obsolete units with modern units. In Poland, this process is ongoing and in the field of coal power is being carried out, inter alia, through replacing obsolete 200 MW class units with modern, large units featuring supercritical parameters. Compared to the obsolete units, the efficient modern units—such as unit 14 at Bełchatów power plant and unit 10 at Łagisza power plant—can work at a higher load range (from 40%), have higher starting speeds (power increase to 5%/min) and deal readily with daily shutdowns and startups.

The poor regulatory properties of 200 MW coal-fired units contribute to the need for frequent shutdowns, which are observed mainly during the weekend valleys. An example of power characteristics for the annual working period of one of the units operating in the Polish power plant is shown in Fig. 2. In this case, in addition to the planned repairs, there were 21 forced withdrawals imposed by the need to reduce the power delivered to the power system. In power plants during the weekends there are situations where the low energy requirements in the system mean that only one unit is working. Although its work is not economically

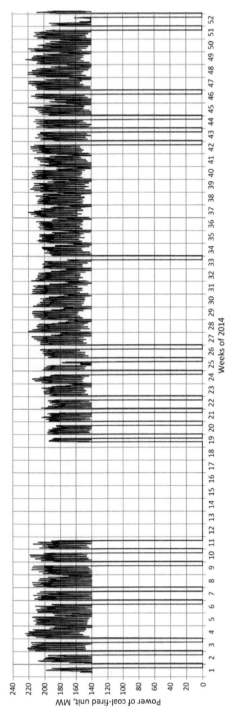

Fig. 2: The power of unit class 200 MW within annual work cycle.

justified, the unit is not switched off due to the need to power up the other units during their startups after the weekend valley. In order to enable all units of the power plant to power off, investments are now considered in steam generators, which will be used only to startup production purposes after weekend valleys. In order to emphasize the essence of the problem of running coal-fired power stations, Fig. 3 shows the time shares of shutdowns (planned and forced) and the levels of load that took place in 2013 and 2014, respectively, for 5 selected national units. Regardless of the considered year, forced withdrawal times should be considered too high. However, as can be observed, the forced withdrawal times for individual units are significantly longer in 2014 than in 2013. It should be noted that the installed capacity of wind farms in 2013 and 2014 were respectively 3389.5 MW and 3833.8 MW (an increase of 13.11%). Increasing the flexibility of power units will contribute to increased energy security and will also reduce the number of unit outages, which will be beneficial due to the economics of their operation. It must be taken into account that the estimated startup cost of a 200 MW unit from hot state (withdrawal time less than 8 hours) is about €20,000, from warm state (withdrawal time less than 60 h) €23,000 and from cold state (withdrawal time over 60 h) €25,000.

Another European country with a large share of coal in electricity production is Germany. Here, renewable energy sources are developing far faster than in Poland. The installed capacity of renewable reached the level of 50% (in 2014) of total sources in the power grid (see Fig. 4). At this time, wind and solar sources were responsible for almost 90 GW installed

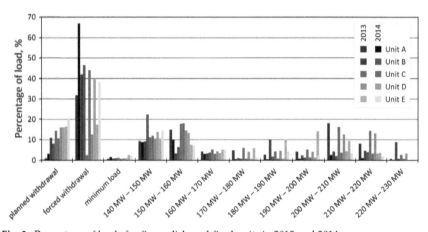

Fig. 3: Percentage of loads for five polish coal-fired units in 2013 and 2014.

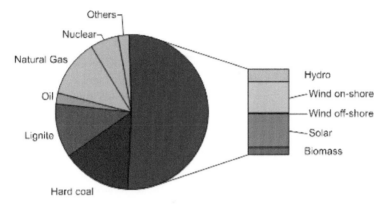

Fig. 4: Installed capacity of power sources in Germany, in 2014 (Quinkertz 2015).

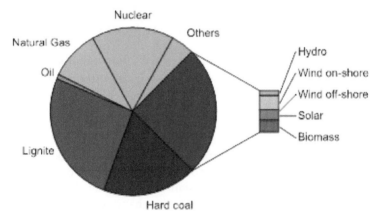

Fig. 5: Electricity generation in various energy sources in Germany in 2014 (Quinkertz 2015).

capacity with the system of total capacity being ca. 184 GW. Installed capacity does not directly translate into energy generation, as solar and wind power sources are radically affected by weather conditions. For example, in Germany in 2014 renewable power sources generated 26% of the total energy produced, but wind and solar sources only 8.6% and 5.8%, respectively (see Fig. 5). This means that 50% of installed power plants generate 25% of energy.

This has a major impact on load distribution and electric network operation. Fossil fuel power plants (lignite, hard coal, and natural gas) and nuclear plants were designed to operate as base load stations, delivering high power generation with small deviations (see Fig. 6).

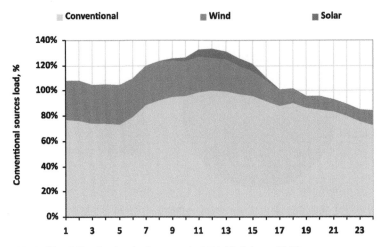

Fig. 6: Typical load distribution in Germany in 2011 (Quinkertz 2015).

Current trends in power sources distribution will force the base load power plant to fulfill a different role in the system–adapting to the changing consumption vs. generation ratio. The highly likely scenario for the short term future will require a shift by the base load power plant from a stable minimum of 40% to 10% or even less. The long term scenario is even more drastic, and will require all fossil fuel power plants in the grid to shut down/standby and all surplus renewable energy to be stored (see Fig. 7). This will require a rethinking of the role and operation of the heavy duty coal-fired steam turbine power plants. Intermediate load/shutdown is the expected future for them. On the other hand, there will be periods in a day or year in which renewable power sources will be unable to cover the demand for electricity. During these periods, fossil fuel/nuclear power plants will have be in operation. Natural gas power plants face an uncertain future, because if lignite/hard coal/nuclear power plants fulfill the role of intermediate load or even peak load there is no place for fuel expensive gas turbine systems. Thus, renewable energy sources will require flexible fossil fired power plants, as this is the only strategy which allows them to survive.

Flexible operation of the power plant is defined by several factors which can be agglomerated under the following time periods (see Fig. 8):

1. Fast Startup
2. Primary Frequency Response

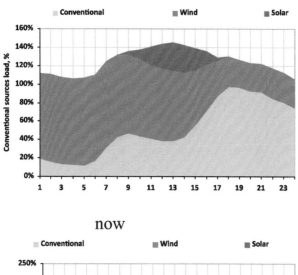

now

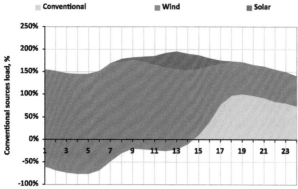

future

Fig. 7: Probable scenarios for future load distribution and load requirements under the "right of way" for energy from renewables (Quinkertz 2015).

3. Secondary Frequency Response
4. Peak Power and Off-Frequency
5. Part load
6. Fast Shutdown
7. Prepared for Startup

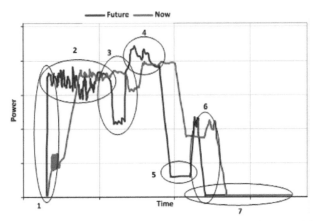

Fig. 8: Requirements for transient operation with the market dominated by renewable energy power sources, 1-fast startup, 2-wider flexibility, 3-lower temporary minimum, 4-higher overpower, 5-lower permanent minimum, faster shutdown (Quinkertz 2015).

For each of these periods, a significant increase in flexibility will be required. Fast Startup can be shortened by avoiding some preliminary tests, controlling thermal stresses in the critical parts of the steam boiler during high load gradient, and speeding up the electric synchronization procedure. Primary and Secondary Frequency Responses can be upgraded by lowering available minimum and rising available overloads. Peak Power and Off-Frequency mode will need to be prepared for extra power delivery and off-frequency operation. Part Load operation can be upgraded by a lower minimum (mainly steam boiler bottlenecked) from 40% to 10–20%. Fast Shutdown will follow similar procedures as Fast Startup by applying adequate procedures for high unload gradients. Key to keeping the system ready for the next startup is implementing optimized preservation and heating concepts.

The startup time for a steam turbine can be reduced in a number of ways. Firstly, faster startup can be achieved by keeping the system in hot state; this measure gives a 20% startup time reduction. This requires reducing the main and re-heating steam temperatures prior to starting the system, resulting in lower efficiency but at the same time preventing excessive thermal stresses during startup. If more advanced solutions are applied, like full optimization of the startup procedure and significant improvement in roll-off of the steam turbine procedure, a 40% shorter startup may eventually be expected. For faster load increase (> 4% P_{nom}/ min) without significant energy losses, it is necessary to improve the fuel supply dynamics. This goal can be achieved through the use of firing torches, which unfortunately may result in a deterioration of economic efficiency. An alternative is to use intermediate carbon dust trays to optimize the operation and maintenance of the mills. Due to the higher concentration of coal dust directed from the intermediate carbon dust tray to the boiler, the minimum load can be reduced to just 15% of the nominal load.

Flexible operation of the steam turbine power plant will adversely impact its life time, reduce its availability and increase maintenance costs. There are various options as regards accounting for this issue, one of which is to change the way Equivalent Operating Hours (EOHs) are counted. Currently, only ordinary startup and shutdown are taken into account, but with the advent of flexible operation the counter should be updated for low load (cool down) and nominal load (heat up) transient periods. Increased flexibility will translate into higher operating costs and replacement costs due to the need for earlier replacement of components. Frequent starts and low loads will cause a decrease in performance of the plant. To minimize the consequences of enhanced thermal flexibility, measures can be introduced such as adaptation of control and monitoring systems, optimization of startup and power change processes, maintenance of optimum thermal state of the turbine components (heating) and adaptation of new techniques and schedules of diagnostic tests.

A number of measures can be taken to improve the flexibility of the steam turbine power plant (see Fig. 9). They act in a range of time frames and have various impacts on system response, power and efficiency.

The most natural actuator of steam turbine power is a governing system based on a set of valves. Depending on the solution used, there can be a single valve (throttling) or group governing. Valves deliver the fastest system response, but there are some time dependent limits (especially with group governing). This can be countered by adding an extra valve to bypass the governing control valves, providing additional steam to the turbine at a time when main valves are still not fully open (see Fig. 10). Additional steam is admitted to some stages downstream of the 1st blade row, increasing the swallowing capacity of the turbine. The turbine is operated with full arc admission at all loads and bypass steam is admitted uniformly around the full arc. This avoids non-uniform mechanical loading of downstream blade rows. The HP stage bypass provides a solution with the best heat rate from minimum stable boiler load up to nominal load with the possibility of frequency response over the entire range.

Steam turbine power can also be increased through bypassing the whole HP part of the turbine, with additional water spraying in the bypass

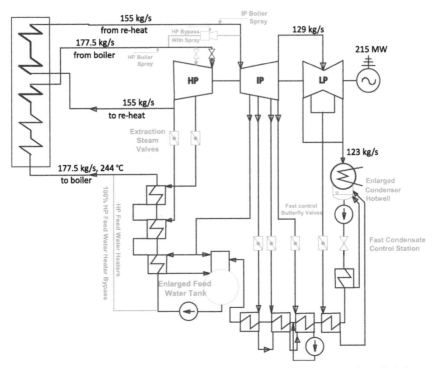

Fig. 9: Control measures for flexibility of a 200 MW steam turbine power plant (Quinkertz 2015).

(see Fig. 11). The measure is obtained by opening the HP bypass valve and spraying water into live steam as well as re-heated steam. This increases steam mass flow and decreases its parameters.

The second measure which can be applied here is to control the condensate in the system. This can be done either by slightly upgrading the condenser and deaerator to enlarge their condensate storage capability or by attaching an additional hot water storage tank (Schuele et al. 2012). The fast condensate control system acts through the condensate valve and bleeders butterfly valves—the control system needs to be modified to achieve a fast, automatic response. By closing all those valves, the bleeders will stop steam from escaping, thereby delivering increased power. This measure stops the condensate flow and accumulates it in the condenser tank. The time response of this measure is limited by the storage capacity of the condenser. The same measure can be used in high pressure regeneration heat exchangers, with the small difference that the feedwater

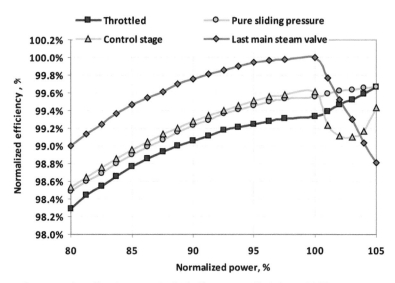

Fig. 10: Increase of swallowing capacity by boiler storage (Quinkertz 2015).

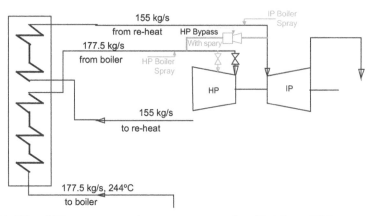

Fig. 11: HP and LP water spray to increase steam mass flow (Quinkertz 2015).

needs to be stored in the deaerator or/and bypass the high pressure heat recovery system.

The standard fuel increase measure is typically implemented, with the slowest time response. It should be noted that all those measures decrease system efficiency under transient response.

All of the measures referred to above have various transient influence on the power changes, thus they should be implemented in the correct

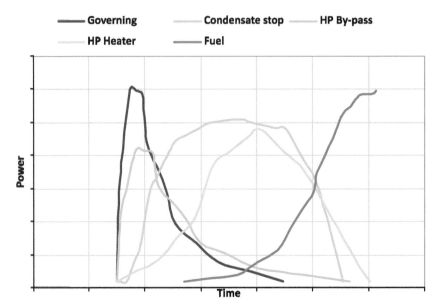

Fig. 12: Time response of the measures (Quinkertz 2015).

order. Figure 12 shows the sequence which gives the most stable turbine power rise for the longest time period possible (to reach the goal by fuel feeding). Taking the efficiency effect into account, the following procedure seems the most suitable:

1. Start increasing firing (very slow response) and opening all the throttled governing valves (very fast response)
2. Opening control/HP part by-pass valve(s)
3. Stop condensate
4. Increase of water spray to live steam and re-heated steam flows
5. Bypass of high pressure heat recuperation by closing high pressure steam bleedings

A power increase of 15% can be obtained in 40 seconds of turbine operation and can be kept on the level for the next 120 seconds. After this time, condensate cannot be accumulated and the power will fall to about 10% above the base power. Then only spraying water and increasing fuel supply will force the power to higher loads.

If all those measures are still too slow, there is also an option to decrease the rotational speed of the turbine shaft—a reduction from 50 Hz to 49.5 Hz will deliver a 10% increase in power.

This chapter presents the main ways of enhancing the flexibility of coal-fired units that can be applied to existing generating units. These measures are particularly important for countries obliged to increase the share of renewable energy sources in the context of energy systems which are based mainly on coal-fired plant. A better solution for improving energy security, but one which needs time to implement and large financial outlay is the construction of modern, highly elastic units for supercritical parameters. An interesting innovation is the concept of the duo-unit, where two steam boilers work for one steam cycle. This increases the range of load, from as low as 10% of nominal load. Another way to ensure energy security is to introduce more flexible technology, i.e., gas turbine units. Gas turbines can be integrated with a coal-fired unit, leading to the construction of a multi-fuel hybrid unit. In addition to increased flexibility, this provides a better energy mix within a single unit, thereby reducing the risk of price fluctuations in the fuel market. A synergies effect can be achieved in hybrid systems, delivering highly efficient fuel conversion without high financial costs. In addition to investing in alternative production technologies, it is also very important to introduce energy storage systems, as they draw power during off-peak demand periods and produce during peak periods.

Acknowledgments

Dr. Rainer Quinkertz, Siemens AG, for consultation.

References

Bartela, Ł., J. Kotowicz and K. Dubiel. 2016. Technical—economic comparative analysis of the energy storage systems equipped with the hydrogen generation installation. Journal of Power Technologies 96(2): 92–100.

Bertsch, J., Ch. Growitsch, S. Lorenczik and S. Nagl. 2016. Flexibility in Europe's power sector—An additional requirement or an automatic complement? Energy Economics 53: 118–131.

Brouwer, A.S., M. van den Broek, A. Seebregts and A. Faaij. 2015. Operational flexibility and economics of power plants in future low-carbon power systems. Applied Energy 156: 107–128.

Després, J., S. Mima, A. Kitous, P. Crique, N. Hadjsaid and I. Noirt. 2016. Storage as a flexibility option in power systems with high shares of variable renewable energy sources: a POLES-based analysis. Energy Econ. http://dx.doi.org/10.1016/j.eneco.2016.03.006.

Eurelectric. 2011. Flexible generation: backing up renewables. Report D/2011/12.105/47. Brussels.

Eurostat. http://ec.europa.eu/eurostat.

Gawlik, L., A. Szurlej and A. Wyrwa. 2015. The impact of the long-term EU target for renewables on the structure of electricity production in Poland. Energy 92: 172–178.

Huber, M., D. Dimkova and T. Hamacher. 2014. Integration of wind and solar power in Europe: assessment of flexibility requirements. Energy 69: 236–246.

Kotowicz, J., Ł. Bartela, D. Węcel and K. Dubiel. 2017. Hydrogen generator characteristics for storage of renewably-generated energy. Energy 118: 156–171.

Kupecki, J., J. Milewski, K. Badyda and J. Jewulski. 2013. Evaluation of sensitivity of a micro-CHP unit performance to SOFC parameters. ECS Transactions 51: 107–116.

Milewski, J., M. Wołowicz, A. Miller and R. Bernat. 2013. A reduced order model of molten carbonate fuel cell: A proposal. International Journal of Hydrogen Energy 38(26): 11565–11575.

Milewski, J., G. Guandalini and S. Campanari. 2014. Modeling an alkaline electrolysis cell through reduced-order and loss-estimate approaches. Journal of Power Sources 269: 203–211.

Papaefthymiou, G. and Ken Dragoon. 2016. Towards 100% renewable energy systems: Uncapping power system flexibility. Energy Policy 92: 69–82.

Polskie Sieci Elektroenergetyczne http://www.pse.pl/.

Quinkertz, R. 2015. Flexible operation of large steam power plants—balancing grids with growing share of renewable. Research and Development in Power Engineering Conference 2015; Plenary Lecture.

Schuele, V., D. Renjewski, F. Bierewirtz and O. Clément. 2012. Hybrid or flexible—integrated approach for renewables integration. International Conference on Power Plants, Zlatibor, Serbia, 30 October–2 November, 2012. Society of Thermal Engineers, Serbia.

Szablowski, Ł., J. Milewski and J. Kuta. 2011. Control strategy of a natural gas fuelled Piston engine working in distributed generation system. Rynek Energii 3: 33–40.

Walker, S.B., D. van Lanen, M. Fowler and U. Mukherjee. 2016. Economic analysis with respect to Power-to-Gas energy storage with consideration of various market mechanisms. International Journal of Hydrogen Energy 41: 7754–7765.

CHAPTER 3

Supercritical Carbon Dioxide Power Cycles

Timothy J. Held

1. Introduction and Background

The primary purpose of a "fluid power cycle" is to convert thermal energy into mechanical or electrical power through use of a working fluid. The idealized Carnot cycle is one in which the fluid undergoes a series of operations, including compression, heat addition, expansion and heat rejection. The two most common practical implementations of fluid power cycles are referred to as the Rankine cycle, in which the fluid is compressed in a liquid state, and the Brayton cycle, where the fluid is compressed in a gaseous state. In the late 1960's, two researchers (Angelino 1968, Feher 1968) proposed alternative fluid power cycles where the fluid was in a supercritical state during the compression phase. As carbon dioxide (CO_2) has a relatively low critical pressure (7.38 MPa), high thermal stability, and low cost, it was selected as a fluid of interest for further investigation.

Beyond the theoretical investigations of these early works, little additional activity was published over the next several decades, other than a study of a supercritical carbon dioxide (sCO_2) power cycle for shipboard exhaust heat recovery (Combs 1977). In the early 2000's, the concept of utilizing an sCO_2 power cycle in connection with advanced nuclear reactor technology (Dostal et al. 2004) created new interest in the cycle. The topic has since gained significant interest in the technical community, and

Echogen Power Systems, 365 Water Street, Akron, Ohio 44308.
Email: theld@echogen.com

has resulted in numerous sCO_2-focused symposia, and a newly-formed technical committee within the International Gas Turbine Institute. Within the last several years, multiple research-scale test loops and component test facilities have been developed, and the first commercial-scale sCO_2 power cycle has undergone factory testing. This chapter provides an overview of the sCO_2 power cycle, and the current state of technology development and implementation.

2. Cycle Basics

A thermodynamic fluid power cycle consists of four basic processes: compression, heat addition, expansion, and heat rejection (Fig. 1). Supercritical CO_2 cycles follow these same processes, but the exact arrangement of the equipment and subsystems used will vary depending on the requirements of the application.

The pressures and temperatures over which the cycle operates are generally constrained by a combination of factors, including material limits, environment temperature, heat source temperature, and performance optimization. Typically, sCO_2 cycles operate between pressures near the critical pressure (6–10 MPa) and 20–30 MPa. The "low-side" pressure (between the expander outlet and compressor inlet) is governed by the temperature to which heat is being rejected, while the "high-side" pressure is limited by either mechanical constraints or a maximum in cycle performance. At these relatively low pressure ratio values (3–5), the fluid exiting the expansion process retains a significant amount of residual enthalpy, which can be recovered through recuperation, where this enthalpy is transferred to the relatively cold, high pressure fluid exiting the compressor. Thus, nearly all practical sCO_2 cycles employ some amount and form of recuperation.

The simplest form of a recuperated cycle is shown schematically in Fig. 2, with the corresponding pressure-enthalpy (PH) diagram in Fig. 3. An environmental heat sink of 25°C, and heat source of 500°C are assumed. The cycle begins at the heat rejection heat exchanger (HRHX) outlet and compressor inlet, where the CO_2 fluid is at its lowest temperature and pressure. The compressor increases the fluid pressure. Heat is transferred to the fluid in the recuperator (RC), and subsequently in the primary heat exchanger (PHX). The high temperature, high pressure fluid is expanded back to the low-pressure condition. Residual enthalpy is transferred to the cold, high-pressure fluid in the recuperator, and the remaining enthalpy is rejected to the environment by the HRHX.

The idealized process (isentropic compression and expansion, constant-pressure heat transfer) is shown as the solid lines in Fig. 3, while the more physically realistic process is shown as the dashed lines. Several

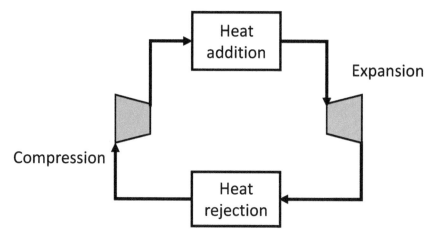

Fig. 1: Generic thermodynamic cycle.

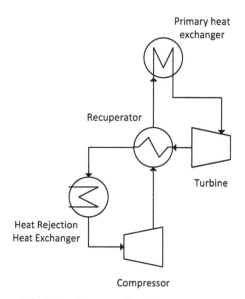

Fig. 2: Simple recuperated sCO_2 cycle process flow diagram.

important general points about sCO_2 power cycles can be made, even on the basis of this simplistic configuration.

The specific net power of a cycle can be defined as the work output per unit mass flow rate of fluid circulated through the system. Due to the low pressure ratio of sCO_2 cycles, this value is rather low compared to a steam Rankine system. In other words, for a given output power, the CO_2 mass flow rate is much higher than for a steam system. To some extent, this difference is offset by the higher molecular weight of CO_2 relative to

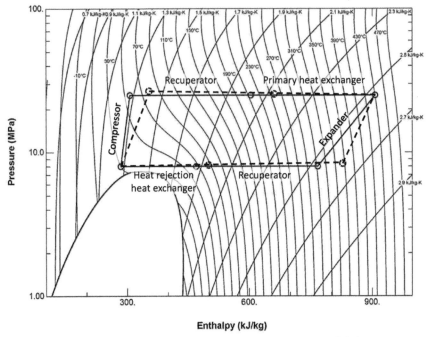

Fig. 3: Pressure-enthalpy diagram for simple recuperated cycle. Solid lines represent isentropic ideal cycle, dashed lines represent more realistic cycle, with non-isentropic compression and expansion processes, and heat exchanger pressure drops.

water, bringing the volumetric flow rates closer together. However, it is to be noted that a drawback of sCO_2 power cycles is the relatively larger piping required for a given power output system.

The specific compression power (relative to net power output) for an sCO_2 cycle is approximately 30%, considerably higher than that for a classical steam Rankine system (typically 1–2%). This difference is primarily driven by the low specific power output, thus requiring larger CO_2 mass flow rate. The somewhat lower density of CO_2 at the compressor inlet condition (typically 600–800 kg/m³) also contributes to the higher compression power requirement. On the other hand, a gas Brayton cycle requires a much higher specific compression power, often as much as 80%.

2.1 Thermodynamic Background

2.1.1 Supercritical Fluids

One of the distinguishing characteristics of the sCO_2 power cycle is its use of a fluid that exists throughout most, if not all, of the cycle in the supercritical state. The critical point (or "critical state") of a fluid is defined as the condition at which the gas and liquid phases have the

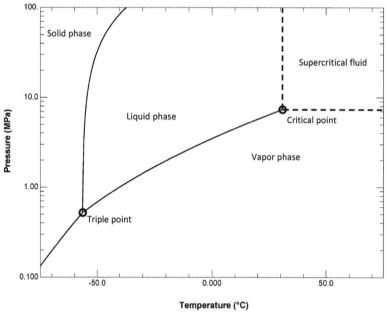

Fig. 4: CO$_2$ phase diagram.

same density, or as the end point of the liquid/vapor equilibrium curve (Fig. 4) (Borgnakke and Sonntag 2014). A supercritical fluid is defined as one where the temperature and pressure both exceed the critical point. While these definitions are important, from the cycle perspective the most important parameter is the critical pressure. Above that pressure, as heat is added to the fluid, the temperature increases and density decreases continuously, with no distinct phase change between a liquid and vapor. This behavior exists whether the initial state of the fluid is above or below the critical temperature. This continuous increase in temperature with heat addition is an important characteristic of sCO$_2$ cycles, and is at least partially responsible for the improved thermodynamic performance of these cycles relative to more conventional Rankine cycles, where heat addition is accompanied by a constant-temperature phase change process.

2.1.2 Supercritical Carbon Dioxide

Carbon dioxide has several properties that make it an attractive working fluid for power cycles. First, its critical pressure is relatively low (7.38 MPa), allowing for cycle operation well above the critical pressure, while maintaining pressures within the range of commonly-used industrial equipment. Second, its thermal stability allows for operation at reasonably high temperatures without risk of thermal breakdown. It is non-flammable, which permits operation in hazardous environments without added fire risk. Its freezing point at typical system operating and stand-still pressures is low (–56°C) allowing operation at extreme climate conditions. And carbon dioxide is readily available at low cost and high purity due to its global use in food and beverage applications.

Carbon dioxide is slightly toxic, with an OSHA 8-hour exposure limit of 5,000 ppm, and a short-term limit of 30,000 ppm (Department of Health and Human Services 2007). In addition, the high molecular weight of CO_2 (44 g/mol) relative to air (29 g/mol) and low temperatures resulting from expansion can result in high concentrations of CO_2 settling in low areas. The potential of solid CO_2 formation during intentional and relief valve venting also complicates the design of safety systems. Many of the safety systems and practices developed for CO_2 pipelines and services (Harper et al. 2011) can be used as a basis for sCO_2 power cycle safety practices.

2.1.3 CO_2 Equation of State and Transport Properties

The design of sCO_2 power cycles requires operation of components throughout conditions where CO_2 can range across vapor, liquid, two-phase and supercritical states. The proper design, operation and control of such a system require an accurate equation of state (EOS), which for a simple compressible substance, relates all thermodynamic properties as a function of two other properties (e.g., density as a function of temperature and pressure). Presently, the majority of the technical community uses a well-established EOS (Span and Wagner 1996) as implemented in REFPROP (Lemmon et al. 2013), or other similar software. The accuracy of this EOS is generally considered acceptable for pure CO_2, although some concerns exist for mixtures of CO_2 and other fluids (Ridens and Brun 2014).

Accurate transport properties (thermal conductivity and viscosity) are needed for design of components such as heat exchangers. REFPROP is a commonly-used source of transport properties, and derives its values from the work of Vesovic and others (Vesovic et al. 1990, Fenghour et al. 1998). More recent work (Harvey et al. 2015) has indicated that improvements can be made by utilizing new data and the Span and Wagner EOS for calculation of derivatives.

2.2 Cycle Architectures

2.2.1 Cycle Basics

Closed-loop thermodynamic cycles can be described by their "architecture", the arrangement of heat exchangers and turbomachinery within the cycle. The simplest architecture for a closed-loop sCO_2 cycle would consist of four primary components:

- Compressor (or pump): Increases the pressure of the working fluid through mechanical work.
- Primary Heat Exchanger (PHX): Transfers heat from the external source to the working fluid.
- Expander: Converts the enthalpy of the high-pressure, high-temperature working fluid to mechanical work while decreasing the pressure of the working fluid.
- Heat Rejection Heat Exchanger (HRHX): Transfers residual heat from the working fluid to the heat sink, generally the environment.

The simple architecture described above is not generally considered for practical sCO_2 applications, due to their low overall pressure ratio (OPR). OPR is the ratio of the "high-side" (compressor discharge through turbine inlet) pressure to the "low-side" (turbine discharge through compressor inlet) pressure. The OPR is an important parameter for determining cycle performance, but for sCO_2 cycles, is limited by practical issues to a narrow range, as described previously. The consequence of this low OPR is that the temperature exiting the turbine is still quite high, only around 100–200°C lower than the turbine inlet temperature. This high temperature, but relatively low-pressure fluid represents unutilized heat that still has substantial thermodynamic value. Thus, most sCO_2 power cycles utilize recuperation, or internal transfer of heat from this low-pressure, high-temperature fluid to lower-temperature, high-pressure fluid within the cycle.

2.2.2 The Simple Recuperated (SR) Cycle

By adding a single heat exchanger (recuperator) between the turbine discharge fluid and the compressor discharge fluid, a large fraction of the unutilized heat can be internally recovered within the cycle, thereby improving the cycle efficiency substantially. The recuperator has an added benefit—the temperature of the fluid entering the HRHX is lower, reducing thermal stresses and potential fouling concerns with water-cooled heat exchangers. The temperature of the working fluid entering the PHX is higher, which reduces the average temperature differential between the heat source and the working fluid. Thus, the conductance (UA) of the PHX needs to be larger for the same amount of heat transferred.

2.2.3 RCB Cycle and Variants

While the SR cycle offers a marked improvement in efficiency over the simple cycle, other architectures offer yet greater improvements at the cost of some added complexity. The most commonly-studied cycle has come to be called the "Recompression Brayton Cycle", although an alternate, more descriptive title would be the "Recuperator Bypass Cycle". The two main differences between the SR and RCB cycles are the division of the recuperator duties into two heat exchangers, and the addition of a high-temperature compressor that bypasses a portion of the high-pressure flow around the low-temperature recuperator (Fig. 5).

The improved efficiency of the RCB cycle over the SR cycle can be best understood by examining plots of the fluid temperature versus heat transferred (called "TQ plots") for the recuperators of both cycles (Fig. 6). For the SR recuperator, a large mismatch can be seen between the slopes of the hot, low-pressure and cold, low-pressure fluids. These

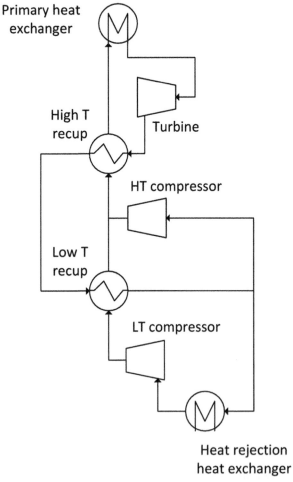

Fig. 5: Recuperator bypass cycle process flow diagram.

slopes are inversely proportional to the product of the mass flow rate and isobaric heat capacity of the fluid streams ($\dot{m}\, c_p)^{-1}$. For the SR cycle, the mass flow rates are equal, and the heat capacity of the cold, high-pressure fluid is much higher than the hot fluid. A simple exergy analysis of the heat exchanger shows that exergy destruction is minimized when the two curves are parallel. The RCB cycle achieves this slope matching by intentionally bypassing a portion of the working fluid around the low temperature recuperator by using a high-temperature compressor (HTC) to compress about 30% of the fluid directly from the low-pressure outlet of the recuperator and reintroducing it downstream of the recuperator high-pressure outlet (thus bypassing the recuperator, giving justification to the "Recuperator Bypass Cycle" nomenclature). The benefits of the LTR temperature slope match outweigh the compression work penalty due to the low fluid density at the HTC inlet.

2.3.4 Heat Recovery Cycles

A side effect of the improved recuperation effectiveness of the RCB cycle is an elevated fluid temperature entering the PHX. Thus, the heat source temperature can only be reduced to a limited degree by this preheated fluid. While this result is acceptable for recirculated heat sources, where unused heat is recycled back to the heat source, certain classes of applications, termed "heat recovery", do not permit this recirculation. Unrecovered heat is lost to the environment, limiting the power output of the system, even if the thermodynamic efficiency of the cycle is high.

Heat recovery cycles are designed to maximize the power output of a non-recirculated heat source by balancing the thermodynamic efficiency of the cycle with the amount of heat extracted from the source. In many cases, this is accomplished by directing some amount of the working fluid directly from the compressor discharge to the PHX. This allows the temperature of the heat source to be reduced far more than the RCB cycle could. This lower-temperature heat addition results in lower thermodynamic efficiency than the RCB cycle, but higher power output due to the larger quantity of heat recovered from the source.

Several varieties of heat recovery cycle have been discussed and studied in the literature. Two examples are shown in Fig. 7. The "cascade" cycle directs a portion of the compressor flow through the primary heat exchanger, where it is raised to its highest temperature. Following expansion through the high temperature turbine, the residual fluid heat is transferred to the remainder of the compressor flow in a high-temperature recuperator (HTR). This preheated fluid is then expanded through a second, low-temperature turbine (LTT). The LTT exhaust is combined with the HTR low-pressure outlet fluid, and again passes through a LTR for additional preheating of the compressor discharge.

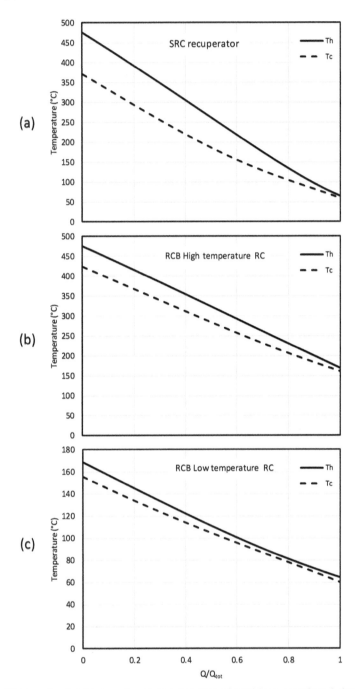

Fig. 6: Recuperator fluid temperatures as a function of total heat transferred, for simple recuperated cycle (a) and RCB cycle (b & c).

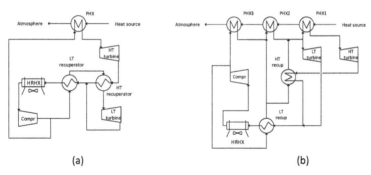

Fig. 7: Heat recovery cycle examples. (a) Cascade cycle and (b) Dual rail cycle.

A second form of heat recovery cycle is the "dual rail" (DR) cycle. Again, the compressor discharge fluid is split between two "rails", one proceeding to a set of PHX coils, the other to the recuperators. As the fluids are heated in parallel, flow is diverted from the recuperator rail to the PHX rail in several steps. This incremental approach allows better matching between the $\dot{m}\,c_p$ of each heat exchanger than does the cascade cycle, thereby attaining higher overall power output for the same application.

2.3 Applications

Supercritical CO_2 power cycles have been studied and proposed for numerous applications, as befitting a general-purpose closed-loop power cycle. Indeed, all these applications are currently most frequently associated with closed-loop steam Rankine cycles. The stated advantages of sCO_2 power cycles generally include improved thermodynamic efficiency, reduced turbomachinery size (and therefore power block cost), and water-free operation.

2.3.1 Nuclear

The re-emergence of sCO_2 power cycles in the early 2000's can be traced to studies of advanced nuclear reactor technology, where the higher heat source temperature ($> 550°C$) relative to pressurized water reactors (PWRs, 300°C) couple well with the improved cycle performance of sCO_2 cycles at high temperatures. In these applications, high temperature reactor coolant, such as molten salt, liquid metal, or gas, is circulated to a primary heat exchanger (PHX). Because the heat transfer fluid is recirculated back to the reactor, the small temperature differential of the RCB cycle is not a disadvantage.

2.3.2 *Concentrating Solar Power (CSP)*

The recent dramatic reduction in price of solar photovoltaic (PV) power initially appeared to render solar thermal, or Concentrating Solar Power (CSP) obsolete before it became a significant factor in utility-scale power production. However, when combined with thermal storage, CSP has recently gained interest as a means to provide dispatchable power to compensate for the intermittency of PV power. Today, several current high-temperature CSP plants utilize molten-salt thermal storage in combination with steam Rankine cycles for power production. Current-generation nitrate salts are limited to approximately 565°C, a temperature at which sCO_2 and steam power cycles yield similar performance. Next-generation (Gen3) CSP technology development is largely centered on advanced, high-temperature (> 700°C) thermal storage media, in combination with sCO_2 power cycles due to their greater thermodynamic efficiency (Mehos et al. 2017).

Advanced thermal storage can take several forms. Sensible heat may be collected and stored by either advanced chloride- or carbonate-based salts or solid particles. Alternatively, thermal energy may be stored in the form of phase-change materials, or thermochemical reactions. In all these cases, the interaction between the sCO_2 power cycle and the storage medium must be considered in the overall technoeconomic selection and optimization process. Sensible heat storage requires the medium to cycle between two states—"cold" and "hot". The narrow temperature range of the RCB cycle constraints the differential between the hot and cold storage states to be smaller than for instance, a steam Rankine cycle might. Thus, a larger mass of the storage medium is required for the same thermal storage capacity, resulting in higher medium costs and storage tank costs. In addition, the flow rate of the storage medium also increases as the temperature differential decreases, causing increased medium transport costs and auxiliary power loads.

Phase change and thermochemical storage utilize very narrow temperature ranges by their nature, thus coupling well with the RCB and similar cycles. However, the readiness level of these technologies is still low, and considerable research and development will be required to fully integrate these solutions with sCO_2 power cycles.

2.3.3 *Primary Indirect-Fired*

The most common application of steam Rankine cycles has historically been power generation from an indirectly-fired source, such as coal combustion. While the combustion temperature is quite high, in excess of 1300°C, the working fluid temperature is limited by the materials of construction of the primary heat exchanger (i.e., the boiler for a steam cycle). Today's supercritical steam power plants operate at steam conditions of 30 MPa and 600°C, with advanced ultrasupercritical (AUSC) power plants being considered at temperatures from 700–760°C, and pressures in excess of 38 MPa (Marion et al. 2014).

Due to their superior thermodynamic efficiency, sCO_2 power cycles are also being considered for these and similar indirectly-fired applications. Recent integration studies of sCO_2 with oxy-coal combustion power plants support the higher potential plant efficiency, and point out some of the challenges for sCO_2 in this application (Shelton et al. 2016, Miller et al. 2017). Again, the relatively narrow temperature range of the RCB cycle limits the temperature to which the flue gas can be reduced to approximately 200°C below the turbine inlet temperature. To achieve high plant efficiency, air preheating is routinely utilized in today's power plants. However, the higher exit temperature of the sCO_2 primary heat exchanger requires a greater degree of air preheating at higher temperature than does the steam cycle. Secondly, although the maximum conditions of the sCO_2 cycle are less severe than the AUSC steam Rankine cycle, the volumetric flow rate of sCO_2 is considerably higher, and the lower pressure ratio of the sCO_2 cycle makes it more sensitive to pressure losses in the high pressure primary heat exchanger. These factors place greater constraints upon the PHX design. The improvements in cycle efficiency, smaller footprint, and lower projected maintenance will be weighed against the PHX design challenges.

2.3.4 Primary Direct-Fired

Unlike the other applications discussed here, the primary direct-fired, or "Allam" cycle (Allam et al. 2014) is unique in that it is only a partially closed-loop, and heat is added internally to the working fluid through a combustion process. The basic cycle architecture is a simple recuperated cycle, with several stages of intercooled compression. Heat is added by burning a fuel, such as natural gas, with oxygen highly diluted by circulated CO_2. Combustion temperatures are 1100–1150°C, at a cycle pressure of 30 MPa. Power is generated by expansion through a turbine to a pressure below the critical state—this higher expansion ratio is necessary to reduce the turbine exit temperature below 750°C due to material limitations in the downstream piping and recuperator. After recuperation to preheat the compressed CO_2, the water produced in the combustion process is separated, and a portion of the CO_2 is extracted from the loop to maintain the overall cycle mass balance. The water is a waste product, and the extracted CO_2 will be sent into a pipeline to either sell to the EOR market, or sequester. One advantage of the cycle is that the CO_2 product is already compressed, and can be dropped directly into the pipeline.

This power cycle is intended to compete against natural gas combined-cycle gas turbines with carbon capture (NGC[4]) for utility power generation. Numerical studies show that the Allam cycle can exceed the overall efficiency of state-of-the-art NGC[4] power plants (Allam et al. 2014, Penkuhn and Tsatsaronis 2016), and a pilot plant is currently under construction (Allam et al. 2017).

2.3.5 *Heat Recovery*

Industrial processes and certain power generation technologies emit excess heat at temperatures high enough to make recovery and conversion to electrical power economically viable (for sCO_2, roughly > 300°C). These applications are distinct from the primary indirect-fired cases described earlier in that heat that is not utilized by the heat recovery power cycle is lost to the environment for either practical or economic reasons. For instance, air preheating of a gas turbine engine would result in lost power due to reduced inlet density, resulting in lower mass flowrate through the engine. For these types of applications, the thermodynamic efficiency of the recovery cycle is less important than the generated output power. The RCB cycle, for instance, has a higher thermodynamic efficiency than does the simple recuperated cycle at the same turbine inlet temperature. However, the higher degree of recuperation in the RCB cycle increases the sCO_2 temperature entering the primary heat exchanger, which then limits the quantity of heat that can be extracted from the exhaust. Thus, the lower-efficiency SR cycle, which has a lower sCO_2 temperature entering the PHX, generates a higher power output.

This feature of heat recovery applications, whereby both the quantity of recovered heat and the cycle efficiency are both important, has given rise to several alternative cycle architectures, including so-called "cascade" and "parallel" cycles (Fig. 7). In all cases, the arrangement of heat exchangers and turbomachinery has been designed to improve the utilization of both externally-supplied heat (from the PHX) and internally-transferred (RHXs) heat. These applications have given rise to the first commercially-available sCO_2 power cycle system (Held 2014).

3. Equipment

The physical components that comprise an sCO_2 power cycle are similar to those used in other applications, but with several features that are related to the high pressure and density of sCO_2. A summary of the major components, and how they apply to sCO_2 systems is given in this section. The physical components of a power cycle can be divided into three categories: rotating equipment, heat exchangers, and piping/instrumentation/controls.

3.1 Rotating Equipment

The task of converting between mechanical and fluid energy is performed by two major types of rotating equipment, expanders and compressors. An expander converts high-temperature fluid enthalpy to mechanical work by reducing its pressure, while compressors use mechanical energy to increase the pressure of the working fluid. The configuration of these devices can vary depending on the size and operating conditions of the application.

3.1.1 Expanders

Two main types of expanders have been used in sCO_2 thermodynamic cycles. Positive-displacement (PD) expanders, such as scroll (Kohsokabe et al. 2008) or piston expanders (Yang et al. 2010), have been used in CO_2 refrigeration cycles at small (kW) scales. Turbine expanders are more common for power cycle applications and are the primary types used and anticipated in current-generation sCO_2 power cycle systems. Turbine expanders again can be divided into two primary subtypes, based on the general flow path. Radial turbines, where the initial flow direction is oriented radially to the shaft centerline, are presently most common, having been used in several test and commercial systems.

A common classical method for the conceptual design of turbomachinery follows a specific speed/specific diameter methodology, in which geometric similarity is used to define a non-dimensional representation of a turbine. For conceptual design purposes, turbine isentropic efficiency can be defined as a two-dimensional function of specific speed and diameter, $\eta = f(N_s, D_s)$ (Baljé 1962). Presuming that turbine will be sized for optimal efficiency at the design point, a specific speed and diameter of 0.55 and 3.5 respectively define the maximum efficiency point on the generalized N_s-D_s chart. By selecting inlet conditions and outlet pressure, an optimal physical speed and diameter can be defined as a function of size for a single-stage turbine. Similarly, by utilizing similar processes for a multistage turbine, where the isentropic enthalpy drop is replaced by a "per-stage" value, an optimal speed and size curve versus the number of stages can be generated (Fig. 8).

The balance between size, efficiency and mechanical complexity will drive the selection of radial versus axial geometry. Of primary importance is the matching of turbine speed with load capability. Below 10 MW, direct drive of permanent magnet or induction generators at high speed (collectively termed "High Speed Alternators" (HSA)) using a single-stage

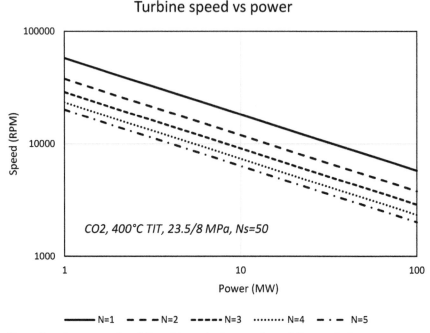

Fig. 8: Speed vs. power for sCO$_2$ turbines, single-stage through 5-stage designs.

turbine is possible. The currently-available high-speed alternators are shown in Fig. 9, with maximum power output plotted versus maximum speed. The curve of optimal turbine speed versus power output is also shown for a single-stage turbine—clearly, today's HSAs are not well-matched to the needs of single-stage sCO$_2$ turbomachinery. Since the HSA maximum speed is a technology limit, the turbine speed must be reduced below the optimal value to match. Due to the need to offset compression work, a single point in lost turbine efficiency will reduce the electrical output power of the sCO$_2$ system by approximately 1.3%. An additional challenge with HSAs is the aerodynamic drag, or "windage loss" created by the spinning rotor. As this drag is proportional to the fluid density in which it is immersed, significant losses and rotor heating can result from residing within the CO$_2$ boundary. To reduce these issues, methods to reduce the rotor cavity pressure are used, such as the "Haskel pumps" described by Wright et al. (2009).

At larger scales (1–10 MW+), HSAs are less available, and the most common power generation method is synchronous generators, operating at 1500/1800 RPM or 3000/3600 RPM depending on the number of generator poles (4 or 2) and grid line frequency (50/60 Hz). Below about 50

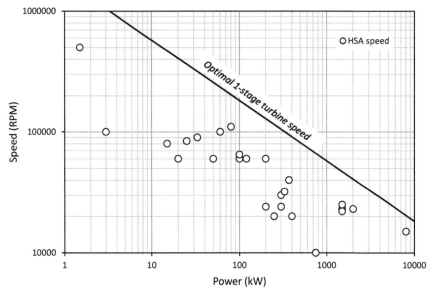

Fig. 9: Published speed vs power for research and commercial high-speed alternators.

MW, gearboxes can be used to match the turbine and generator speeds—however, in contrast to HSAs, once the generator is synchronized to the power grid, its speed is fixed. The part-load characteristics of fixed-speed turbines are substantially different from a variable-speed turbine, which should be accounted for in cycle modeling.

At even larger scales (50 MW+), gearboxes become less available, and 2-pole (3000/3600 RPM) synchronous generators become more common. Therefore, direct driving of the generator becomes necessary at fixed speed. Given the sCO$_2$ cycle characteristics, constant 3000/3600 RPM operation will require a multistage expander to maintain good aerodynamic efficiency. Depending on specific cycle design conditions, a minimum of 3 expander stages is likely necessary. Aerodynamic blade loading and other design constraints could result in a turbine configuration with as many as 12 stages.

3.1.2 Compressors and Pumps

The other major category of rotating equipment in a sCO$_2$ power cycle includes compressors and pumps. The distinction between the two is rather blurred, as pumps are typically classified as increasing the pressure of a liquid, while a compressor increases the pressure of a gas. For sCO$_2$ power cycles, the fluid state at the compressor inlet is near the critical pressure, and the temperature can range from subcritical to supercritical. For the purposes of this chapter, we refer to the device as a compressor, although the characteristics of the low temperature compressor for the RCB cycle and the sole compressor for the HR cycles are more characteristic of a liquid pump, due to the liquid-like properties of the fluid.

Similar sizing exercises can be carried out for compressor aerodynamic design as was described above for expanders, using non-dimensionalized efficiency maps. Again, speed-matching to the drive is an important design constraint. For the compressors, either turbine-driven or motor-driven options are possible. The turbine-driven option is more efficient, as the electromechanical efficiency of the gearbox, generator and compressor motor reduce the net efficiency of a motor-driven compressor. However, the motor-driven option allows more direct and independent control over the system flow rate(s), making it a more flexible option.

3.1.3 Seals

Rotating equipment requires both internal and external seals. Internal seals are primarily required to reduce inter-stage leakage, and can take the form of labyrinth, hole-pattern or honeycomb seals. External seals are necessary to reduce overboard leakage of CO_2, which must be recovered and/or replenished to maintain the total closed-loop fluid inventory. The current state-of-the-art is the dry gas seal, which operate with micron-level running gaps that are created by hydrodynamic forces. These were developed originally for compressor service, but their adaptation to turbine seals is relatively straightforward. The main areas of design concern are thermal management in the higher-temperature turbine environment due to secondary elastomer seals, and fluid conditioning and buffering to prevent contamination and condensation in the leak gap.

3.1.4 Bearings

The selection of fluid bearings is important in the design of the overall turbomachinery system, and has major impact on the remainder of the turbomachinery design. Conventional oil-lubricated bearings are acceptable outside the working fluid boundary, but their location outside the sealing system increases rotor length, which is detrimental to rotor stability. Moving the bearing system inside the working fluid boundary reduces the rotor length, but requires that the bearings be compatible with the working fluid. Potential candidate technologies include CO_2-lubricated hydrodynamic (e.g., "air-foil"), hydrostatic, or magnetic bearings.

3.2 Internal Heat Exchangers (Recuperators)

Because sCO$_2$ power cycles operate over a relatively small pressure ratio (3–6), the expansion process through the turbine only reduces the fluid temperature by about 100–150°C. This residual enthalpy is used to preheat the sCO$_2$ prior to entering the primary heat exchanger, by employing one or more internal heat exchangers, termed "recuperators". These heat exchangers need to be able to withstand the full operating pressure of the power cycle, typically in the range of 20–30 MPa, and temperatures that can range up to near the turbine inlet temperature.

Another key characteristic of recuperators for sCO$_2$ power cycles is their combination of high effectiveness and near-unity heat capacity ratio. For instance, in the RCB cycle, approximately 30–35% of the flow bypasses the low temperature recuperator to optimize cycle thermodynamic efficiency. This flow distribution not coincidentally results in the two sides of the recuperator having nearly identical heat capacities ($C = \dot{m}c_p$). Again, in the interests of cycle performance, typical cycle designs use recuperator effectiveness values in excess of 95%. This combination of high effectiveness at a near-unity heat capacity ratio ($C_r = C_{min}/C_{max}$) requires heat exchanger performance that closely approaches pure counter-flow (Incropera and DeWitt 2007).

The only two commercially-available heat exchanger technologies that are capable of this type of operation are the "Printed Circuit Heat Exchanger" (PCHE, discussed below) and Shell & Tube (S&T) configurations. While S&T heat exchangers can operate at these conditions, their large size and cost are prohibitive for sCO$_2$ power cycle use. In addition, the combination of effectiveness and $Cr = 1$ performance is very difficult to meet with S&T configurations due to their mixture of cross-flow and counter-flow characteristics, thus requiring multiple heat exchangers in series.

While microtube heat exchangers are in development (Chordia 2015, Kelly 2017), they have not yet attained commercial readiness for CO$_2$ systems. Their heat transfer performance will likely be intermediate between PCHE and S&T heat exchangers.

3.2.1 Printed Circuit Heat Exchangers

The PCHE is a highly-compact, high-pressure-capable heat exchanger. It has been used extensively in the oil and gas industry for over 30 years as a compressed-gas aftercooler, and is robust and well-suited to sCO$_2$ power cycle applications. The heat exchanger core is assembled from stacks of thin sheet metal plates that have been chemically-etched with flow passages. This stack is formed into a monolithic block by a diffusion-bonding process, which results in a structure that has comparable strength to a single block of parent material. The use of relatively small (1–3 mm) passages and thin sheet metal plates allows for low thermal convective and conductive resistance, providing a very compact device. These small passages also increase the propensity of the heat exchanger for plugging and fouling with debris, which must be managed carefully in a closed-loop system. Although questions have been raised about PCHE durability and resistance to thermal cycling during process upset conditions, detailed analysis of thermal stresses under these transient conditions has not yet been completed, although initial modeling and data comparison efforts are underway (Held and Avadhanula 2017). However, the balanced flow through the two sides of the recuperator, in which it is difficult to conceive of a situation where flow would be interrupted through only one side of the heat exchanger, limit the severity of the thermal transients. In addition, the extensive field experience of the PCHE has shown that under severe conditions, thermal stresses are an unlikely failure mode.

3.3 External Heat Exchangers

Beyond the recuperators, the remaining heat exchangers in the sCO_2 power cycle are responsible for transferring heat into and out of the working fluid through interaction with the environment. The primary heat exchanger (PHX) is responsible for transferring heat into the power cycle, while the heat rejection heat exchanger (HRHX) transfers residual heat to a heat sink, typically either to water or air.

3.3.1 Primary Heat Exchanger

The PHX is highly application-specific. Its configuration will depend primarily on the heat source characteristics. Two examples are given here:

Hot gas source: In a gas turbine combined-cycle application, the heat source is a near-atmospheric-pressure exhaust gas. The allowable pressure drop through the exhaust-side of the heat exchanger is generally quite low (1000–2500 Pa) to avoid creating a significant performance penalty on the gas turbine, and limiting the pressure-containment requirements on the ductwork transitioning from the gas turbine exhaust to the PHX inlet. For this type of application, the most common heat exchanger configuration is the finned-tube design, similar to superheater sections of a HRSG. Due to the low exhaust-side heat transfer coefficient, significantly more heat transfer area is required on that side of the heat exchanger. To meet the low pressure-drop requirement, an enlarged face (cross-sectional) area is used to reduce the average gas velocity. These characteristics are common for all exhaust heat exchangers, whether the working fluid is single-phase, supercritical or two-phase. However, because the CO_2 in the PHX is well above the critical pressure, the PHX configuration is relatively simple compared to a steam-based HRSG, where multiple pressures, steam drums, and separate superheater, boiler and economizer sections are required for each pressure.

Molten salt: For certain advanced nuclear power applications, and concentrating solar power (CSP) with thermal energy storage (TES), the primary heat source for the power cycle can be molten salt. The only current commercial use of molten salts is for CSP/TES, using nitrate salts, which have a maximum utilization temperature of approximately 565°C. Advanced chloride, fluoride or carbonate-based salts are possible future developments, which could allow operation at temperatures exceeding 700°C.

A molten salt-to-CO_2 heat exchanger would need to withstand the full heat source temperature and high-side sCO_2 pressure simultaneously. The high-pressure requirement again limits the current choices of heat exchanger technology to S&T and PCHE, or the related H²X hybrid configuration (Southall 2009) if a larger flow area is required for the molten salt. The simultaneous high-temperature and pressure requirements combined with the corrosivity of molten salts will place extreme challenges on material selection and cost of the PHX for this application.

3.3.2 Heat Rejection Heat Exchanger

As with any closed-loop thermodynamic cycle, residual enthalpy needs to be rejected to the environment. As sCO_2 power cycles can operate in either condensing or fully-supercritical mode, the heat exchanger used to reject this heat can be called either a condenser or a cooler. To avoid confusion, we refer to this device simply as a Heat Rejection Heat Exchanger (HRHX), thus encompassing both applications.

It is important to note that the distinction between the two modes of operation is somewhat arbitrary, as the same device can function in either mode, provided proper attention is given to ensuring that phase separation is managed properly. Even in a fully supercritical mode of operation, the density ratio between the outlet and inlet states is around 3–4, while in condensing mode, the same ratio is approximately 5–6. Therefore, although two phases exist during condensing operation, the mass velocity found in typical HRHX applications (500–700 kg/s·m²) is high enough that the flow will largely follow relatively well-mixed flow patterns (Cheng et al. 2008). In practice, very little effect of transitioning between subcritical and supercritical operation has been observed experimentally (Held and Avadhanula 2017).

The specific design of the HRHX will depend on whether the coolant is air or water. For water-cooled applications, the high CO_2 pressure of even the low-pressure side of the system leads to selection of either a PCHE (Held 2014) or S&T (Fourspring and Nehrbauer 2015) heat exchanger design. The relative cost and size advantages of PCHEs are also clear in HRHX applications, but the propensity of fouling leads the manufacturer to recommend against their use with "open cooling water" sources, such as cooling towers. Thus, an intermediate fluid loop is necessary to provide a clean heat transfer fluid to the PCHE, negatively impacting cost and performance. The S&T heat exchanger can be more easily disassembled for maintenance, and therefore can use lower quality cooling water, but at the cost of larger size and cost.

The high density of CO_2 in the HRHX allows for a relatively compact air-cooled design, which follow commonly-used air-cooled gas cooler and condenser commercial practices. Typical designs feature several banks of parallel finned tubes, with air cross-flowing over the tube banks driven by large fans. Both flat and V-shaped configurations are commonly available. Air-cooled sCO_2 applications are particularly advantageous, as the combination of a dry working fluid and cooling technology allows for a completely water-free, low maintenance installation, with minimal cost or performance degradation compared to water-cooled installations (Held et al. 2016).

4. Materials

Material selection for sCO_2 power cycles is governed by strength and material compatibility and corrosion considerations. Due to the high pressures involved in the power cycle, systems should be constructed to applicable codes and standards, such as ASME B31.1, ASME B31.3 and/or Section VIII BPV. These codes provide the designer guidance on acceptable practices and material selection from the perspective of yield strength and creep. However, it is important to recognize that these codes generally do not address material compatibility issues such as corrosion resistance. An acceptable design must consider these issues as well—fortunately, a large body of work has been created over the past decade to understand common piping material compatibility with dry CO_2. Some recent studies of moderate to high-temperature material compatibility with pure CO_2 can be found in the following references: (Roman et al. 2013, Pint et al. 2016a), while the effects of impurities have been investigated in the following: (Mahaffey et al. 2016, Pint et al. 2016b).

In addition to metals used for pressure containment and flowpath definition, flexible materials are commonly used for seals. Carbon dioxide has a well-known propensity to damage these types of materials through a process known as "rapid gas decompression" (RGD) or "explosive decompression", where the fluid dissolves into the material at high pressure. When the pressure is released, the dissolved CO_2 outgasses rapidly, but due to the limited permeability of the material, this process results in physical damage to the material (Briscoe et al. 1994). The resistance of various materials to RGD damage is advertised by suppliers, but the actual performance of the material can vary substantially due to seal gland design, temperature and rate of depressurization. Currently, no universal guidelines for the successful design and material selection of a flexible seal exists; typical design processes are empirical in nature and application-specific.

5. Operation, Instrumentation and Controls

At present, several operational sCO_2 power cycle loops have been tested, ranging from 10–100 kW laboratory-scale systems to 8 MW systems.

Proper control of a power cycle requires the ability to maintain stability at steady-state conditions, to maneuver between different steady-state conditions—for instance, as power demand, heat source or heat sink conditions change—and to transition between stopped and started conditions in both normal and emergency situations. Depending on the application, the heat source conditions may or may not be under the control of the power cycle system. For instance, in a nuclear power plant, the reactor state can be controlled, but for a waste heat recovery application, the bottoming cycle will have no control over the heat source. In general, whether the cooling source is air or water, the temperature of the cooling source will be governed by the ambient state. And for non-base-load applications, the power cycle will need to be able to deliver a variable output in response to a demand setpoint. If the plant is not connected to a larger electrical grid, it may also be responsible for maintaining frequency control as well.

Due to the broad variety of application-specific requirements, a single control strategy cannot be defined. However, several key points can be made regarding basic control methodologies.

Compressor operation: One of the key design decisions in the overall cycle layout process is whether the compressor(s) will be driven by a constant-speed turbine (e.g., the "power" turbine, which also drives a synchronous generator), a variable-speed turbine (either by a power turbine driving a non-synchronous generator or an independent compressor-drive turbine), or a fully independent drive (e.g., a variable-speed electric motor). This decision has a major impact on the control methodology, as compressor speed can be a powerful control variable. If compressor speed cannot be varied, other compressor control methods can be employed, such as variable inlet guide vanes.

Compressor operability range is limited by surge at high pressure-rise, low flow conditions, and by choke at low pressure-rise, high flow conditions. Methods of maintaining compressor operating margin over variable operating conditions include variable geometry, speed, or throttling and/or flow bypass. All these strategies impact cycle performance to some degree, making their selection and employment part of the overall series of design compromises.

Turbine control: The power turbine creates shaft power, which drives the electrical generator (or a mechanical load in certain cases). For most larger-scale applications, the electrical generator will be connected to a large-scale electrical grid, which provides frequency (or shaft speed, for a synchronous generator) stability. However, even in this case, turbine speed control needs to be provided during the grid synchronization process. Once power is being generated, the system output will need to be controlled either for demand matching, to optimize the power cycle efficiency, or during the process of increasing power from an idle state to full power. Several options for turbine control include throttling, fluid bypass, variable turbine geometry, or variable inlet pressure. As with compressor control, selection of these options also impacts cycle performance.

Startup and shutdown: Starting a sCO_2 system from full rest conditions offers some unique challenges. First, the system rest state will depend on the ambient temperature of the system. For a subcritical ambient temperature (below 31°C), CO_2 condensation will occur within the system. Therefore, provision must be made to either preheat the system, or to manage the two-phase state of the system prior to starting.

For a normal shutdown, compressor speed can be reduced by either bypassing or throttling, and system pressure ratio allowed to drop. However, management of system residual heat should be done with care to avoid system over-pressurization. This can be especially challenging in cases where system power has been interrupted, and cooling capacity has been lost.

Emergency shutdown scenarios frequently require rapid shutoff of turbine power to avoid overspeed, particularly in cases where the electrical load is suddenly removed. The typical approach used in steam power plants is a fast-acting, close-coupled turbine stop valve. A similar approach is expected to be needed for sCO_2 power plants. The stop valve closing time requirement can be calculated through a simplified transient analysis, comparing the residual energy in the fluid system downstream of the valve to the acceleration rate of the power turbine/gearbox/generator assembly when load is removed. Although the small size of sCO$_2$ turbomachinery will reduce the polar moment of inertia of the assembly, the generator inertia dominates the assembly total inertia. Typical system calculations result in stop valve closing time requirements in the 0.1–0.5 second range, which is within the capability of commercial steam turbine stop valves. The transient response of the system following the rapid closing of the turbine stop valve, including the action of bypass valves, maintenance of compressor surge margin, and pressure equalization, requires system-level transient model evaluation.

6. Summary

Supercritical CO$_2$ power cycles have clear potential advantages in thermodynamic performance over steam- and organic fluid-based Rankine cycles. System architecture and heat source integration flexibility allow for a broad range of applications to be served. Initial commercial developments are directed toward heat recovery cycles, with current research and development activities focused on higher-temperature concentrating solar power, indirect-fired fossil and advanced nuclear applications that require higher turbine inlet temperature and the RCB cycle.

References

Allam, R., S. Martin, B. Forrest, J. Fetvedt, X. Lu, D. Freed et al. 2017. Demonstration of the Allam Cycle: An update on the development status of a high efficiency supercritical carbon dioxide power process employing full carbon capture Energy Procedia 114: 5948–5966.

Allam, R.J., J.E. Fetvedt, B.A. Forrest and D.A. Freed. 2014. The oxy-fuel, supercritical CO$_2$ Allam Cycle: New cycle developments to produce even lower-cost electricity from fossil fuels without atmospheric emissions. *In* ASME Turbo Expo 2014: Turbine Technical Conference and Exposition GT2014–6952. Düsseldorf, Germany.

Angelino, G. 1968. Carbon dioxide condensation cycles for power production. ASME J. Eng. Power 90: 287–296.

Baljé, O.E. 1962. A study on design criteria and matching of turbomachines: Part A— Similarity relations and design criteria of turbines. ASME. J. Eng. Power 84: 83–102.

Borgnakke, C. and R.E. Sonntag. 2014. Fundamentals of Thermodynamics. Wiley, New York.

Briscoe, B.J., T. Savvas and C.T. Kelly. 1994. Explosive decompression failure of rubbers: A review of the origins of pneumatic stress induced rupture in elastomers. Rubber Chem. Technol. 67: 384–416.

Cheng, L., G. Ribatski, J. Moreno Quibén and J.R. Thome. 2008. New prediction methods for CO$_2$ evaporation inside tubes: Part I—A two-phase flow pattern map and a flow pattern based phenomenological model for two-phase flow frictional pressure drops. Int. J. Heat Mass Transf. 51: 111–124.

Chordia, L. 2015. Thar Energy—Manufacturer of heat exchangers for sCO_2 power cycles. *In* University Turbine Systems Research Workshop. Atlanta, Georgia.

Combs, O.V.J. 1977. An investigation of the supercritical CO_2 cycle for shipboard application. M.S. Thesis, Massachusetts Institute of Technology, Cambridge, Massachusetts.

Department of Health and Human Services. 2007. NIOSH pocket guide to chemical hazards, DHHS (NIOSH) Publication No. 2005–149.

Dostal, V., M.J. Driscoll and P. Hejzlar. 2004. A supercritical carbon dioxide cycle for next generation nuclear reactors. Technical Report MIT-ANP-TR-100. Massachusetts Institute of Technology.

Feher, E.G. 1968. The supercritical thermodynamic power cycle. Energy Convers. 8: 85–90.

Fenghour, A., W.A. Wakeham and V. Vesovic. 1998. The viscosity of carbon dioxide. AIP Publishing. J. Phys. Chem. Ref. Data 27: 31–44.

Fourspring, P.M. and J.P. Nehrbauer. 2015. Performance testing of the 100 kW shell-and-tube heat exchanger using low-finned tubes with supercritical carbon dioxide on the shell side and water on the tube side. *In* ASME Turbo Expo 2015: Turbine Technical Conference and Exposition GT2015–42245. Montréal, Canada.

Harper, P., J. Wilday and M. Bilio. 2011. Assessment of the major hazard potential of carbon dioxide (CO_2). Heal. Saf. Exec.: 1–28.

Harvey, A.H., M.L. Huber, A. Laesecke, C.D. Muzny and R.A. Perkins. 2015. Progress toward new reference correlations for the transport properties of carbon dioxide. *In* The 4th International Symposium—Supercritical CO_2 Power Cycles. Pittsburgh, Pennsylvania.

Held, T.J. 2014. Initial test results of a megawatt-class supercritical CO_2 heat engine. 4th Int. Symp.—Supercrit. CO_2 Power Cycles. Pittsburgh, Pennsylvania.

Held, T.J. and V.K. Avadhanula. 2017. Printed circuit heat exchanger steady-state, off-design and transient performance modeling in a supercritical CO_2 power cycle. *In* Clearwater Clean Energy Conference.

Held, T.J., J. Miller and D.J. Buckmaster. 2016. A comparative study of heat rejection systems for sCO_2 power cycles. *In* 5th International Symposium—Supercritical CO_2 Power Cycles. San Antonio, Texas.

Incropera, F.P. and D.P. DeWitt. 2007. Fundamentals of Heat and Mass Transfer. John Wiley & Sons.

Kelly, K. 2017. Mezzo Technologies. <https://mezzotech.com/> (6 January 2017).

Kohsokabe, H., M. Koyama, K. Tojo, M. Matsunaga and S. Nakayama. 2008. Performance characteristics of scroll expander for CO_2 refrigeration cycles. *In* International Compressor Engineering Conference Paper 1847.

Lemmon, E.W., M.L. Huber and M.O. McLinden. 2013. NIST standard reference database 23: Reference fluid thermodynamic and transport properties—REFPROP, Version 9.1. National Institute of Standards and Technology, Gaithersburg, Maryland.

Mahaffey, J., D. Adam, M. Anderson and K. Sridharan. 2016. Effect of oxygen impurity on corrosion in supercritical CO_2 environments. *In* 5th International Symposium—Supercritical CO_2 Power Cycles. San Antonio, Texas.

Marion, J., F. Kluger, M. Sell and A. Skea. 2014. Advanced ultra-supercritical steam power plants. *In* POWER-GEN Asia. Kuala Lumpur, Malaysia.

Mehos, M., C. Turchi, J. Vidal, M. Wagner, Z. Ma, C. Ho, W. Kolb, C. andraka and A. Kurizenga. 2017. Concentrating solar power Gen3 demonstration roadmap. NREL/TP-5500–67464.

Miller, J.D., D.J. Buckmaster, K. Hart, T.J. Held, D. Thimsen, A. Maxson, J.N. Phillips and S. Hume. 2017. Comparison of supercritical CO_2 power cycles to steam Rankine cycles in coal-fired applications. *In* ASME Turbo Expo 2017: Turbomachinery Technical Conference and Exposition GT2017–64933. Charlotte, North Carolina.

Penkuhn, M. and G. Tsatsaronis. 2016. Exergy analysis of the Allam cycle. San Antonio, Texas.

Pint, B.A., R.G. Brese and J.R. Keiser. 2016a. Supercritical CO_2 compatibility of structural alloys at 400°–750°C. *In* 5th International Symposium—Supercritical CO_2 Power Cycles Paper No. 7747. San Antonio, Texas.

Pint, B.A., R.G. Brese and J.R. Keiser. 2016b. The Effect of O_2 and H_2O on oxidation in CO_2 at 700°–800°C. San Antonio, Texas.

Ridens, B.L. and K. Brun. 2014. High pressure thermophysical gas property testing, uncertainty analyses, and equation of state comparison for supercritical CO_2 compression applications. *In* 4th International Symposium—Supercritical CO_2 Power Cycles.

Roman, P.J., K. Sridharan, T.R. Allen, J.J. Jelinek, G. Cao and M. Anderson. 2013. Corrosion study of candidate alloys in high temperature, high pressure supercritical carbon dioxide for brayton cycle applications. *In* Corrosion 2013 Conference & Expo.

Shelton, W.W., N. Weiland, C. White, J. Plunkett and D. Gray. 2016. Oxy-coal-fired circulating fluid bed combustion with a commercial utility-size supercritical CO_2 power cycle. *In* 5th International Symposium—Supercritical CO_2 Power Cycles. San Antonio, Texas.

Southall, D. 2009. Diffusion bonding in compact heat exchangers. *In* $SCCO_2$ Power Cycle Symposium. Troy, New York.

Span, R. and Wagner, W. 1996. A new EOS for CO_2 covering the fluid region from the triple point temperature to 1100 K at pressures up to 800 MPa. J. Phys. Chem. Ref. Data 25: 1509–1596.

Vesovic, V., W.A. Wakeham, G.A. Olchowy, J.V. Sengers, J.T.R. Watson and J. Millat. 1990. The transport properties of carbon dioxide. J. Phys. Chem. Ref. Data 19: 763–808.

Wright, S.A., P.S. Pickard, R. Fuller, R.F. Radel and M.E. Vernon. 2009. Supercritical CO_2 compression loop operation and test results. *In* Supercritical CO_2 Power Cycle Symposium.

Yang, J., L. Zhang and H. Yuan Li. 2010. Development of a two-cylinder rolling piston CO_2 expander. *In* International Compressor Engineering Conference, Purdue University USA 2022.

CHAPTER 4

Nuclear Power

M. Cumo and *R. Gatto**

1. Introduction

The demonstration of the feasibility of producing nuclear energy in a controlled manner, which materialized with the Chicago pile built by the Italian physicist Enrico Fermi and his co-workers in December 1942, can certainly be considered one of the major scientific and technological achievements of last century, opening up the way to the pacific utilization of nuclear energy. This discovery has acquired even more value since the emergence of concerns about the negative effects of the anthropogenic release of CO_2 in the atmosphere, the source of which is mostly due to fossil fuel fired power plants. However, even though today there are about 450 nuclear power plants operating around the world, present-day public perception of nuclear energy is controversial: on the one hand, as an economical form of bulk energy that can offer a decisive relief from global warming, and, on the other hand, by a sense of distrust in the aftermath of the 2011 Fukushima-Daiichi accident. This chapter presents a comprehensive introduction to nuclear power generation and to the characteristics of the new generation of reactors currently under construction, focusing on their innovative aspects. The latter introduce substantial improvements on those problematics (safety, sustainability,

Sapienza University of Rome, Corso Vittorio Emanuele II 244, 00186 Rome, Italy.
Email: maurizio.cumo@uniroma1.it
* Corresponding author: renato.gatto@uniroma1.it

economics, and proliferation resistance) the satisfactory solution of which is required before Fermi's achievement could turn into a widely accepted form of safe and sustainable energy production.

2. Nuclear Power

Current nuclear power originates from the energy released in nuclear reactions in which a heavy nucleus, like ^{233}U, ^{235}U or ^{239}Pu, absorbs a neutron and splits, or "fissions", into two lighter nuclei, the fission fragment (FPs), each one with a mass of the order of half the mass of the original heavy nucleus, and with the simultaneous release of $\upsilon = 2$–3 additional neutrons. In a fission reaction, the rest-mass energy of the final products is smaller than the rest-mass energy of the initial reactants (exothermic reaction), and the final products emerge with a large amount of kinetic energy, in accord with Einstein energy-mass relation. In a power nuclear reactor, fission reactions must continuously take place in such a way to have steady-state energy production. For this to occur, one of the emerging fission neutrons must induce a new fission reaction, while the remaining fission neutrons are lost by parasitic absorption (e.g., in structural materials) or by leakage from the system. A reactor core is said to be "critical" when the fission chain reaction proceeds at a steady rate, and the power generated is constant. In term of the "multiplication factor" k, defined as the ratio of the number of fissions (or fission neutrons) in one neutron generation and the number of fissions (or fission neutrons) in the preceding generation, criticality is obtained when k = 1. A useful related quantity is the "reactivity" $\rho = (k-1)/k$, which denotes the reactor's relative departure from criticality. A positive (negative) deviation of the reactivity from zero leads to an increasing (decreasing) rate of fission reactions, and thus of reactor's power. Changes of reactivity are planned, such as when a variation of the power is desired or when reactivity must be introduced to make up for fuel burning, or unplanned as a consequence of abnormal conditions. Planned reactivity changes are usually performed by small variations of the degree of insertion of control rods in the reactor core (cylinders made of neutron absorber materials). An abrupt insertion of the entire system of control rods leads to a rapid ending of the chain reaction, as needed for example when potentially dangerous abnormal conditions suggest the immediate shut-down of the reactor's power.

In the Chicago Pile, the first nuclear assembly that under the leadership of Italian physicist Enrico Fermi in 1942 proved the feasibility of a chain reaction, effectively opening the way to civilian nuclear power, the nuclear fuel was natural uranium embedded in a graphite matrix, and control rods were made of cadmium, a very efficient neutron absorber. A typical fission reaction occurring in such a case is the following: $n + {}^{235}U \rightarrow {}^{147}La + {}^{87}Br + 2n$. The Q-value of the reaction, where Q is defined as the difference in the rest-mass energies between the reactants and the products of one fission reaction, is about 200 MeV ($\sim 3.2 \times 10^{-11}$ J). A simple order-of-magnitude calculation shows that the energy produced by the burning of one gram of ^{235}U is about 8.3×10^{10} J, three times the energy produced by one ton of coal. This realization motivated in the post second world-war period an intense physical and technological research activity, the result of which was the construction of the first electricity-generating nuclear powers plant in the mid-1950s. As of today, 30 Countries worldwide are operating about 450 nuclear reactors for electricity generation. Installed nuclear capacity is the highest that it has ever been at 397 gigawatts electrical. Twenty new reactors were connected to the grid in the last two years, and sixty power reactors are being built around the world (of which 20 in China, 7 in Russia and 5 in India) (IAEA Annual Report 2016).

Except for the core of the nuclear reactor, i.e., the region where the nuclear chain reaction takes place, a nuclear power plant for electricity generation is very similar to a conventional power plants using fossil fuel. The presence of the nuclear part, however, and the totally new kind of accidents that this presence could lead to, as well as the technologically very challenging environment in which the nuclear section of the plant operates (not last the presence of a high energy neutron flux), has led to the development of new sophisticated approaches to the assessment of the operational risks in power stations, and has raised to a new level the technological properties required by the materials employed in the construction of power plants.

Besides electricity generation, nuclear reactors are used today for a variety of other purposes. The nuclear industry provides the radioactive materials employed in medical diagnosis and the treatment of cancer and contrast agents (radio-pharmaceuticals). For example, iodine and barium are the most common types of contrast medium for enhancing X-ray-based imaging methods, while gadolinium is used in magnetic resonance imaging. All these radioisotopes are produced in nuclear reactors, either large plants whose principal goal is energy production, or smaller reactors used also for research activities. Beside from nuclear medicine, nuclear and isotope techniques are employed to improve health and safety of nutrition. Finally, high-temperature nuclear reactors could be utilized for thermal power production, as well as for water desalination and hydrogen production. These latter applications of nuclear energy are a key goal of next generation nuclear reactors.

3. Present-day Power Plants (Generation II)

Going from a critical system, i.e., a system that is capable of realizing a self-sustaining neutron fission chain reaction, like Fermi's pile, to a commercial power plant has revealed to be a challenging but not a particularly difficult task. After only 12 years from Fermi's achievement, the first nuclear power plant connected to the electrical grid was realized. In 2015, the world gross electrical energy production by nuclear source amounted to 10.6% (39.3% by coal, 22.9% by natural gas), increasing to 18% in OECD Countries (27.9% by coal, 27.7% by natural gas) (International Energy Agency 2017).

A basic distinction between types of nuclear reactors is the average neutron energy in the core. Reactors with a thermal neutron spectrum, in which the neutron population is in thermal equilibrium with the surrounding medium, with average temperature of fractions of eV, are referred to as "thermal" reactors (TRs), while a "fast" reactor (FR) denotes a system in which the average energy is increased to hundreds of eV. The occurrence of these two kind of reactors arises naturally in uranium fueled reactors due to the different nuclear properties of the two main isotopes of uranium, ^{235}U (0.72% in natural uranium) and ^{238}U (99.27%), and therefore the different role that they can play. Figure 1 shows the incident neutron kinetic energy dependence of the fission cross section[1] of the two isotopes (Krane 1988). An isotope like ^{235}U which has a large probability of fissioning at low energies is termed "fissile" (beside ^{235}U, other fissile isotopes employed in the nuclear industry are ^{233}U, ^{239}Pu and ^{241}Pu). Differently, due to the threshold-type cross section of ^{238}U, the fissioning of this isotope can occur only with very high energy neutrons. An isotope with this fission characteristic is termed "fissionable". Due to the very high value of the fission cross section of ^{235}U at low energy, it is natural to think about reaching criticality in a thermal reactor, as was the case with Fermi's pile. To accomplish this, fission neutrons which are born in the MeV range must be slowed down. This is typically done introducing in the core a moderating material, i.e., a light element material such as water or graphite. Due to the small quantity of ^{235}U in nature, uranium fuel for thermal reactor must be artificially enriched, even though a small percentage of the order of 2–5% is sufficient to reach criticality (LEU: low enriched uranium)[2] in a well-designed nuclear reactor core. The fission cross section of ^{238}U is practically zero below 1 MeV, and doesn't rise to very high values for higher incident neutron energies. Even though ^{238}U is

---

[1] The nuclear cross section of a nucleus quantifies the probability that a nuclear reaction will occur. The concept of nuclear cross section can be interpreted physically in terms of "effective area", where a larger effective area means a larger probability of interaction. The standard unit for measuring a nuclear cross section (conventionally denoted by σ) is the barn, which is equal to 10–24 cm².

[2] Nuclear fuel in thermal reactors is usually in the form of uranium oxide UO_2 (UOx).

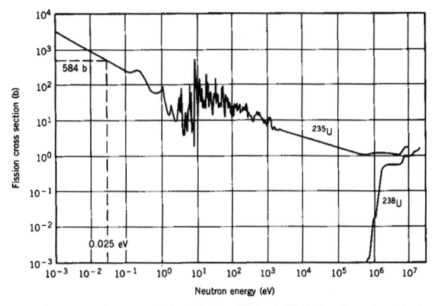

Fig. 1: Cross section for neutron-induced fission of ^{235}U and ^{238}U (Both scales are logarithmic. 1 barn = 10^{-24} cm^2). The energy of fission neutrons is in the high energy range (~ 2 MeV on average).

present in large quantities in natural uranium, the behavior of its fission cross section would suggest no clear advantage in trying to operate a fast reactor. However, ^{238}U has a very important property, that is, a significant probability to absorb low energy neutrons so to initiate a chain of reactions and decays that ultimately leads to the formation of ^{239}Pu, a fissile isotope: $^{238}_{92}$U(n,γ)$^{239}_{92}$U → $^{239}_{93}$Np → $^{239}_{94}$Pu, where arrows indicate β$^-$ decays.

The process of creating a fissile isotope (in this case, ^{239}Pu) from a fertile isotope (in this case, ^{238}U) is called "conversion" (or "breeding" if the mass of fissile material produced by fertile neutron captures is higher than the mass of fissile material consumed). The ability to breed a fissile isotope, which then contributes to the sustainment of the chain reaction in the same way as ^{235}U does, is a very important property of ^{238}U, which is termed for this reason a "fertile" isotope.[3] And it is this very property that has led to the design of uranium-fueled fast reactors, in which the sustainment of the chain reaction is due to fission reactions of ^{235}U and of the bred ^{239}Pu. In Fig. 2 (Nuclear Power) we present typical neutron spectra in the core of a thermal and a fast reactor, showing how the peak

---

[3] This nuclear sequence is the backbone of the uranium-plutonium (U-Pu) fuel cycle. An alternative cycle involves the other important fertile isotope, ^{232}Th, and the ^{232}Th-^{233}U cycle that originates from it. We will discuss the Th-U cycle later on when describing high-temperature gas reactors, since the latter are very well suited to employ such a fuel cycle.

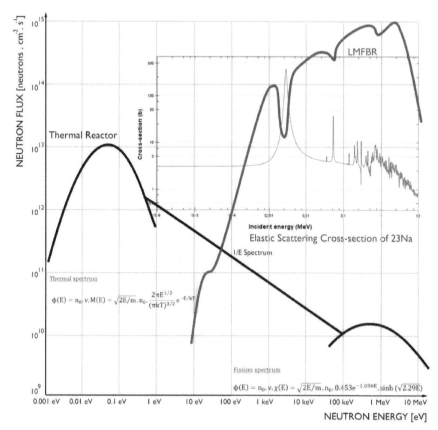

Fig. 2: Neutron flux in the core of a thermal reactor (blue) and liquid metal fast breeder reactor (red). The dips in the flux of fast reactors are due to peaks in the neutron absorption cross section of the liquid sodium used as cooling vector. Analytical models of the flux dependence on energy are also reported.

of neutron flux (neutrons crossing the unit surface in the unit time) occurs at low and high energies, respectively. The fresh fuel in fast reactors is either made of highly enriched uranium (HEU, with enrichment > 20%), so that the large quantity of fissile isotopes compensates for the low cross section at high energy, or a mixture of LEU (or even natural uranium) and plutonium,[4] taking advantage of the larger average number of fission neutrons produced per neutron absorbed when ^{239}Pu fissions (~ 3 for ^{239}Pu versus ~ 2.4 for ^{235}U for 1 MeV incident neutrons).

---

[4] Nuclear fuel for fast reactors is usually in the form of mixed oxide (U,Pu)O$_2$ (said MOx), with a content of PuO$_2$ that varies from 1.5 wt.% to 25–30 wt. %, or in the form of metallic fuel, such as uranium-aluminum or uranium-zirconium. The plutonium used to fabricate fresh fuel elements derives from used fuel (which contains bred plutonium), or from Pu-rich fuel of disposed nuclear weapons.

Two main advantages are related to fast reactors. First, there is no need to include the core moderator material since fission neutrons do not have to be slowed down. This reduces in principle the size and the complexity of the core. Second, the surplus of neutrons from the fission of ^{239}Pu favors the self-generation of fissile material while in operation and therefore leads to a more efficient utilization of uranium resources. Moreover, following optimized fuel cycle strategies, the fast neutron spectrum is more apt at minimizing the radio-toxicity of spent fuel, in particular by reducing the amount of highly radio-toxic minor actinide elements by transmuting them into less toxic elements. The technological challenges associated with fast reactors are however higher than those confronting thermal reactors (briefly: higher power densities and higher neutron fluxes, and thus higher thermo-mechanical stresses of structural materials; greater nuclear proliferation and security issues; more costly to build and operate), and this is the reason why, as of today essentially all commercial nuclear power plants worldwide are thermal reactors, while only a few experimental/prototype fast reactors have been built. The contribution from commercial fast reactors should however become important, and eventually predominant, starting from next generation power plants, principally in the name of nuclear energy sustainability.

Today, the most widely spread thermal nuclear power reactor is the pressurized water reactor (PWR), covering over 60% of the installed nuclear capacity (generation II (in short, Gen II) reactors), and the larger part of the reactor under construction (generation III/III+ (in short, Gen III/III+) reactors). We will give therefore a rather detailed description of characteristics of a nuclear power plant referring to the Gen II PWR concept, a "loop-type" reactor (the primary circuit is a loop, as opposed to a pool as in many fast reactor designs) in which pressurized water acts both as moderator and coolant medium. Further, we provide more succinct descriptions of the other kinds of nuclear reactors that are currently in operation.

Pressurized Water Reactor

In Fig. 3 (US Nuclear Regulatory Commission) we present a simplified schematic of a typical pressurized water reactor (PWR) plant. In describing its main features, we will refer in particular to the Westinghouse PWR concept (Westinghouse Electric Corporation, Water Reactor Division 1984).

The "nuclear" part of the plant is shown on the left-hand side and is enclosed by the reactor building, housing the reactor vessel confining the reactor core, where the chain fission reaction takes place, and the other components of the primary loop (pressurizer, steam generator, and coolant pumps). The remaining part is "conventional", with devices found in any fossil fuel power plant: the turbine building contains the turbine-generator system constituting the secondary loop, and a tertiary

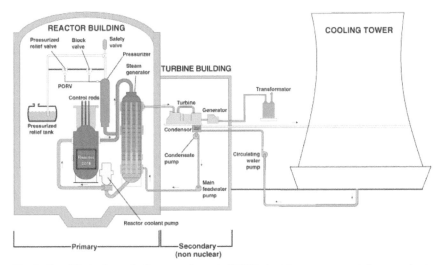

Fig. 3: Simplified schematic diagram of a typical PWR plant. Large commercial power plants have up to four primary loops, each one comprised of a steam generator, a reactor coolant pump, and interconnecting piping.

(or auxiliary) loop comprises the cooling tower required to disperse the disposal heat to the environment.[5] The basic fuel element is the fuel pellet, a small cylinder of uranium dioxide (UOx) powder with a diameter of 0.78 cm and a length of few cm, shown in Fig. 4 (US Nuclear Regulatory Commission, Mitsubishi Nuclear Fuel). These pellets are then inserted one on top of the other inside the fuel rod made of Zircaloy-4[6] and with a length of 365 cm. Pellets are treated so to be able to retain fission products and to resist deterioration caused by high temperature water in case of an accidental breach of the cladding. A square array of 17 × 17 fuel rods structurally bound together constitutes a fuel assembly. Control rod guide thimbles replace fuel rods at selected spaces in the array. A large number of fuel assemblies (193) are then assembled inside a core baffle to form the reactor core. The pressure vessel of a PWR, shown on the left side of Fig. 5 (U.S. Energy Information Agency), is a carbon steel cylindrical container with all wet surfaces made of stainless steel to limit corrosion. The right side of the figure (Westinghouse Electric Corporation, Water Reactor Division 1984) shows a cross-section of a typical four-primary-loop core.

The power of the reactor, proportional to the neutron flux in the core, is regulated by a cluster of control rods in the fuel assemblies, a neutron absorber (boric acid) dissolved in the reactor coolant, and proper

---

[5] Alternatively, the residual heat can be dispersed using a heat sink such as a river, a lake or the sea.

[6] Zircaloy is used because it absorbs relatively few neutrons and has good heat transfer properties.

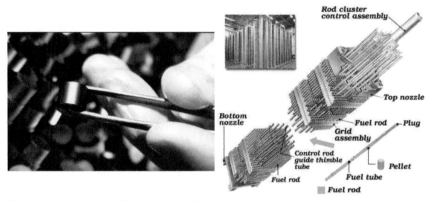

Fig. 4: Left: a fuel pellet. Right: a fuel rod and a fuel assembly.

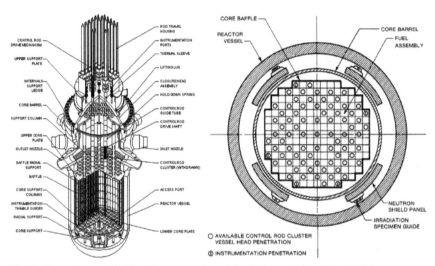

Fig. 5: Pressure vessel (left) and cross-section of the fuel assembly region (right).

allowance for physical phenomena which influence the core neutron balance.[7] The heat generated in the reactor core, originating from the slowing down of the fission products (FPs) inside the fuel, is removed by the flow of pressurized, under-cooled water in the primary loop. After increasing its enthalpy without reaching a full boiling state, the water enters the containment side of the steam generator with a few degrees of

---

[7] Various physical phenomena can lead to positive or negative insertion of reactivity. For example, the reactivity associated with an incidental increase of the void fraction in the moderating/cooling water (very small in normal operating condition) is negative. Physically, more void in the water leads to an hardening of the neutron spectrum, and therefore to an increase in neutron leakage and a decrease in fission reactions. The effect of void therefore acts against power increase and contributes to the reactor stability.

sub-cooling. The task of maintaining the pressure above the saturation pressure so that bulk boiling does not occur is taken by the pressurizer connected to one of the coolant hot legs. The need to have steam with features not too distant compared to those of modern conventional thermal plants, namely steam at ~ 70 bar, dry saturated at ~ 285°C, compared to, e.g., steam at 180 bar overheated to 540°C of modern fossil fuel plants, requires, given the inevitable thermal degradation due to heat transfer, to operate in the primary loop at ~ 150 bar. The heat acquired from the core is then transferred to the secondary loop (water-steam), undergoing a cooling of the order of 30°C. Besides the steam generator and the turbine-alternator system, the secondary loop includes a condenser, feed pumps, pipes and regulating organs, and additional components having the goal of improving turbine operation and the efficiency of the thermodynamic cycle (e.g., superheaters, condensate separators, regenerators). The tertiary loop is the heat rejection loop where the latent heat of vaporization is rejected to the environment through the condenser cooling water. Depending on the specific site, this heat is released directly to a river, a lake, a sea, or by means of cooling tower systems.

The single reactor containment building of a PWR is made of a continuous reinforced concrete structure with a steel liner covering its inner surface in order to ensure water tightness. Its principal function is that of containing the mass and energy of the reactor coolant in the postulated event of a rupture in the reactor coolant piping (i.e., an assumed loss-of-coolant accident, or LOCA,[8] which could lead to peak internal pressures up to 4–5 bar). The design of the reactor building is made taking into account normal loads, overloads and stresses that can arise as consequences of more severe conditions. In particular, critical entries for the dimensioning of the structure are the stresses induced by seismic events and by incidental sequences. Note that in the PWR concept the use of a steam generator to separate the primary loop from a secondary loop largely confines the radioactive material to a single reactor building during normal power operation.

Safety systems designed to protect the plant from internal origin events are designed bearing in mind the principle of separation and redundancy in the implementation logic, power supplies and components cooling loops. All modern PWRs may be cooled, in emergency conditions, even with the "feed and bleed" method, using the emergency core cooling systems (ECCS) which inject water into the core and download an equal flow rate by the pressurizer relief valves. A simplified scheme of the ECCS is shown

---

[8] Loss of Coolant Accident (LOCA) is defined as an accident in which reactor coolant pressure boundary breaks to freely discharge reactor coolant. LOCA which is caused by a large break in the primary coolant system is a design basis accident (DBA) for PWRs.

in Fig. 6 (US Nuclear Regulatory Commission). Besides providing core cooling to minimize fuel damage following a LOCA, ECCS deliver extra neutron poisons to ensure that the reactor remains shutdown following the cool-down associated with a main steam line rupture (US Nuclear Regulatory Commission). In the Westinghouse PWR the ECCS consist of four separate systems. The first system is the high pressure injection system, which takes borated water from the refueling water storage tank (RWST) and pumps it into the reactor coolant system, during emergencies in which the reactor coolant system pressure remains relatively high (such as small break in the reactor coolant system). Similarly, the intermediate system and the low pressure injection system (also called the residual heat removal (RHR) system) inject water from the RWST to the reactor coolant system in case of emergencies in which the pressure of the primary remains moderately high (intermediate system), or drops to very low values, such as in the case of large primary breaks (RHR system). In addition, the RHR system has a feature that allows it to take water from the containment sump, pump it through an auxiliary heat exchanger for cooling, and then send the cooled water back to the reactor for core cooling. This is the method of cooling that is employed when the RWST goes empty after a large primary system break. This cooling system is thus designed for long

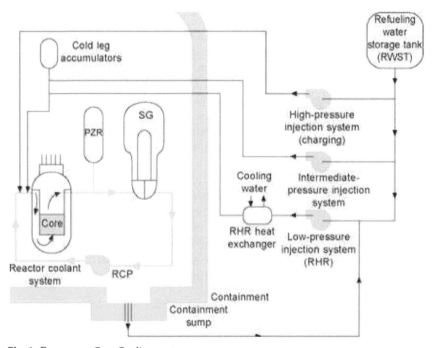

Fig. 6: Emergency Core Cooling systems.

term cooling. The last system of the ECCS is the cold leg accumulators, i.e., tanks containing large amounts of borated water with pressurized nitrogen gas in the top. If the pressure of the primary system drops low enough, the nitrogen will force the borated water out of the tank into the reactor coolant system. These tanks are designed to provide water to the reactor coolant system during emergencies in which the pressure of the primary drops very rapidly, such as large primary breaks.

Besides core overheating, accidents can lead to a dangerous increase of pressure and temperature inside the containment building. In fact, upon the occurrence of either a primary or secondary break inside the containment building, the containment atmosphere would become filled with steam. To cope with such a situation, two systems have been designed: the fan cooler system, which circulates the air through heat exchangers to accomplish the cooling, and the containment spray system. This reduces pressure and temperature of the building by taking water from the RWST and pumping it into spray rings located in the upper part of the containment. The water droplets, being colder than steam, will remove heat from the steam, which will cause the steam to condense. This will cause a reduction in pressure of the building and will also reduce the temperature of the containment atmosphere. Like the RHR system, the containment spray system has the capability to take water from the containment sump if the RWST goes empty.

It is important to note that Gen II power plants have adopted safety systems that are mostly "active", meaning that they need external power (either form the normal electrical grid, or emergency power suppliers such as diesel generators or electrical batteries) in order to operate. One of the main evolutionary feature adopted in many Gen III/III+ power plants (those presently under construction) with respect to safety consists in adopting passive safety systems, i.e., systems the activation of which relies on the laws of physics, with no need of external power (e.g., natural circulation of coolant). An example of this rather substantial evolution is represented by the AP1000 reactor, described later on.

We now provide very succinct overviews of the other three thermal reactor concepts presently in operation worldwide—the boiling-water reactor and the pressurized heavy-water reactor, using respectively light and heavy water, and the high-temperature gas-cooled reactor, using either helium or carbon dioxide as coolant.

Boiling Water Reactor

Both PWRs and boiling water reactors (BWRs) are thermal light-water reactors (LWRs) which use enriched uranium as fuel and water as both coolant and neutron moderator. The major difference between these two type of reactors is that while in a PWR the primary circuit is under high pressure and steam is produced in the steam generator of a secondary circuit, in a BWR the water boils while passing through the reactor core, producing steam at the reactor vessel output line. While the direct production of steam simplifies the overall plant scheme due to the lack of a secondary circuit, it introduces the additional problem of induced radioactivity of the turbine due to the activated steam. A typical BWR pressure vessel is shown in Fig. 7.

The steam separators required to separate water from steam before sending the latter to the turbine are located in the upper reactor shell. As a consequence, control rods are normally inserted from the bottom. Thanks to the lack of a separate steam generator, the overall thermal efficiency of a BWR is greater than that of a PWR. The design of the reactor vessel itself is however much more complex for a BWR because of the presence of a large amount of steam inside it. The design of the containment building of a BWR needs to consider the possibility of an accidental sequence that releases inside it a large amount of high temperature steam. Usually this requires suppression pool water that can absorb the energy released and avoid the build-up of high pressure inside the primary containment. As it is not possible to introduce neutron poison in the primary loop due to the changing state of the coolant, reactor power is controlled by the movement

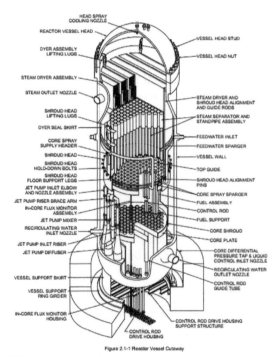

Figure 2.1-1 Reactor Vessel Cutaway

Fig. 7: Typical BWR pressure vessel.

of control rods and by changing the flow of water through the core. If more water is pumped through the core and more steam is generated, more power is produced.

Heavy Water Reactor

Pressurized heavy water reactors (PHWRs) use heavy water as moderator, and possibly as primary coolant. Since heavy water has a parasite neutron capture cross section that is 1/600 of that of ordinary light water, the core of a PHWR can be designed to reach criticality using natural uranium UOx. This is obviously a major advantage for countries that do not have their own enrichment facilities. Alternatively, the plutonium produced during irradiation can be recycled, as done in LWRs, reprocessing the spent fuel without need of any enrichment plant, and building fuel elements as MOx. The favorable neutron economy of these reactors allows the obtainment of average conversion factors of the order of 0.7–0.8, against the 0.5–0.6 of LWRs. For this reason the heavy water reactor (HWR) is referred to as "advanced converter": aside from the net production of plutonium, a considerable amount of energy (about half) is obtained by burning *"in situ"*. Different types of PHWRs have been built around the world since the 1960s, but the only type that has reached an industrial maturity and that is still interesting and development worthy is the heavy water moderated and heavy water cooled reactor (PHWHW, or CANDU (CANada Deuterium Uranium) reactor).[9] A simplified scheme of a CANDU reactor is shown in Fig. 8 (International Atomic Energy Agency).

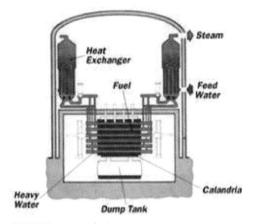

Fig. 8: PHWHW, or CANDU, reactor design.

---

[9] The heavy water moderated reactor has been developed mainly in Canada. Presently, there are about 20 CANDU reactors operating in Canada. Few other units are located in Argentina, China, India, Pakistan, Romania and South Korea.

A primary cooling loop, with pressurized D_2O (~ 100 bar), is coupled to a secondary loop, with steam generator and turbine (similarly to a PWR). Heavy water temperatures at the inlet and outlet of the reactor are around 250 and 300°C, and this low maximum temperature limits the thermodynamic efficiency (~ 30%). The moderator is D_2O at atmospheric pressure and low temperature. An original solution adopted in CANDU is the replacement of the conventional pressure vessel with pressure tubes containing the fuel elements, with the consequence of simplifying the procedures and reducing costs. The container of the moderator (calandria) is not pressurized and can be prefabricated. A very large quantity of heavy water needs to be immobilized, for a commercial power plant, of the order of hundreds of tons. Heavy water has a very high unit cost, and for this reason the plant cost of a HWR tends to be higher than the plant cost of a LWR. Due to the large operating experience of HWRs, the development of this reactor type is only a refinement, as its performance and reliability have already been amply demonstrated. In later advanced models developed for the Canadian industry, denoted by the initials ACR700 and ACR1000, the use of heavy water is limited to only moderate (reducing its amount and therefore the cost), using light water as the heat transfer fluid.

High Temperature Gas Reactor

High temperature gas-cooled reactors (HTGRs) are thermal reactors which have achieved a very promising development stage, are beginning to acquire commercial space, and represent one of the sectors with prospects for the future. They are all characterized by ceramic material cores (e.g., graphite as the only structural material) and gas helium as coolant, both for the excellent chemical compatibility with structural materials and for good thermal conductivity. Thanks to these favorable properties of helium and its gas state, a HTGR reactor is made of only a primary loop, with consequent simplification of the overall plant design. The absence of parasitic absorbers in the core such as steels, allows the use of almost every possible combination of the three fissile isotopes (^{233}U, ^{235}U, ^{239}Pu) and two fertile isotopes (^{232}Th, ^{238}U), in the form of oxides or carbides. Experimental reactors have shown the ability to achieve helium temperatures of 850–1000°C, with consequent high cycle efficiency (over 40%), opening the way to a whole host of interesting prospects (hydrogen production, coal gasification, water desalination, and in general heat for industrial processes).

The possibility to achieve very high temperature, thanks to the use of ceramic materials and a non-corrosive coolant medium, has led to a unique concept for the fuel, which is characterized by microspheres with a diameter of the order of one millimeter, consisting of an inner core of fuel (dioxides or carbides) covered with concentric layers of pyrolytic carbon and silicon carbide. The microspheres can be either embedded in a spherical graphite matrix, forming spherical fuel elements with diameter of about 6 cm (pebbles), or compacted to form cylindrical fuel elements, which in turn are used to form fuel assemblies resembling those of LWRs, and then to form a prismatic core. Initially, it was thought a reactor concept that fully transferred the volatile FPs into the coolant of the primary loop, so you can reach, with a suitable continuous purification of primary coolant, extremely high burn-up rates.[10] It was found, in spite of everything, a good degree of FP retention from fuel spherules (initially without any protective layers, then greatly improved), while reaching very high burn-up, which reversed the project philosophy, focusing on high confinement of fission products (as in the other reactor concepts).

The feasibility of HTGRs was first demonstrated by various experimental/prototype HTGR plants, such as Dragon (operation period 1964–1975, UK), Peach Bottom (1966–1974, USA), AVR (1967–1988, FRG), THTR (1985–1991, FRG), Fort St. Vrain (1976–1989, USA), and is being advanced in concepts such as the HTR-PM (evolution of HTR-10, China) and NGNP (USA). Two HTGRs are currently operating in the world, both using uranium fuel: the HTTR (Japan, prismatic core, first criticality 1999), the first reactor to demonstrate continuous H_2 production by sulfur/iodine process, and the HTR-10 (China, pebble bed core, first criticality 2000), which has demonstrated an inherent safety performance with electricity production and co-generation at a power level of 10 MWt.

For illustrative purpose, we analyze more in detail the core of AVR, the first pebble bed HTR worldwide, operated in Jülich (Germany) during the period 1967–1988. The analysis of this system gives us the opportunity to present the mechanism of operation of the pebble bed core, a concept radically different from conventional LWR cores, and the rationale for using thorium as nuclear fuel in a thermal reactor, an option alternative to uranium, that is being intensively studied, especially by those countries (e.g., India)[11] with abundant thorium reserves and little uranium availability. In Fig. 9 we show a sketch of the AVR (Moormann 2009), and a picture of the pebble bed core (Thorium High Temperature Reactor).

The graphite-moderated AVR core, generating 46 MWt/15 MWe, consists of about 100,000 matrix graphite pebbles located in a cylindrical vessel made of graphite. The vessel has a diameter of 3 m and a fuel element packing height of 2.8 m, and serves also as a neutron reflector. The helium heated in the pebble bed (pressure 10 bar) is delivered by two cooling gas blowers through channels in the reflector head to the steam generator

---

[10] Burn-up, or fuel utilization, is a measure of how much energy is extracted from a primary nuclear fuel source. One way to quantify it is by giving the actual energy released per mass of initial fuel in gigawatt-days per metric ton of heavy metal, GWd/tHM.

[11] India has only around 1–2% of the global uranium reserves, but one of the largest shares of global thorium reserves at about 25% of the world's known reserves.

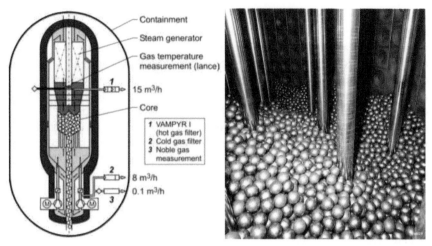

Fig. 9: AVR scheme (left) and core interior (right).

installed above the reactor core (Theenhaus and Storch 1990). Pebbles are inserted in the core from the top, residing in it for an average period of 6–8 months, before reaching the defueling tube at the core bottom and being re-fed to the core top. This cycle is repeated until the desired burn-up is met and then the fuel element is replaced by a fresh one. The continuously moving active core is a unique feature of pebble bed reactors, and allows minimizing the amount of reactivity stored in the core. In certain phases of the AVR life the outlet helium temperature reached 950°C. Since these high temperatures are suitable for process heat applications, the pebble bed technology finds an interest worldwide.

Most of the fuel pebbles used in the course of AVR operation contained 1 gr of ^{235}U, but the reactor experimented also with thorium, and with $(Th,U)O_2$ fuel elements with a larger heavy metal content (from 6 to 11 g). Indeed, one of the missions of the AVR was experimenting with the Th-U cycle, as an alternative to the conventional U-Pu cycle. Thorium in nature is constituted mainly by the fertile isotope ^{232}Th (99.98%), with an alpha-decay half-life of 1.4×10^{10} years, and therefore is still present in large quantities in the Earth's crust. When this isotope is exposed to a neutron flux, radiative neutron capture leads to a chain of beta decays that eventually terminate with the production of the fissile isotope ^{233}U: $^{232}_{90}Th(n,\gamma)^{233}_{90}Th \rightarrow ^{233}_{91}Pa \rightarrow ^{233}_{92}U$. There is however an important difference between the Th-U cycle and the U-Pu cycle. In order to start the Th-U cycles, a source of neutrons is required, meaning that there must be also an initial amount of fissile isotope in the fuel element (the "driver" fuel: uranium enriched in ^{235}U, ^{233}U, or ^{239}Pu). Theoretically, while the fresh fuel elements must be enriched in the driver fuel, it is conceivable that after enough ^{233}U has been produced, the replacing fuel element could be enriched with the self-produced ^{233}U, therefore achieving a "closed cycle".[12]

The reason for experimentation with thorium is that this material is 3 to 4 times more abundant in nature compared to uranium and is widely distributed in nature as an easily exploitable resource in many Countries. Other reasons to fuel the interest in the Th-U cycle are: the intrinsic proliferation resistance of thorium fuel cycle due to the presence of ^{232}U and its strong gamma emitting daughter products; better thermophysical properties (higher thermal conductivity and lower coefficient of thermal expansion) and better chemical stability of ThO_2 compared with UO_2, which ensures better in-pile performance and a more stable waste form; less plutonium and long lived MA production compared to the traditional uranium fuel cycle; superior plutonium incineration in (Th,U) O_2 fuel as compared to $(U,Pu)O_2$.[13] However, there are several challenges in the front and back end of the thorium fuel cycle. The melting point of ThO_2 (3350°C) is much higher compared to that of UO_2 (2800°C). Hence, a much higher sintering temperature is required to produce high density ThO_2 and ThO_2-based MOx fuel. Irradiated ThO_2 and spent ThO_2-based fuels are relatively inert and difficult to dissolve in HNO_3 because of the inertness of RhO_2. The high gamma radiation associated with the short lived daughter products of ^{232}U (in particular, ^{208}Ti, ^{212}Bi), which is always associated with ^{233}U, necessitates remote reprocessing and re-fabrication of fuel.[14] In the back end of the thorium cycle, there are other radionuclides such as ^{231}Pa, ^{229}Th and ^{230}U which may have long term radiological impact, a problem which needs to be suitably resolved.

Fast Reactor

The characterizing difference between thermal and fast reactors (FRs) is the lack of neutron moderator in the latter. As a consequence, the average neutron kinetic energy rises from the electronvolt in thermal cores to hundreds of eV, an energy region where the fission cross section of ^{235}U and ^{239}Pu is much reduced (the energy dependence of the fission cross section of these two fissile isotopes is very similar). Restricting ourselves to the cycle U-Pu, who has found widespread achievement in fast breeder reactors (FBRs),[15] and

---

[12] This equilibrium thorium fuel cycle can obviously be operated only after a sufficient amount of ^{233}U has been built-up. For practical systems this timescale is very long (especially in a thermal spectrum), and this has to be taken into account when considering a possible transition to a thorium fuel cycle.

[13] Plutonium is not bred in the once-through thorium cycle (contrary to what happens in the U-Pu cycle).

[14] ^{232}U forms via (n,2n) reactions with ^{232}Th, ^{233}Pa and ^{233}U. The half-life of ^{232}U is only 73.6 years and daughter products have a very short half-life.

[15] So-called because the high neutron energy favors the U-Pu sequence and thus conversion/breeding.

currently established technology, we can briefly say that the central region of the core (the "seed", where most of fissions are taking place) is formed from U and Pu dioxide with high enrichment (in initial loads it employs ^{235}U or ^{239}Pu obtained from other reactors, e.g., thermal). Order of enrichments of 30–40% is possible, according to the particular core optimization project. During operation, the hard spectrum favors neutron capture in ^{238}U (in depleted UO_2 fuel elements usually located in the outer region of the core, the "mantle" or "blanket"), with formation of ^{239}Pu. The conversion ratio that can be achieved in a FBR is of the order of 1.2. Moreover, the high energy spectrum transmutes MAs at a higher rate than in a thermal reactor. These favorable neutronic characteristics of fast reactors explain the predominant role that FBRs occupy in the research for the next generation of nuclear reactors. By producing more fissile than they consume, FBRs aim at increasing a hundred times the energy obtainable from uranium with multiple recycles.

Because of their high technological requirements, much fewer fast reactors have being built and operated so far compared to thermal reactors and most of them with an experimental or prototype status. As an example of a fast reactor, we describe the French reactor Phénix, a prototype sodium-cooled reactor with nominal power of 560 MWt/250 MWe, which operated commercially in the period 1974–2009. The schematic of the plant circuit of the reactor, presented in Fig. 10 (Aoto et al. 2014), shows the "pool-type" (or integrated) design of the vessel, a novelty with respect to the loop-type design adopted by all thermal reactors (and few fast reactors as well).

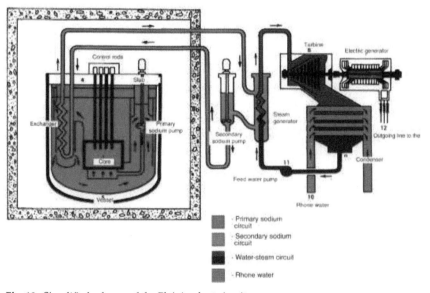

Fig. 10: Simplified scheme of the Phénix plant circuits.

The main primary circuit, with its three pumps and six intermediate heat exchangers (IHXs) arranged around the core, is contained completely within a stainless steel vessel filled with 800 tons of sodium. The core, located in the lower region of the primary reactor vessel, is cooled by the flowing of the sodium driven by the primary pumps. The sodium exits the core (mean inlet temperature of 360°C and mean outlet temperature of 435°C for the 120 MWt initial state), and transfers its heat in the shell side of the IHXs to the sodium circulating in the three secondary loops (each with 2 IHXs). Finally, three tertiary loops with water/steam as a working fluid generate electricity.

The adoption of the pool-type layout containing the entire primary loop leads to several advantages, the main one being the long thermal response time associated with a higher thermal inertia of sodium. In case of a drop in the rate of circulation of primary sodium in the core, the temperature of coolant due to the ensuing transient will increase slowly due to higher heat capacity of sodium. Another advantage associated with the pool-type configuration is the reduction in the amount of external piping, and the elimination of LOCA due to a leak in the primary loop, a feared occurrence in loop-type reactors.

The long period of Phénix operations (and of the other few experimental fast reactors around the world) have brought significant contributions to the development of fast reactors, demonstrating the viability of SFRs, while conducting a wide range of irradiation experiments. Among the most significant achievement there is the reaching of a burn-up exceeding 17% HA in experimental pins, and a measured conversion ratio of 1.16 (breeding). The U-Pu fuel cycle, based on mixed oxide fuel and PUREX[16] reprocessing, has been closed and the first fuel subassembly made with reprocessed plutonium was loaded in the reactor in January 1980. In total, five of Phénix cores (~ 25 tons) were reprocessed (51 cycles). The successful and regular operation at the highest temperatures and nominal power resulted in validation of the pool concept option and gained much knowledge regarding the high temperature design and structural material of fast reactors. During periods of stabilized operations with nominal parameters the plant reached a gross/net plant's thermal efficiency of 45.3/42.3%.

Various experimental campaigns have been carried out also on the study of MAs transmutation. The first experiment, called SUPERFACT, led to the incineration of MAs (neptunium and americium). Subsequently, this programme was further strengthened involving transmutation of MAs and long-lived FPs.

---

[16] Essentially all nuclear fuel recycling is performed using the PUREX (plutonium uranium redox extraction) process, based on liquid extraction ion-exchange, which was initially developed for extracting pure plutonium for nuclear weapons.

4. Nuclear Fuel and Nuclear Cycle

Contrary to conventional fossil-fueled power plants, in which the fuel (coal, or natural gas) is essentially unchanged from plant to plant, and what differentiates a plant design is the way the steam is utilized for electricity production, for a nuclear power plant the kind of nuclear fuel, and the way it is utilized, i.e., the adopted fuel cycle, is a major defining characteristic. In fact, one of the most important innovation introduced by next generation reactors resides in the adoption of new strategies of fuel utilization and reprocessing, i.e., in advanced fuel cycles.

In general, the isotopes that are relevant for nuclear power production are the three fissile isotopes ^{233}U, ^{235}U and ^{239}Pu, and the two fertile isotopes ^{232}Th and ^{238}U. Most present-day LWRs (boiling and pressurized type) utilize UOx enriched in ^{235}U, possibly containing also ^{239}Pu in MOx fuel. These two fissile isotopes have very large fission cross sections at low neutron energies.

During normal operation, the fraction of fissile isotope in the fuel elements progressively diminishes, down to a point in which these elements need to be extracted and replaced with fresh ones with the required enrichment. Used, or "spent", nuclear fuel (UNF or SNF) element differs from fresh one not only in the quantity of original fissile isotope present, which is obviously reduced with respect to the initial enrichment, but also on the presence of FPs accumulated during operation. Moreover, neutron capture on ^{238}U (higher in the low and intermediate energy range) leads eventually to the production of isotopes of plutonium, as well as other transuranics (TRUs, i.e., elements heavier than uranium), in particular Am and Cu.

The presence of radioactive FPs and TRUs constitutes the problem of "radioactive waste", one of the most negative byproduct of fission energy production. In Fig. 11 we specify the relative presence of radioisotopes in a spent fuel element of a PWR reactor, initially composed by ^{235}U and ^{238}U in the percentage of, respectively, 4.2% and 95%, after a specific burn-up of 50 GWd/ton heavy atoms, and after four years from the moment of extraction from the reactor core. FPs and TRUs are radioactive, emitting particles and/or radiation of various energies with a mean life which span an enormous range, from few seconds to more than 10^6 years. From the waste disposal point of view, the most dangerous isotopes are those with high fission yields and medium/long half-life, $T_{1/2}$. Among the FPs, we mention ^{90}Sr and ^{137}Cs which have $T_{1/2} \sim 39$ years, and the isotopes of iodine (^{129}I, ^{131}I) which although have either small yield or short decay time, represent a relevant biohazard because they concentrate in the thyroid gland. Other FP with long half-lives (up to 15.7×10^6 yr) are ^{99}Tc, ^{93}Zr, and ^{135}Cs. Among the TRUs: $T_{1/2}$ [^{237}Np] $= 2.1 \times 10^6$ yr, $T_{1/2}$ [^{239}Pu] $= 2.4 \times 10^4$ yr, $T_{1/2}$ [^{242}Pu] $= 3.7 \times 10^5$ yr, $T_{1/2}$ [^{243}Am] $= 7.3 \times 10^3$ yr, $T_{1/2}$

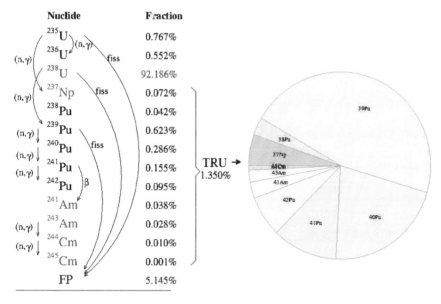

Nuclide	Fraction	
^{235}U	0.767%	
^{236}U	0.552%	
^{238}U	92.186%	
^{237}Np	0.072%	
^{238}Pu	0.042%	
^{239}Pu	0.623%	
^{240}Pu	0.286%	
^{241}Pu	0.155%	TRU → 1.350%
^{242}Pu	0.095%	
^{241}Am	0.038%	
^{243}Am	0.028%	
^{244}Cm	0.010%	
^{245}Cm	0.001%	
FP	5.145%	

$MA\ (Np,Am,Cm) = 0.149\%$

Fig. 11: Composition (left) and TRU content (right) of the spent fuel of a standard PWR with enriched uranium fuel irradiated up to a specific burn-up of 50 GWd/ton, and after four years of cooling. Arrows indicate how most isotopes are formed.

[^{245}Cm] = 8.5 × 10^3 yr. As these numbers show, SNF represents a hazard to life forms when released into environment, and its disposal requires isolation from the biosphere in stable deep geological formations for a very long period of time (hundreds of thousand years). To quantify the hazard associated with radioactive nuclides, what is relevant is not simply the rate of disintegration, identified by the activity A (in unit of becquerel, 1 Bq = one disintegration per second), but the health effects of the released ionizing radiation on human body. This effect is usually quantified in terms of "radio-toxicity", or equivalent dose: introducing the "dose coefficient" ε quantifying the "dangerousness" of particular isotope (and relative to ingestion or inhalation),[17] with units of Sv/Bq, the radio-toxicity relative to isotope j is given by $\varepsilon_j A_j$ in units of Sv. Figure 12 (OECD-NEA 2006) shows the level of radio-toxicity of uranium spent fuel, compared to the radio-toxicity of uranium ore, which is taken as a "natural" point of reference. The radio-toxicity of the FPs is seen to be relevant during the first 100 years, while the long-term radio-toxicity is solely associated to plutonium and americium isotopes. At present, the problem of nuclear waste remains open, and one of the main drives behind the research effort on next generation reactors.

---

[17] Dose coefficients are tabulated.

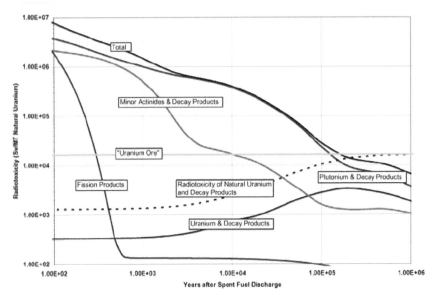

Fig. 12: Radio-toxicity of spent fuel as a function of time after discharge.

In present-day light-water thermal reactors, two fuel utilization strategies are adopted. The simplest one is the so called one-through cycle or "open cycle" strategy: it consists in fabricating nuclear fuel made from naturally-occurring material, using it once in a nuclear reactor, and after a short period of interim storage in cooling ponds and/or dry-casks located at reactor site to partially reduce the radio-toxicity level (10–40 yr), sending it to permanent disposal, typically in a deep geological repository. With this simple strategy, the fissile isotopes still contained in spent nuclear fuel are wasted, and moreover the toxicity level of the stored fuel is very high,[18] requiring their isolation from the biosphere for a period as long as 250,000 years to allow time for radioactive decay to reduce the radiological hazard. A scheme of the open fuel cycle (which is not a cycle per se) is presented in Fig. 13 (U.S. Department of Energy 2005). Once-through fuel cycles can be based both on the fissile isotope ^{235}U, or the fertile isotope ^{232}Th which under irradiation generates the fissile isotope ^{233}U. Currently, the once-through fuel cycle is the strategy pursued in most nuclear Countries, including the United States.

Direct disposal of SNF which has been irradiated only once is however problematic, due to the very long time duration in which performance of the repository is required and the large uncertainties concerning the potential for release of radioactive material. Even though the permanent

---

[18] The fission product and actinide content of spent nuclear fuel are primarily affected by the type of fuel and the discharge burn-up, i.e., the extent to which the fuel has been consumed.

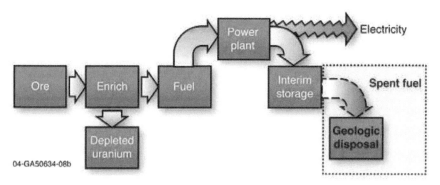

Fig. 13: Schematic of the open, or once-through, nuclear fuel cycle using enriched uranium.

disposal of radioactive nuclear waste in a geological formation could be developed on convincing scientific basis, the nuclear industry based on the utilization of uranium fuel needs to cope with the problem of limited resources. Since current utilization for all nuclear reactors is around 57,000 metric tons of natural uranium per year, and the total identified uranium resources are ~ 5.7 Mtons (NEA-IAEA 2016), the open cycle strategy would leave us with a uranium supply lasting not more than additional 100 years.[19] A way to improve this situation comes from the observation that SNF contains useful fissile isotopes: e.g., for UOx fuel, the remaining quantity of unburned ^{235}U and the generated quantity of ^{239}Pu and ^{241}Pu. A step of reprocessing can be thus inserted before sending the SNF to geological disposal, with the aim of separating the fissile uranium and plutonium isotopes. The latter would be then used to fabricate fresh MOx fuel elements to be recycled in the reactor. Such an improved fuel strategy is referred to as "closing the fuel cycle". The scheme of a closed fuel cycle is presented in Fig. 14 (U.S. Department of Energy 2005). Since the direct disposal of spent fuel indicates the limit of the recycling activities, such closed cycle is termed "limited". Even though this limited closed fuel cycle is beneficial with respect to efficient fuel utilization, it still produces SNF with a high level of radio-toxicity due to the presence of those MAs which have not been transmuted during the recycling.[20] In fact, the residual spent fuel still needs to be isolated for about 100,000 years before equating the radio-toxicity of natural uranium ore. This time period, although more than halved that required in the open cycle case, is still very high, and unsustainable.

---

[19] Since about additionally 10.5 Mtons remain undiscovered, this period of time can become of the order of 280 years. Further exploration and improvements in extraction technology are likely to at least double this estimate over time (assuming the current level of utilization remains unchanged).

[20] Assuming that all uranium and plutonium have been recycled.

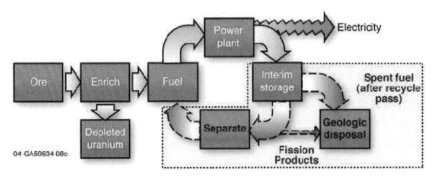

Fig. 14: A closed fuel cycle with a limited recycle strategy using enriched uranium fuel.

The limited closed fuel cycle is presently employed only in few nuclear countries with consolidated nuclear programs, including France, Japan and U.K. In these countries, after a series of recycles, used fuel is temporarily stored pending its use in future generation fast breeders reactors which would adopt more advanced fuel cycles able to fission all plutonium and transmute MAs.

5. Evolutive Solutions Adopted for Generation III/III+ Power Plants

One of the aspect on which the nuclear community is focusing its research effort is that of safe operation and effective accident mitigation. Studies have compared radiation doses from airborne effluents of model coal-fired and nuclear power plants, concluding that population doses from the coal plant are typically higher than those from light water reactors that meet government regulations (McBride et al. 1978). The potential dangerousness of the two kind of power generators in case of accident is however very different, since the core of a nuclear power plant contains a large amount of highly radioactive material, very well contained during normal operation but which, if put in direct contact with the environment, can lead to catastrophic consequences involving not only the population living nearby the plant, but also people living in an area that might span countries, as the Chernobyl disaster in 1986 demonstrated. Among OECD countries (i.e., not considering the Russian Federation and the Chernobyl event), nuclear power plant safe operation has a pretty good track record, since only the recent accident in Japan—which was related to an estimated, and not considered possible, tsunami event, induced by the most powerful earthquake ever registered in Japan—has led to serious, widespread consequences. This major accident, together with a series of human errors and technical malfunctions that caused minor incidents has led to a series of improvements so that the power plants currently under construction, Gen III/III+ plants, have safety characteristics that are substantially improved with respect to the ones present in power plants built in the 1950s and 1960s.[21] From the safety point of view, Gen III+ power plants have design features that should avoid any evacuation of the population around these plants, even under the worst-case conditions of core melt-down.

Besides safety, Gen III/III+ plants show improvements on performance and economics, since in order to be attractive nuclear energy has to be competitive with other forms of energy production.

5.1 Lessons Learnt from Nuclear Accidents and Terrorist Attacks

With respect to safety, the analysis of the three major nuclear accidents to date which occurred to civilian power plants, i.e., Three Mile Island (USA) accident in 1979, Chernobyl (USSR) in 1986, and the more recent Fukushima-Daiichi (Japan) accident in 2011, together with the fear of possible terrorist attacks of the kind occurred against the World Trade Center in New York City in 2001, has provided important lessons which induced nuclear safety authorities worldwide to introduce recommendations aimed at improving the safety of currently operating nuclear plants, and to guide the design of new generation reactors.

In particular, the 9/11 terrorist attacks introduced the novel threat of the intentional impact of a civilian airliner full of kerosene in the hand of a terrorist-suicide group of people on the containment vessel of a nuclear plant. For example, US Nuclear Regulatory Commission (NRC) published a final resolution in 2009 requiring all new nuclear power plants to incorporate design features that would ensure that, in the event of a crash by an aircraft, the reactor core would remain cooled and the reactor containment would remain intact, so that radioactive release would not occur from spent fuel storage pools.

The Three Mile Island and Fukushima-Daiichi events[22] have shown that the main causes of nuclear accidents are technical malfunctioning of major components of the plants (e.g., primary pumps), a plant design that is not able to cope with the occurrence of every possible natural event

---

[21] Gen III nuclear reactor are essentially Gen II reactors with evolutionary, state-of-the-art design improvements in the areas of fuel technology, thermal efficiency, modularized construction, safety systems, and standardized design. Gen III+ reactor designs are an evolutionary development of Gen III reactors, offering significant improvements in safety (Goldberg and Rosner 2011).

[22] The Chernobyl accident occurred in a reactor of very different characteristics compared to occidental-type reactors, and was induced by a sequence of human errors in handling a reactor experiment. For these reasons, the lessons learnt from Chernobyl fell short of requiring immediate changes in the occidental nuclear regulatory authorities.

and that relies in a determinant way on power sources external and/or internal to the plant, and finally human errors in handling the various phases of an accident. Given that it is impossible to exclude the occurrence of a unforeseen relevant accident, a statement that holds true for any type of industrial activity, nuclear power plant included, the long-term safety goal for next-generation nuclear facilities can be summarized in three major objectives. First, to improve the safety and reliability of plants. This can be accomplished by increasing the quality control of all components of the plants, from a simple valve to the pressure vessel, and to provide adequate monitoring of the main components during reactor operation. In this respect, replacement of the analogue control instrumentation with a fully digital system has provided a major improvement. Training of control room personnel can also be improved, and strategies to maintain high controller's state of alert can be applied, e.g., creating frequent simulated malfunctioning under the control of a supervisor. Second, to lessen the possibility of significant damage (in particular, partial or total core meltdown) during accidents. To this end, the main design approach of Gen II plants based on redundancy and separation of active emergency systems must be raised to a new level. Alternatively, or in synergy, safety systems can be designed so that they rely in the least possible way on human action and on a source of power. That is, emergency cooling systems must be passive, meaning that they do no rely on external or internal power sources and operate thanks only to laws of physics. In both cases of redundancy or passive control, a major aim is that of simplifying as much as possible the design of the plant, and of the various safety systems in particular. Focus is also given to the effectiveness of the systems dedicated to remove the heat generated by the radioactive irradiated fuel elements over long periods, adopting the redundancy/separation or/and a passive design approach. Additionally, more attention has to be given to hydrogen production and accumulation inside the primary containment building, to avoid chemical explosions which can damage the reactor building and lead to radioactive leaks into the environment. The third major objective consists in minimizing the potential consequences of any accident that does occur (in particular, the release of radioactivity). To accomplish this goal, the containment building already present in all OECD power plants in operation today, can be made more effective in containing radioactive material, and can be surrounded by a second containment structure, designed also to resist the deliberate crash of a large commercial airliner plane full of kerosene.

5.2 Generation III/III+ Objectives

Besides being safer in comparison to Gen II plans, Gen III/III+ power plants are more reliable, have better economic return, and consume fuel in a more efficient way. The main objectives that underlined the design of Gen III/III+ LWRs can be summarized as follows:

- Base-load generation of electricity (hydrogen and other process heat applications still not emphasized)
- Improved plant safety and reliability
  - Reduced need for operator action
  - More redundancy and/or passive control
  - Improved protection against natural events such as earthquakes and tsunamis, as well as terrorist attacks
  - Core Damage Frequency (CDF)[23] less than 10^{-4}/ry, and probability of a large radiation release to values less than 10^{-6}/ry (values once required by the US NRC for licensing)
- Improved economics
  - Increased plant design life (60 years)
  - Shorter construction schedule (36 months from first concrete to fuel loading)
  - Low overnight capital cost (~ $1000/kWe)
  - Low Operation and Maintenance cost of electricity (1¢/kWh).

In the following two sections we will present an overview of the two most widespread Gen III/III+ reactors under construction. The first is the European Pressurized Reactor (EPR), developed by the Franco-German cooperation Framatome-Siemens, now Areva NP, and the second is the Advanced Passive 1000 (AP1000) reactor, developed by Westinghouse in USA. Both reactors represent the development line called "evolutive", i.e., they are evolutionary designs that ensure continuity in the mastery of PWR technology, minimizing the risks associated with the introduction of more drastic changes. Building up on the valuable experience gained in the operation of several reactors per years of LWR operation worldwide, these evolutive designs increase the level of safety and improve plant performances with minor changes. As opposed, the "innovative" line of

---

[23] The CDF relative to a nuclear reactor design is a single number that quantifies the likelihood that, given the way a reactor is designed and operated, an accident could cause the fuel in reactor to be damaged. This event is considered extremely serious because severe damage to fuel in the core might prevent adequate heat removal or even safe shutdown, which can lead to a nuclear meltdown. The CDF is expressed in the units of reactor year: for example, a CDF = 1 x 10^{-4} ry means that one core damage incident is likely to occur in one of 10,000 reactors operating for a year. Assuming there are 500 reactors in use in the world, the above CDF means that, statistically, one core damage incident would be expected to occur somewhere in the world every 20 years. The CDF, which therefore quantifies the strengths and weaknesses of the design and operation of a nuclear power plant, is the most basic level of assessment estimate provided by the probabilistic risk assessment (PRA) methodology.

development, i.e., the next generation nuclear plants still under research phase (described in the next chapter of this book), is characterized by fundamental innovations compared to existing designs, and thus it will need to go through the realization of experimental devices and reactor prototypes before becoming operative, very likely not before 2040/2050.

5.3 *The EPR Reactor*

Areva NP is currently building its first Gen III+ EPRs in Finland (Olkiluoto 3), France (Flamanville 3) and China (Taishan 1&2). The EPR is a PWR designed for the high capacity range of 1500 to 1800 MWe, with a pressurized vessel connected with four primary loops. It is a design based on tried-and-tested technologies and principles. Contrary to previous generations, its design is characterized by a higher level of safety, by the economic savings that it achieves, and by a more sustainable and flexible fuel cycle. From a safety point of view, the EPR basic design philosophy has been that of simplification of safety systems, their redundancy (both mechanical and electric) and their physical separation, so as to ensure prevention of severe accidents, as well as mitigation of their consequences on the environment. Thanks to these improvements, the CDF value for this reactor has been decreased to 10^{-6}/ry, well below the initial value of 10^{-4}/ry required by the US NRC for licensing. Figure 15 (World Nuclear Association) shows a cross-section of the nuclear island emphasizing the main safety features.

To cope with an intentional direct impact of an airliner full of kerosene, or other man-made hazards (e.g., explosion pressure waves), and to strongly limit the radioactive releases that such events could induce, the reactor building of the EPR is based on double containment, the conventional inner wall of prestressed concrete to resist to every possible internal pressure peaks and withstand hydrogen explosions (in case of core meltdown and its leaking from the primary loop with reaction with the oxygen of the air), and an additional external wall, of reinforced concrete, to face every conceivable external attacks of natural and anthropogenic origin. Any radioactive leakage from the primary containment entering the annulus between the two containment shells is passed through a filter system before being discharged to the plant environs.

The EPR safety systems for emergency cooling of the core are designed as an extension of previous generation plants, with mainly active actuation. Simplification is pursued when possible to increase availability and reduce costs, and strong emphasis is posed on redundancy and separation.

The Containment Heat Removal System (CHRS) consists of a spray system which injects water taken from the in-containment refueling water storage tank (IRWST) to the upper part of the containment building, and is designed both to remove heat and to control pressure buildup due

Main EPR Safety Features

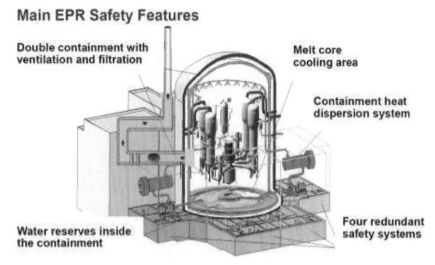

Double containment with ventilation and filtration

Melt core cooling area

Containment heat dispersion system

Water reserves inside the containment

Four redundant safety systems

Fig. 15: Main EPR safety features. Clearly visible is the double containment, the water reserve (IRWST), the spreading area where the melted core is isolated and cooled, and two of the four CHRS systems, each one housed in a different safeguard building.

to an accident. While the design philosophy of these safety systems is similar to that of present-day reactors, in the EPR there are four different safeguard buildings (each containing a CHRS) according to the principle of redundancy and physical separation safety criteria.[24] Two of these safeguard buildings are protected by the same reinforced concrete outer shell that contains the reactor building, the control room and the spent fuel building, so to have protection against natural or external man-made hazards. The remaining two safeguard buildings are located at opposite sides of the reactor building so that only one would be destroyed in case of hazard.

The mitigation of loss-of-coolant accidents of all sizes, specific non-LOCA events, such as main steam line breaks and sequences leading to feed and bleed is accomplished by the Safety Injection Systems (SIS). These systems ensure heat removal, coolant inventory and reactivity control. They consist of four trains, like the CHRS. Each train comprises a medium-head safety injection pump (MHSI) and an accumulator injecting into the reactor primary circuit cold leg, a low-head safety injection pump

---

[24] Safety systems have been designed according to the principle of quadruple redundancy, mechanical and electrical. This means that each system is actually composed by four systems ("trains") each one able to provide its safety function independently from the others. In this way the overall system satisfies the "N+2" failure criterion: it performs its function even though one system is out of order, and another is not available due to maintenance.

(LHSI) injecting into a cold leg and, at a later stage of the accident, also into a corresponding hot leg. A heat exchanger is ensured in the low head injection path, providing cooling for the injection flow (Bonhomme 1999).

The fluid systems of both CHRS and SIS are detailed in Fig. 16 (Teller 2010), together with the Emergency Feedwater System (EFWS) and the Chemical and Volume Control System (CVCS). The EFWS is designed to feed water to the steam generators in case of lack of normal operation water. The system consists of four trains; each one assigned to one steam generator, and consists of a tank, a pump, and piping and valves in a simple linear arrangement. As in any PWR, the CVCS has the main purpose of adjusting the reactor coolant boric acid concentration, and maintaining the proper water inventory in the primary loops in conjunction with the pressurized level control system.

Great attention has been given to the danger of chemical explosions inside the containment building due to accumulation of hydrogen in accidental sequences (e.g., melting and spreading of core). Hydrogen combustion is prevented by reducing the hydrogen concentration in

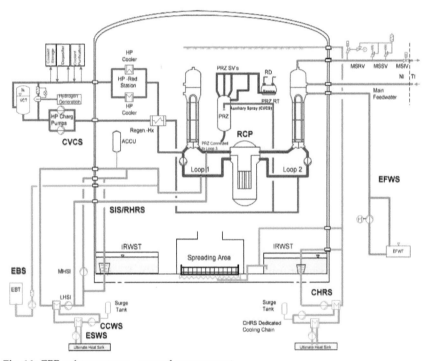

Fig. 16: EPR safety systems—general arrangement.

the containment at an early stage by catalytic hydrogen recombiners (hydrogen concentration limited to 10%).

Two other improvements have been pursued in EPR. First, the water inventories for emergency cooling actions as well as for various components (e.g., pressurizer and steam generator) have been increased in size and capacity, with the goal of smoothing accidental transients and to provide operators an extended grace period for intervention (12–24 hours). Second, creating an optimized man-machine interface based on fully digitalized instrumentation and control systems and status-oriented information supplied by modern operator information systems, with the goal of reducing sensitivity to human error.

Besides reducing the probability of occurrence of core damage states, the EPR is designed to restrict the effects of a postulated core melt accident to the plant itself. The EPR design carries an improvement in this regard with the adoption of a special compartment underneath the nuclear reactor apt to confine and cool the molten core in case of breakage of reactor pressure vessel, thereby preventing the drilling of the reactor building and minimizing both in time and space the release of radioactive pollution in the environment that surrounds the reactor. The basic goal of these safety measures and innovations is to reduce post-accident stringent countermeasures, limiting the relocation or evacuation of population in the immediate vicinity of the plant, and limiting the restriction of the use of cereals and other crops to the first year harvest.

An overall view of the EPR power plant is shown in Fig. 17 (www. modernpowersystems.com). Along with the four safeguard buildings hosting the CHRS (two of them with a double containment shell, and the remaining two located on opposite sides of the reactor building), are the two buildings containing the emergency diesel generators, each one containing two generators, which are located on opposite sides with respect to the main reactor building, according to the principle of physical separation.

In addition to the innovative features associated with its reinforced level of safety, the EPR benefits from many technological innovations, such as a reactor core surrounded by a neutron reflector that improves fuel utilization and protects the pressure vessel against irradiation-related aging phenomena; a pressure vessel made of optimized steel resistant to aging and designed with a reduced number of welds; and a steam generator equipped with an axial economizer, allowing production of high-quality steam (78 bar) and therefore high plant efficiency (36–37%) (Leverenz et al. 2004).

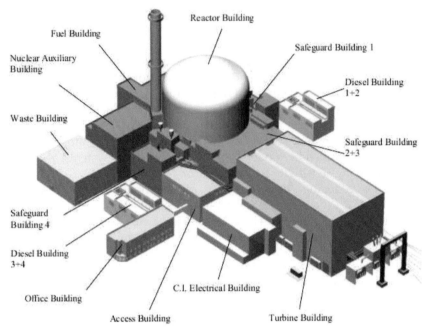

Fig. 17: Overall view of a power plant based on the EPR.

Improvements have been introduced also in the flexibility of operation. The EPR can operate with UOx fuel enriched up to 5%, as well as with reprocessed MOx (UO_2 + PuO_2) fuel, with variable proportion up to 100%, thereby offering the possibility to alleviate the burden of waste disposal. Fuel cycle length varies between 12 and 24 months, allowing for a better management of a power plant fleet. The power output in the 60–100% nominal output range can be adjusted at a rate of 5% nominal power per minute at constant temperature, preserving the service life of components and plant.

The EPR is designed with the ability to withstand an earthquake/tsunami event. In particular, the entire nuclear island stands on a single reinforced concrete basement, more able to resist seismic vibration. Moreover, the height of the buildings has been minimized, and the heaviest components, in particular the water tanks, are located at the lowest possible level.

The high power of the EPR reactor, between 1500 and 1800 MWe, is achieved with a large core, capable of holding more fuel with respect to present-day PWRs, and an advanced nuclear steam supply system comprising 4 primary coolant loops. The level of availability of the plant is 92% thanks to reduced recharge and maintenance periods. The plant life is 60 years, with the claim that the proprietary Company should not make important interventions or replacements of components.

5.4 The AP1000 Reactor

The Westinghouse AP (Advanced Passive) 1000 reactor is another example of a Gen III+ PWR reactor, the first few units of which are currently under construction in the US and China. The AP1000 has the same reactor vessel of a standard Westinghouse three-loop plant, with nozzles adjusted to accommodate the two loops of the new design. The internals are also standard, with minor modification. The core, located low in the vessel to minimize core temperature during LOCAs, is very similar to existing operating PWRs, with a 18-months fuel-cycle. Contrary to previous reactor generation, the AP1000 is characterized by extensive use of "passive" systems which has led to a significant improvement in the safety of the system, its reliability, and simplification of operation. The major advantage of these passive systems is that the long-term accident mitigation can be maintained without the involvement of operators and reliance on off-site or on-site power sources, thanks to reliance on laws of physics (gravity, natural convection, condensation, heat circulation) and/or on inherent characteristics (properties of materials, internally stored energy, etc.) and/or 'intelligent' use of energy (decay heat, chemical reactions, etc.). Probabilistic risk assessment provided a maximum CDF of $\sim 5 \times 10^{-7}$/ry (Health & Safety Executive Nuclear Directorate Assessment Report). The passive safety systems also reduce the capital and operating costs of the plant thanks to the significant reduction in number of pipes, wires, valves, and associated components.

Figure 18 (World Nuclear Association) shows a section of the main reactor building, with its double-containment walls, an inner steel containment and an outer concrete shield building. Like for the EPR reactor, the outer concrete building provides resistance to external hazards, such as missiles, including big aircraft impact. The inner containment is a steel pressure vessel designed to contain radioactivity in case of a reactor accident, to shield the core during normal operation and to provide a removal heat system in case of core melt.

The action of the Passive Containment Cooling water System (PCCS), schematized in Fig. 18, provides a safety-related ultimate heat sink for the plant. When the PCCS is activated, water drains by gravity on the steel containment from the water tank located on the top of the containment shield building. Thanks to the chimney effect, cool air enters inside the outer containment in its lower region, and rises by natural

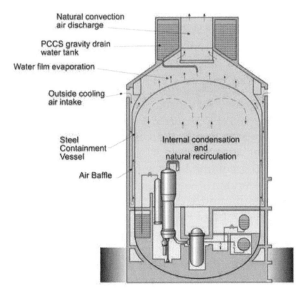

Fig. 18: AP1000 passive containment cooling system (PCCS).

circulation along the external surface of the steel cylinder causing water evaporation and therefore cooling it and decreasing its internal pressure. The containment vessel therefore acts as the heat transfer surface, and the ultimate heat sink is the atmosphere. The system is designed so that even with failure of water drain, air-only cooling is capable of maintaining the containment below a predicted failure pressure. The system, besides being passive, is very simple, not needing all the complex system of pumps and heat exchangers characterizing the emergency cooling systems of the old generation of nuclear plants.

Figure 18 also shows three additional water storage tanks, dedicated to the emergency cooling of the core in case of leaks and ruptures of various sizes in the primary loop. On the left of the figure is the In-Containment Refueling Water Storage Tank (IRWST), located in the containment just above the primary loop, while on the right is the Core Makeup Tank (CMT, upper tanks), and the Accumulator (ACC, lower tanks). These three water tanks are the sources of water used by the Passive Core Cooling System (PXS), and work in synergy to provide core residual heat removal, safety injection, and depressurization. The detailed structure of the PXS is shown in Fig. 19 (http://www.westinghousenuclear.com/New-Plants/AP1000-PWR/Safety/Passive-Safety-Systems). The passive residual heat removal (PRHR) system, also part of the PXS, cools the water of the hot leg of primary loop using one heat exchanger located in the IRWST, and re-inject

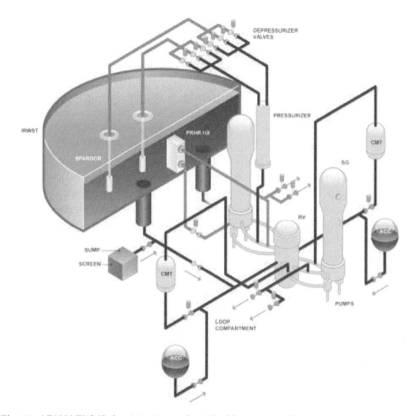

Fig. 19: AP1000 PXS (Safety injection and residual heat removal) system.

water into the steam generator cold leg plenum of one of the two primary loops. This system is postulated for non-LOCA events, where a loss capability to remove core decay heat via the steam generator occurs, and works in a totally passive way (gravity instead of pumped recirculation). The heat can be absorbed by the IRWST for more than one hour before the water begins to boil. Once boiling starts, steam passes to the containment. The steam condenses on the steel containment vessel and, after collection, drains by gravity back into the IRWST. In this way, the system provides indefinite decay heat removal capability with no operator action required (Schulz 2006).

The passive safety injection system (PSIS) is designed to provide sufficient water to the reactor primary loop to rapidly refill the reactor vessel, reflood the core, and continuously remove the core decay heat so as to mitigate the effects of a LOCA. It consists of two ACCs, two CMTs, and the IRWST. The two ACCs act similarly to current power plants, i.e., injecting medium pressure borated water into the primary circuit if the pressure of the primary circuit drops low enough. The ACCs provide a very high flow for a limited duration of minutes. The CMT is a high pressure safety water injection system which enters directly into the pressure vessel, and therefore replaces the high head injection pumps of conventional PWRs. The CMT provides a relatively high flow rate for a longer duration. When the primary loop is nearly depressurized, the water injection is taken over by the IRWST by gravity, which can be seen as replacing low head injection pumps of conventional PWRs. The IRWST provides a lower flow rate, but for a much longer time. The containment is the final long-term source of water. It becomes available following the injection of the other three sources and floodup of containment.

Figure 19 also shows the Automatic Depressurization System (ADS), part of the primary loop, which is a system designed to automatically depressurize the primary loop so that water injection from the safety systems can occur.

A major safety advantage of passive plant AP1000 versus conventional PWRs is that long-term accident mitigation is maintained without operator action and without reliance on off-site or on-site AC power sources. In particular, the PCXS can maintain safe shutdown conditions for several days after an event without operator action and without both non-safety related on-site and off-site power. The use of passive safety-related systems also simplifies the design by eliminating many safety-related components including pumps, valves, pipes and their associated buildings.[25] In addition, many other safety-related components have been reduced to non-safety. These changes result in great simplification in procurement, construction, startup, and operation including inservice inspection/testing and maintenance.

The simplification of design benefits plant costs and construction schedules. In addition, modular construction techniques have been adopted by the AP1000. The number of modules has been maximized in the design. These modules are built in factories and then shipped to the plant, where they are assembled in parallel construction areas and then lifted into the plant as needed.

6. Small Modular Reactors

Notable interest is being posed on Gen III/III+ designs that are smaller in scale than current designs (generating less than 300 MWt) and make

---

[25] Compared to currently operating PWRs: 50% fewer valves, 35% fewer safety grade pumps, 80% less pipes, 45% less seismic building volume, and 85% less cables (Schulz 2006).

extensive use of modular construction techniques (Goldberg and Rosner 2011, OECD-NEA 2016).

Modularity is a key feature of the concept. The modules, which are intended to be produced in factory conditions, could even be complete reactor units. Other equipment such as turbine-generator, condenser, the cooling system, etc., could also be produced as modules. These stand-alone modules could then be transported to the construction site (which could also involve factory-produced structures) and installed.

These reactors, referred to as small modular reactors (SMRs), could be built quickly and installed at existing nuclear sites, or they could replace existing coal-fired plants in order to comply with more stringent emission standards. Moreover, SMRs are the logical choice for smaller countries, or countries with a limited electrical grid. An interesting aspect of these small reactors is their ability to stretch out refueling schedules, from 18 months to possibly 3–5 years and potentially as long as 10 years. Because of these attractive features, combined with their simplicity, enhanced safety, and relatively limited financial resources required to build them, SMRs are currently in different stages of developments throughout the world, with two reactors being under constructions: the CAREM-25 (a prototype) in Argentina, and the KLT-40S in the Russian Federation. Whether SMRs will acquire a significant portion of the market depends in a determinant way on their economics, which at present has to be developed.

7. Summary

This chapter provides an introduction to the field of nuclear energy production, illustrates the main features of nuclear reactors currently in operation (generation II), and gives an overview of two reactors (EPR and AP1000) presently under construction. These new reactors are classified as generation III/III+ designs, and present evolutionary state-of-the-art improvements over generation II reactors in the areas of safety systems, fuel technology, thermal efficiency, modularized construction, and standardized design. With respect to safety, in particular, both the EPR and the AP1000 design, although using two different approaches, respectively based on the concept of redundancy and on the adoption of passive systems, ensure a core damage frequency value that is well below that of currently operating plants. The nuclear industry is also looking at a new type of small reactor based on generation III/III+ design, generating a power less than 300 MWt and making extensive use of modular construction techniques. These small modular reactors substantially reduce construction times, and may represent the logical choice for smaller countries or countries with a limited electrical grid.

References

Aoto, K., P. Dufour, Y. Hongyi, J.P. Glatz, Y.-l. Kim, Y. Ashurko, R. Hill and N. Uto et al. 2014. A summary of sodium-cooled fast reactor development. Prog. Nucl. Energ. 77: 247–265.

Bonhomme, N. 1999. System organization for the European pressurized water reactor (EPR). Nucl. Eng. Des. 187: 71–78.

From: http://www.westinghousenuclear.com/New-Plants/AP1000-PWR/Safety/Passive-Safety-Systems.

From: www.modernpowersystems.com/features/featuresite-work-underway-on-finland-s-1600-mwe-epr.

Goldberg, S.M. and R. Rosner. 2011. Nuclear reactors: generation to generation. American Academy of Arts and Sciences.

Health & Safety Executive Nuclear Directorate Assessment Report. Westinghouse AP1000 Step 2 PSA Assessment.

International Atomic Energy Agency, https://nucleus.iaea.org.

International Atomic Energy Agency Annual Report 2016.

International Energy Agency. 2017. Electricity information: overview.

Krane, K.S. 1988. Introductory Nuclear Physics. John Wiley & Sons.

Leverenz, R., L. Gerhard and A. Goebel. 2004. The European Pressurized Water Reactor: a safe and competitive solution for future energy needs. International Conference on Nuclear Energy for New Europe, Portoroz, Slovenia.

McBride, J.P., R.E. Moore, J.P. Witherspoon and R.E. Blanco. 1978. Radiological impact of airborne effluents of coal and nuclear plants. Science 202: 1045–1050.

Mitsubishi Nuclear Fuel, Co., http://www.mnf.co.jp.

NEA-IAEA. 2016. Report: Uranium 2016: Resources, Production and Demand.

Nuclear Power, http://www.nuclear-power.net.

OECD-NEA. 2006. Physics and Safety of Transmutation Systems. A Status Report. No. 6090.

OECD-NEA. 2016. Small modular reactors: nuclear energy market potential for near-term deployment.

Moormann, R. 2009. AVR prototype pebble bed reactor: a safety re-evaluation of its operation and consequences for future reactors. Kerntechnik 74: 8–21.

Schulz, T.L. 2006. Westinghouse AP1000 advanced passive plant. Nucl. Eng. Des. 236: 1547–1557.

Teller, A. 2010. EPR Safety Systems. Areva.

Theenhaus, R. and S. Storch. 1990. The AVR high-temperature reactor—Operating experience, storage and final disposal of spent fuel elements. CONF-900210-Vol2. Post, R.G. (Ed.). United States.

Thorium High Temperature Reactor, http://www.thtr.de/technik-bau.htm.

U.S. Department of Energy, Office of Nuclear Energy, Science, and Technology. 2005. Report to Congress—Advanced Fuel Cycle Initiative: Objectives, Approach, and Technology Summary.

U.S. Energy Information Agency (EIA). http://www.eia.doe.gov.

US Nuclear Regulatory Commission, https://www.nrc.gov.

US Nuclear Regulatory Commission. Reactor Concepts Manual—Pressurized Water Reactor (PWR) Systems.

Westinghouse Electric Corporation, Water Reactor Division. 1984. The Westinghouse pressurized water reactor nuclear power plant.

World Nuclear Association. Nuclear reactors. Information library.

CHAPTER 5

Next Generation Nuclear Reactors[1]

M. Cumo and *R. Gatto**

1. Introduction

Commercial nuclear power reactors conform to OECD safety standards have shown a very good safety record since the beginning of the civilian nuclear power era more than 60 years ago, with only one event leading to a major release of radioactivity, the Fukushima-Daiichi accident. Despite this, the consequences of this event (long term evacuation of population living nearby the plant, long term contamination of harvest, long term health effects associated with radiation) have spread a sense of distrust toward nuclear energy in part of the public opinion worldwide. Due to the advantageous aspects of nuclear energy (no CO_2 release, economical production of bulk electrical energy as well as of hydrogen and high quality heat for industrial processes), this form of energy is however still pursued in many OECD Countries, and is expanding in new powers such as China and India. In this context, the ongoing research effort on next generation nuclear reactors, an overview of which is presented in this chapter, aims at resolving in a fully satisfactory way the problematic aspects of present-day nuclear energy (safety and reliability, sustainability,

Sapienza University of Rome, Corso Vittorio Emanuele II 244, 00186 Rome, Italy.
Email: maurizio.cumo@uniroma1.it
* Corresponding author: renato.gatto@uniroma1.it

[1] This chapter is intended as a sequential reading of the preceding chapter in "Nuclear power", and many concepts and definitions relative to nuclear energy production and nuclear reactors presented in the preceding chapter are not repeated here.

economics, and proliferation resistance), with the goal of turning both public opinion and national orientations toward a renewed acceptance and promotion of this form of energy production.

2. Innovative Generation IV Reactors

Thanks to passive safety systems and/or extensive redundancy and physical separation, evolutive Generation III/III+ (Gen III/III+) reactors, such as the EPR and AP1000 that are presently under construction, have a lower probability of suffering a severe accident leading to damage of the nuclear core (core damage factor, or CDF, of the order of ~ 10^{-6}/ry) compared to present-day generation II (Gen II) plants. Moreover, Gen III/III+ reactors adopt thermal cycles with improved efficiency, and reduce construction costs. A nuclear energy production system based on such reactors, however, would not be acceptable as a long term solution. The principal reason for this resides in the adopted fuel cycle, which does not pay due attention to the minimization of long-lived radioactive waste, and to the saving of natural resources of uranium. In one word, Gen III/III+ reactors do not provide a long-term sustainable solution for nuclear energy production.

On the initiative of the U.S. Department of Energy an International Committee, grouping 13 Countries,[2] named GIF (Generation IV International Forum), has been established in January 2000 for the development of new advanced nuclear systems which can be designed, tested and manufactured to prototype level by 2040–2050, when many reactors presently in operation will be at the end of their operating licenses. The Table 1 describes the eight technological goals that have been defined for generation IV (Gen IV) systems, which are grouped into four broad areas: sustainability, economics, safety and reliability, proliferation resistance and physical protection (Kelly 2014). These areas can thus be identified as the four pillars for future nuclear energy.

In general terms, the way to reach these objectives have been identified as follows:

- Sustainability can be accomplished by a better use of fuel resources, and in particular with the adoption of new fuel cycles that increase as much as possible the utilization of both uranium and plutonium, and that are able to minimize the residual waste radio-toxicity by destroying also the actinides by nuclear fissions to produce useful energy;

---

[2] Argentina, Brazil, Canada, France, Japan, the Republic of Korea, the Republic of South Africa, the United Kingdom, and the United States signed the GIF Charter in July 2001. It was subsequently signed by Switzerland in 2002, Euratom in 2003, and the People's Republic of China and Russian Federation in 2006.

Table 1: Technological goals for generation IV (Gen IV) systems.

Goals	Description
Sustainability	(1) Gen IV nuclear energy systems will provide sustainable energy generation that meets clean air objectives and provides long-term availability of systems and effective fuel utilization for worldwide energy production.
	(2) Gen IV nuclear energy systems will minimize their nuclear waste and notably reduce the long-term stewardship burden, improving protection for public health and environment.
Economics	(3) Gen IV nuclear energy systems will have a clear life-cycle cost advantage over other energy sources. (4) Gen IV nuclear energy systems will have a level of financial risk comparable to other energy projects.
Safety and Reliability	(5) Gen IV nuclear energy systems operations will excel in safety and reliability. (6) Gen IV nuclear energy systems will have a very low likelihood and degree of reactor core damage. (7) Gen IV nuclear energy systems will eliminate the need for off-site emergency response.
Proliferation Resistance and Physical Protection	(8) Gen IV nuclear energy systems will increase the assurance that they are very unattractive and least desirable route for diversification or theft of weapons-usable materials, and provide increased physical protection against acts of terrorism.

- Long-term economic competitiveness can be accomplished by lower initial capital cost of the plant, by providing certainty in the construction times using modularity of components, and improving the thermodynamic efficiency of the plant.

- Safety and reliability can be accomplished following the direction already undertaken with Gen III+ reactors, therefore introducing simplification, redundancy, spatial separation of systems, passive emergency system, and possibly incorporating inherent safety characteristics.[3]

- Proliferation resistance and physical protection can be obtained by cautious reprocessing activities and increasing the resistance of reactor building so to withstand acts of terrorisms, in particular the intentional crash of a big civilian plane full of kerosene.

---

[3] A passive safety system employs components which do not need any external input to operate, i.e., their functions are achieved by means of static or dormant unpowered or self-acting means, based only on the laws of nature. An inherent safety characteristic refers to the elimination of a specified hazard by means of the choice of material and design.

While the objective of safety and economic competitiveness can be achieved with evolutive-kind of steps, as done in Gen III+ reactors, the accomplishing of sustainability requires the adoption of new fuel concepts and advanced fuel cycles and management, therefore giving an innovative footprint to Gen IV plants. Because of the innovative character of many of the adopted solutions, considerable fundamental research is still needed to be fully developed. The way to the first commercial Gen IV plant has to be preceded by the construction of experimental plants and prototypes, and no emergence of a Gen IV commercial reactor is foreseen before the mid of the century.

3. Advanced Fuel Cycles with Partitioning/Transmutation and Breeding

One of the most important goal for Gen IV reactors is better utilization of fuel resources and reduction of nuclear waste quantity and toxicity, in other words, sustainability of nuclear energy. Present-day light-water thermal reactors (LWRs) adopt either an open (one-through) cycle strategy, in which spent fuel is sent directly to permanent storage, or a limited closed fuel cycle with separation and recycling of fissile U and Pu isotopes contained in the irradiated nuclear fuel before the latter is sent to a geological repository. The upper two curves in Fig. 1 (Salvatores 2016) show the radio-toxicity level of spent nuclear fuel in the cases of open fuel cycle, and a limited closed fuel cycle with U and Pu recycling. A storage time of respectively 250,000 and 100,000 years is required in order to reduce the radio-toxicity at the level of the original natural uranium ore used for the LWR fuel, so that man-made impact is eliminated. To improve this situation, the only obvious option is to reprocess (separate and recover) not only the fissile isotopes of uranium and plutonium for further use as fuel, but also the problematic MA isotopes, with the goal of transmuting them into less hazardous or shorter-lived radioactive materials, substantially altering in a positive way the characteristics of the material destined for disposal. A fundamental result of studies on the topic of transmutation is that the latter is best performed in reactor cores characterized by a fast neutron spectrum, both because MAs are fissionable, and because fast spectrum systems have an essential advantage in neutron surplus production available for transmutation when being compared with other spectrum types. A closed cycle with MAs partition and transmutation (P&T) is referred to as "advanced cycle". Ideally, all the MA elements separated from the used fuel and re-introduced in the core would fission after several recycles (with opportune recycle strategies),

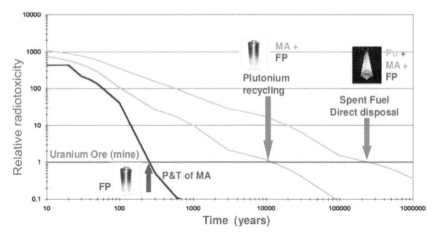

Fig. 1: Radio-toxicity level of nuclear waste. The three curves indicate the time evolution of radio-toxicity level of spent fuel in case of (from upper to lower curve): direct disposal of spent fuel, elimination of Pu by recycling, elimination of Pu and MAs by P&T.

and what is left as nuclear waste are the fission products (FPs). In this case, as shown by the lower curve in Fig. 1, the ore-level of radio-toxicity would be reached after a storage time of few hundred years, a time-span that should be acceptable by the public opinion.

An example of implementation of an advanced fuel cycle is presented in Fig. 2, in which all the MAs are fissioned while all the fissile isotopes of U and Pu are recycled (Salvatores 2006). Besides the Gen II nuclear plant producing highly radio-toxic spent fuel, the implementation of the cycle requires a fast reactor, a fuel element production facility, and reprocessing (separation of Pu and MAs) facilities. The advantages of such an advanced cycle is that energy is produced while the amount of Pu and MAs is reduced. Studies have however shown that a prerequisite of these radio-toxicity reduction is a nearly complete fissioning of the actinides, for which multi-recycling is a requirement. Also, losses during reprocessing and re-fabrication must be well below 1% and probably in the region of 0.1%.

A beneficial side-effect of P&T strategies is the reduction of the radionuclides masses to be stored, the reduction of their associated residual heat, and as a consequence a reduction of volume and cost of the repository. Moreover, the reduction of masses associated with P&T strategies results in a reduction of the proliferation risk, in particular if transuranic (TRU) elements are not separated from each other.

Even with successful recycling of both plutonium and MAs, the disposal of spent fuel is still problematic due to the radioactivity of FPs, both generated from fissions of the original fuel, and from transmutation

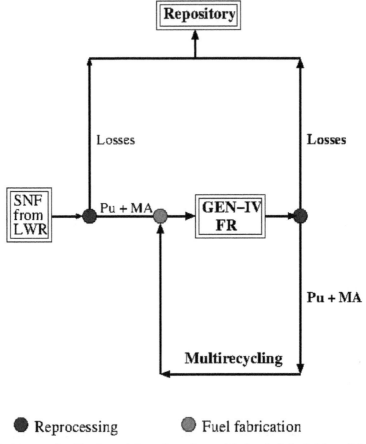

● Reprocessing   ◉ Fuel fabrication

Fig. 2: An example of implementation of an advanced fuel cycle (adapted from Salvatores 2006).

of produced MAs. It is then reasonable to question whether P&T could be conveniently applied also to the most radio-toxic FPs. With respect to FP, transmutation means a sequence of neutron captures which convert them into stable or short-lived nuclides. Attention has been posed on those low-level fission products (LLFPs) for which the rate of neutron absorption during irradiation significantly exceed the rate of natural radioactive decay, so that they can be effectively transmuted: in particular, ^{99}Tc, ^{129}I, ^{135}Cs, ^{93}Zr and ^{107}Pd. While a significant amount of studies on the transmutation of LLFPs in both fast and thermal spectrum systems have been performed, and transmutation strategies have been proposed, as of today it is not clear whether the effort of transmuting LLFPs instead of sending them to permanent disposal is a worthwhile endeavor, considering also that the probability of loss of control in the process of fabrication and irradiation of FPs is higher as compared with keeping the same radioactivity in storage facilities.

Recycle strategies are obviously more complex than once-through strategies, due to the undertaking of the additional steps of spent fuel reprocessing, fabrication of recycled fuel, and the need of irradiation facilities for transmutation of the recovered materials in recycled fuel. For example, the fabrication of recycled fuel containing one or more of the TRU elements is a complex task, requiring shielding and remote handling as determined by the activity of the TRU.[4] Also, the irradiation of a fuel element containing a certain MA must be performed in such a way that its transmutation rate is greater than its production rate. As a general statement it can be said that effective reduction of fuel toxicity after irradiation can be accomplished with a closed fuel cycle with low enough reprocessing losses and high burn-up level (long fuel life). At present, the scientific feasibility of transmutation has been proven by the analysis of a few pins containing minor actinides, irradiated in various reactors (e.g., Phénix in France). But the capacity of a transmuting system to burn all the actinides it produces has not yet been proven, and it is one of the major goals of Gen IV R&D. Together with transmutation in a critical fast reactor, as envisaged in Fig. 2, the possibility to transmute in an accelerator-driven sub-critical fast system[5] is presently studied, as well, due to several advantages of the latter, not least its higher safety in dealing with a core containing significant quantities of MAs.

The conversion aspect of the fuel cycle, a key element of advanced fuel cycles, has different characteristics in the U-Pu and Th-U cycles. In order to have an efficient converter reactor the average number of neutrons produced for each neutron absorbed, denoted by η, must be greater than 2. For the breeding of ^{239}Pu, this occurs only if the neutrons have an energy of at least 100 keV (for example, for 3.5% enriched uranium in a thermal spectrum, η is only ~ 1.85). Therefore, a thermal reactor is not able to breed ^{239}Pu, so a fast reactor is required. Different is the case of thorium cycle,

---

[4] For instance, when Cm produced in the U-Pu cycle is irradiated with neutrons it forms Cf and Fm which undergo spontaneous fission. As a result, the neutron emission from a used fuel element which has included Cm will be much higher, potentially posing a risk to workers at the back end of the cycle unless all reprocessing is done remotely. This could be seen as a disadvantage, but on the other hand it also makes the nuclear material difficult to steal or divert, making it more resistant to nuclear proliferation. France, for example, has decided to reprocess Am but not Cm.

[5] In a sub-critical system, the multiplication factor, indicating the number of neutrons in successive generations, is slightly less than one. To maintain a steady-state fission rate, an external neutron source is required. The latter can be an accelerator producing neutrons via spallation nuclear reactions.

in that for ^{233}U the number of neutrons liberated per neutron absorbed is greater than 2.0 over a wide range of thermal neutron spectrum.[6] Thus, high conversion factors, and even breeding, can occur in thermal spectra.[7] Thorium cycles are categorized by the type of added fissile material, and are also significantly influenced by the way in which the fissile and fertile materials are distributed within the fuel bundle and within the core. The simplest of these fuel cycles are based on a homogeneous thorium fuel design, where the fissile material is mixed uniformly with the fertile thorium. These fuel cycles can be competitive in resource utilization with uranium-based fuel cycles, building up an inventory of ^{233}U in the spent fuel for possible recycle in thermal reactors.

At present thorium fuel has only been used in research reactors, and demonstration irradiations were performed in power reactors. The question on whether the Th-U cycle is more advantageous than the U-Pu cycle in terms of nuclear fuel resources and waste production, as well economics and proliferation resistance, is open. Thorium is more abundant than uranium, and combined with a breeding cycle is potentially a major energy resource.

With regard to radio-toxicity, if a reactor using thorium fuel reaches the long term equilibrium condition with self-sustained ^{233}U recycle, the Th-U long term fuel activity would have only trace quantities of TRUs and therefore lower radio-toxicity after 500 years. In a practical scenario, the reduction in radio-toxicity is more modest than the long term equilibrium would indicate. Another potential advantage of using Th is the larger negative void reactivity coefficient which allows for the use of Th-Pu oxide fuels with a larger quantity of Pu (with respect to U-Pu MOx), giving more flexibility for plutonium re-use in LWRs. With regard to disadvantages, many daughters of ^{232}Th and ^{233}U are either strongly gamma emitters (e.g., ^{208}Tl), therefore making it difficult to reprocess thorium fuel, or are neutrons absorbers (e.g., ^{233}Pa), therefore decreasing the efficiency of the process. Moreover, ^{233}U is a weapons usable material with a low fissile mass. This situation however is ameliorated by the fact that the presence of ^{232}U inhibits fuel diversions. Finally, reprocessing thorium fuel is less straightforward than with the U-Pu cycle. The THOREX process has been demonstrated at small scale, but will require further R&D to achieve its commercial readiness. In summary, the only clear advantage in pursuing thorium fuel is the associated sustainability, and this is an enough

---

[6] Typically, in a thermal spectrum like that of HTRs (and PWRs), the neutron reproduction factor η (the average number of neutrons produced for each neutron absorbed) is 2.29 for ^{233}U, but only 2.05 for ^{235}U, and 1.80 for ^{239}Pu. Thus ^{233}U is by far the best fissile isotope for thermal spectrum reactors.

[7] The possibility of breeding in a thermal reactor with fuel using ^{233}U was demonstrated under experimental conditions in the Shippingport reactor (USA) in the early 1970s.

important drive for those Countries that lack uranium resources and point to a long-term utilization of nuclear energy.

4. Generation IV Reactors

The GIF has identified six innovative reactor systems (reactor and all related fuel cycle facilities): Sodium-cooled Fast Reactor (SFR), Very High Temperature Reactor (VHTR), Gas-cooled Fast Reactor (GFR), Lead-cooled Fast Reactor (LFR), Molten Salt Reactor (MSR), and SuperCritical Water-cooled Reactor (SCWR).

The Table 2 provides a broad classification of these systems according to the fuel cycle/spectrum and coolant type and outlet temperature (CFC = closed fuel cycle, OFC = open fuel cycle, F = fast spectrum, ET = epi-thermal spectrum, T = thermal spectrum):

Table 2: Classification of Gen IV reactor systems.

	SFR	VHTR	GFR	LFR	MSR	SCWR
Neutron spectrum	F	T	F	F	T/ET/F	T
Moderator	-	Graphite	-	-	-	Light water
Core outlet temperature	~ 500°C	1000°C	~ 850°C	~ 500°C	1000°C	625°C
Coolant	Sodium	Helium	Helium	Lead	Molten salt	Light water
Fuel type	MOx or metal allow	Coated particles	Plate or ceramic	MOx or nitrides	Molten salt	UOx
Fuel/Fuel cycle	CFC	OFC	CFC	CFC	CFC	OFC[8]

In order to achieve all goals fixed by GIF and obtain a long term solution, the strategy would be to adopt a synergistic combination of some of the proposed reactors so to form a nuclear energy production system in which each type of reactor will cover its role at the best.

Following the latest technology roadmap update for Gen IV systems available today (OECD Nuclear Energy Agency 2014), we now provide a brief overview of the six Gen IV concepts, putting in evidence the most innovative aspects of each system.

---

[8] Variants of the initial reference design are characterized by fast spectrum and full actinide recycle.

SFR

Liquid sodium is a coolant with several advantageous thermo-physical properties, namely, low melting point and high boiling point, and high heat of vaporization, heat capacity and thermal conductivity. The adoption of liquid sodium therefore allows a low-pressure coolant system and high-power-density operation with low coolant volume fraction in the core. Vapor can be obtain with excellent features (e.g., 180 bar and 510°C), leading to high system thermodynamic efficiency. Figure 3 presents the schematic of Gen IV SFR layout with pool-type reactor. The large thermal inertia in the primary coolant (long thermal response time), the large margin to coolant boiling, a primary system that operates near atmospheric pressure, and an intermediate sodium system between the radioactive sodium in the primary system and the power conversion system, are all important safety features of a sodium-cooled reactor. Due to the low neutron absorption and negligible moderation properties of sodium, its optimal use is in reactors with fast neutron spectrum. The development of a closed fuel recycle system is perhaps the main mission of the SFR, allowing for improved resource utilization, and advanced high-level waste management including plutonium and MAs. Because of these sustainability features, SFR is an attractive energy source for nations

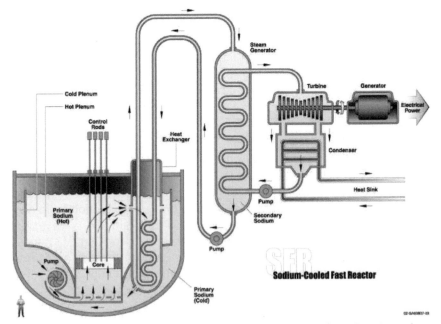

Fig. 3: Gen IV Sodium-Cooled Fast Reactor. The pool-type layout places the primary heat exchanger and pumps for the primary sodium inside the reactor tank.

desiring to make the best use of limited nuclear fuel resources and manage nuclear waste by closing the fuel cycle.

A major drawback is that sodium reacts chemically with air and water[9] and requires a sealed coolant system, and high attention to passive safety features which can prevent, and in case mitigate, the occurrence of sodium-air-water reactions. An intermediate sodium system between the radioactive sodium in the primary system and the power conversion system is therefore required. Corrosion is also a problem with liquid sodium (as with most liquid metal coolants), a problem that can be mitigated by the required oxygen-free environment. Another sodium drawback is associated with the activation of the stable isotope ^{23}Na upon neutron absorption, so that during reactor operation the sodium reaches high levels of radioactivity.

GIF considers three base line operations of fast reactor layout, according to the power level, and the structure of primary circuit:

- A large size (600–1500 MWe) loop-type reactor with mixed uranium-plutonium oxide fuel and potentially MAs, supported by a fuel cycle based upon advanced aqueous processing at a central location serving a number of reactors.

- An intermediate-to-large size (300–1500 MWe) pool-type reactor with oxide or metal fuel.

- A small size (50–150 MWe) modular-type reactor with uranium-plutonium-minor actinide-zirconium metal alloy fuel, supported by a fuel cycle based on pyrometallurgical processing in facilities integrated with the reactor.

Technological bases of SFR have been established by past or existing demonstration and/or prototype SFRs such as PFR (UKAEA), Phénix (France), BN-350 (Kazakhstan), Super Phénix (France), BN-600 (Russia), Monju (Japan). France, Japan and Russia are designing new SFR demonstration units for near-term deployment; China, the Republic of S. Korea and India are also proceeding with their national SFR projects. The SFR technology is thus more mature than other fast reactor technologies, and is deployable in the very near-term for actinides management. With innovation aiming to reduce capital cost, the SFR also aims to be economically competitive in future electricity markets.

R&D performed in the last decade have led to major accomplishments. In the area of safety and operation, activities have been carried out in the field of numeric modeling and multi-dimensional calculation, sodium void reactivity effect, components of decay heat removal systems, in-

---

[9] Sodium reacts violently on water contact, producing hydrogen.

service inspection methodology, and radioactive elements transportation. Experimental studies on sodium boiling, fuel pin failure modes, and analysis of metal fuel pin disruption tests were also performed. In the area of advanced fuels and actinide management, activities have been carried out on fuel irradiation tests and post-irradiation examinations for MA-bearing oxide, metal, nitride and carbide fuels. In the area of component design and balance-of-plant, several technical solutions and improvements have been suggested for steam generators, such as double-walled steam generator tubes, and new safety approaches to the monitoring of sodium/water reaction. As an alternative to the traditional Rankine cycle, the viability of the supercritical CO_2 Brayton energy conversion cycle has been investigated both theoretically and experimentally, and data on the interaction of sodium with CO_2 and on corrosion of austenitic and ferritic steels by supercritical CO_2 have been collected. A major advantage of a SFR with a supercritical CO_2 Brayton energy conversion cycle consists in the elimination of secondary sodium circuits because of no sodium/water reaction. The power generation efficiency of such a system could reach values of 42%, and the size of the turbine and compressors (and thus the volume of the reactor building) would be notably reduced thanks to the high fluid density of supercritical CO_2.

The Advanced Sodium Technical Reactor for Industrial Demonstration (ASTRID) is a reactor proposed as the successor of the Phénix reactor in its role as a French-built SFR within the international Gen IV reactor program. ASTRID is a pool-type sodium-cooled fast reactor with a nominal power of 600 MWe, conceived as a technological (not commercial) reactor for demonstration of the relevancy and performance of innovations in the field of safety and operation. The innovations include (Aoto et al. 2014) a core with an overall negative sodium void effect, specific features to prevent and mitigate severe accidents (core catcher, combination of proved Decay Heat Removal systems and Vessel Natural Air draft cooling), a power conversion system that drastically decreases the sodium-water reaction risk (e.g., innovative sodium leak detection systems), and improvements in the in-service inspection and repair. Coupled with a pilot reprocessing facility, ASTRID has also the objective to test the industrial feasibility of the multi-recycling of plutonium, and to demonstrate the possibility of industrial transmutation of americium.[10] Two concepts will be explored: the transmutation in homogeneous mode, with up to few per cent of Am in the core, allowing the demonstration of break-even-breeding (the quantity of MA that is transmuted equals the quantity that is produced in

---

[10] Some of the isotopes of curium are highly radioactive and have a high thermal power. Moreover, the even isotopes spontaneously fission and are considerable neutron emitter. For these reasons manipulating curium is particularly challenging, and some Country (e.g., France) has opted to transmute only Am.

the core), and transmutation in heterogeneous mode, with 10% to 20% of Am in peripheral blankets (Grouiller et al. 2013).

VHTR

Helium is an inert gas which can be heated to higher temperatures than water, thereby allowing to obtain higher plant efficiency compared to the water cooled design. Similarly to the high-temperature gas-cooled reactor, of which the VHTR is a next evolutionary step, the VHTR is characterized by a thermal neutron spectrum using graphite as moderator, and low power density. Thanks to the capacity of reaching core outlet temperatures of up to 1000°C (much higher than the conventional value of 550°C of the steam produced in a LWR), the VHTR is very well suited for co-generation of electricity and nuclear heat. The latter can be used as process heat for refineries, petrochemistry, metallurgy and hydrogen production, making the HTGR an attractive heat source for large industrial complexes. The production of hydrogen can occur in several ways, from only heat and water by using thermochemical processes (such as the sulfur-iodine process), combined thermochemical and electrolysis (such as the hybrid sulfur process), high temperature steam electrolysis, or from heat, water, and natural gas by applying the steam reformer technology. With respect to electricity generation, two options are considered: a direct cycle with a helium gas turbine system directly placed in the primary coolant loop, or an indirect cycle with a steam generator and a conventional Rankine cycle. Figure 4 presents the general layout of a VHTR power plant.

A major advantage associated to VHTR is the inherent safety features related to the adoption as basic fuel component a coated fuel particle that can survive extremely high temperatures before losing its integrity and releasing radioactive FPs. The version of the coated fuel particle adopted by the Gen IV VHTR, the TRISO (Tristructural Isotropic) particle, is shown on the upper left of Fig. 5 (IAEA 2011, Allen et al. 2010). Each particle, the size of about 0.5 mm, is made of a central kernel of fuel surrounded by several layers of non-fuel material which form a very effective containment structure for the FPs. Experimental tests have already demonstrated this inherent safety in case of extreme accidents such a severe reactivity insertion (as in the Chernobyl case) or residual heat removal failure (as in Fukushima case). The TRISO particles can be arranged in two reactor core configurations, namely the pebble bed type, and the prismatic block type (see Fig. 5). Although the shape of the final fuel element for the two configurations is different, the technical basis for both is the same. The TRISO coated particle fuel can support alternative fuel cycles such as U-Pu, Pu, MOx, and Th-U.

Besides the robustness of TRISO fuel, the VHTR safety features are related to the graphite core, mainly for two reasons. First, the graphite

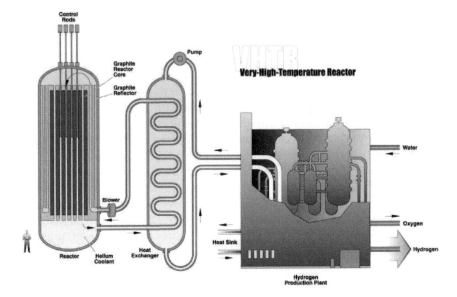

Fig. 4: Gen IV VHTR.

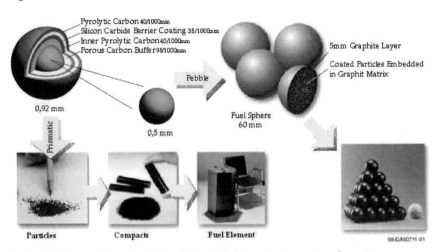

Fig. 5: TRISO coated particle fuel, and its pebble bed and prismatic configuration.

core of HTGRs has a high thermal conductivity, which aids in preventing hot-spots from forming within the core. Second, the graphite core has a high heat capacity. The latter characteristic, combined with the low core power, provides a relatively long delay in the thermal response during LOCAs or reactivity insertions, and produces a reactor concept that does not need off-site power to survive multiple failures or severe natural events. Studies have shown that the maximum fuel temperature in the VHTR graphite core is not expected to occur for several days following the loss of coolant (OECD 2002), providing significant time for operators to take action. Figure 6 shows the peak core temperature during a loss of coolant computer analysis for different VHTR core layouts (Bayless 2003). In each case, the peak temperatures are not reached for approximately 2–3 days. In addition, the helium coolant, graphite moderator, and TRISO fuel combine to give the core a strong negative temperature coefficient for reactivity. This provides power and temperature attenuation during accidents, since the fission reaction rate (i.e., the rate of heat generation) slows as the core temperature increases. These core safety aspects have been confirmed in safety demonstration tests in past and present HTRs (AVR in Germany, HTTR in Japan).

Because of these beneficial features, the major goal of accident mitigation and prevention in VHTR becomes the limitation of maximum core and fuel temperatures. If the ceramic and non-ceramic core components can be designed to have sufficient high-temperature capabilities to preclude structural damage and subsequent release of FPs (the degradation temperature of current TRISO fuel is in the range of 1600–1800°C), VHTRs would be able to prevent radioactive release without operator intervention or active safety system. No external accident management would have to be undertaken outside the plant fence, and no off-site emergency response would be necessary. Ultimately, VHTR could be built at industrial sites in areas with dense population, in order to support process heat applications and reduce carbon emission (Chapin et al. 2004).

R&D activities performed in the last decade have led to major accomplishments, and the accumulation of a large experience and extensive international databases. One example is the AVR reactor, the first pebble bed HTR worldwide, operated in Germany during the period 1967–1988. While AVR operation for 21 year has demonstrated the technical

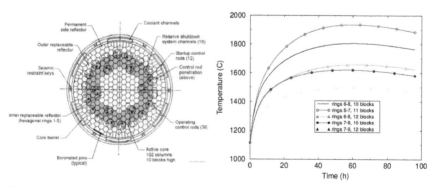

Fig. 6: VHTR reactor vessel cross section (left) and time evolution of peak core temperature during a loss of coolant computer analysis for different core layouts.

feasibility of HTGRs, serious problems have also occurred,[11] implying that a large R&D effort still needs to be performed on the concept. Some of the R&D tasks suggested by the operation of AVR are summarized as follows: the development of a coated fuel particle even more capable of retaining metallic fission products in long term operation; development of reliable quality control for fuel elements; full understanding and reliable modeling of core temperature behavior and of pebble bed mechanics; full understanding of fission product transport in the coolant circuit; development of fast detection system for metallic fission product release from core; material development of process heat components; HTR specific dismantling and disposal items.

GFR

By virtue of its fast spectrum, the SFR is very well suited to enhance uranium utilization via breeding and MA burning, while the very high outlet temperature of the cooling gas of the VHTR allows it to achieve high efficiencies. Replacing the liquid metal coolant of a SFR with a gas results therefore in a system that combines the advantages of fast-spectrum systems for long-term sustainability of uranium resources and waste minimization, with those of high-temperature systems for high thermal efficiency and industrial use of the generated high-quality heat. Moreover a gas coolant is chemically inert (allowing operation without corrosion and coolant radio-toxicity) and single phase (eliminating boiling), and is an inefficient moderator (the void coefficient of reactivity is small).

The absence of graphite (present in VHTRs) and liquid metal coolant (present in SFRs) introduces however some new technological challenges. First and foremost, the low thermal inertia of the gas leads to rapid heat-up of the core following loss of forced cooling. Since GFR has a high power density, removal of the decay heat becomes problematic. Also, gas-coolant density is too low to achieve enough natural convection to cool the core, and the power requirements for the blower are important at low pressure. Lastly, additional consideration will need to be given to the effects of a fast neutron dose on the reactor pressure vessel in the absence of core moderation.

The reference design for the GFR, the plant schematic of which is shown in Fig. 7, is currently based around 2400 MWt, since a lesser power would not allow break-even-breeding. The power plant would adopt an indirect cycle (to lower technological risk and increase flexibility

---

[11] The major problem has been the heavy contamination of the primary circuit with dust bound metallic fission products, in particular ^{90}Sr and ^{137}Cs, due to both inadequate fuel quality and inadmissible high core temperatures. The latter are probably caused by the insufficiently examined pebble bed mechanics.

of working fluid in turbine) with pressurized helium (~ 90 bar) on the primary circuit, exiting the core at ~ 850°C, a Brayton cycle on the secondary circuit, and a steam cycle on the tertiary circuit. The target net efficiency of the current design is 48%. The reference fuel compound is (U,Pu)C/SiC (70/30%)[12] with about 20% Pu content, and with a core volume fraction of

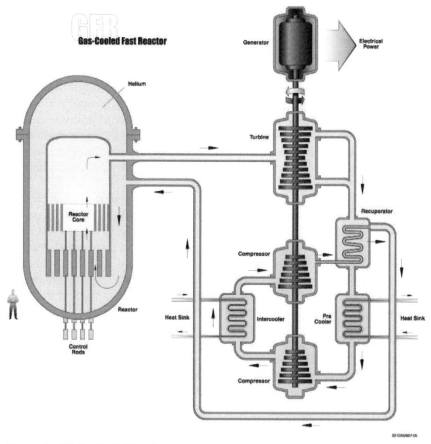

Fig. 7: Gen IV Gas-Cooled Fast Reactor.

---

[12] As a fast reactor fuel, the mixed carbide has many advantages over the oxide fuel. The high metal atom density and the high thermal conductivity imply that the carbide fuel is a better breeder than an oxide fuel. Further, because of the high thermal conductivity, the temperatures as well as the temperature gradients in the carbide fuel are much lower than in the case of oxide fuels, so that the migration of the fuel constituents as well as the FPs is reduced. On the negative aspects, carbide fuel is difficult to fabricate as it is highly pyrophoric and highly susceptible to oxidation and hydrolysis. Moreover, reprocessing spent carbide fuel is difficult as it is difficult to dissolve in nitric acid and leaves behind organic complexes.

fuel/gas/SiC equal to 50/40/10%. The conversion ratio is set to one (self-sufficiency), with burn-up of 5% fissions per initial metal atom.

Since 2006 R&D activity on GFR has been carried out mainly in two areas, conceptual design and safety, and fuel and core materials. Because of the negligible thermal inertia of the gas coolant, ensuring a robust shutdown and decay heat removal (DHR) system without external power input, even under depressurized conditions, is required. The orientation is for electrical (battery) driven blowers to handle depressurized DHR. Attention has to be focused on the integrity of these electrical infrastructures following an extreme event. Work is needed on two fronts in particular, first to reduce the likelihood of full depressurization and second, to increase the autonomy of the DHR system through the use of self-powered systems, i.e., not requiring any external power unit. The development of an acceptable fuel system is also a key viability issue for the GFR system. Target criteria for the cladding are: clad temperature of 1000°C, during normal operation; no FP release for a clad temperature of 1600°C during a few hours; and maintaining the core-cooling capability up to a clad temperature of 2000°C.

Within the research activity on GFR, the experimental reactor project ALLEGRO is currently being undertaken by a consortium of four central-European Countries. This reactor is expected to be built within the next 10–20 years, and it would be the first fast spectrum gas-cooled reactor to be constructed, and the test bed to develop and qualify the high-temperature, high-power density fuel that is required for a commercial-scale high-temperature GFR. A scheme of the reactor is presented in Fig. 8 (Stainsby et al. 2009). The nominal power is 75 MWt, with a primary circuit with helium at 70 bar, and with an inlet/outlet core temperature of 260/530°C. There are two secondary pressurized water loops, and a tertiary atmospheric air (no power conversion).

The main objective of ALLEGRO is the demonstration of the following key GFR technologies: core behavior and control, development of ceramic fuel, helium circuits and components, decay heat removal. The reactor shall be operated with two different cores. The starting core with UOx or MOx fuel in stainless steel cladding will serve as a driving core for six experimental fuel assemblies containing the advanced carbide (ceramic) fuel. The long term core will consist solely of the ceramic fuel and will enable to operate the reactor at the high target temperature of a GFR power plant (increase the inlet/outlet temperature to 400/850°C).

The DHR system consists of three loops (helium/water heat exchanger) located above the core to facilitate natural helium circulation, each one of them being designed to remove 3% of the nominal power. If the blowers are not available, the DHR system should be able to operate in forced and natural circulation.

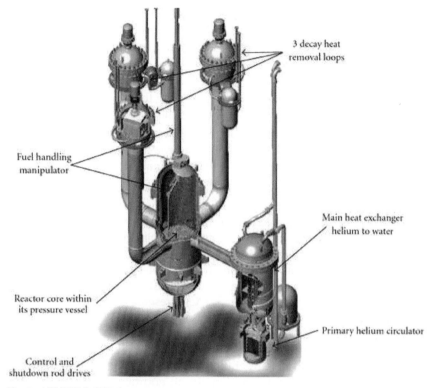

3 decay heat
removal loops

Fuel handling
manipulator

Main heat exchanger
helium to water

Reactor core within
its pressure vessel

Primary helium circulator

Control and
shutdown rod drives

Fig. 8: ALLEGRO GFR demonstrator.

LFR

The SFR (like all fast neutron spectrum reactors) is very well suited to satisfy the requirement of sustainability sought by GIF, in that MAs are fissioned efficiently with high energy neutrons, and breeding is favored. However, SFR has to deal with the problematic aspect of exothermic reaction of sodium with air and water. This leads to the necessity of an intermediate loop, and very stringent requirements on reactor leak tightness. This negative aspect of sodium can be eliminated by substituting it with Pb, or with the Pb-Bi eutectic.[13] Besides the elimination of the intermediate loop, the physical properties of Pb lead to several other advantages:

- The very high boiling point of lead (up to 1743°C) allows operation at low pressure and high temperature, and eliminates the risk of core voiding due to coolant boiling.

---

[13] LBE: 44.5wt.%Pb-55.5wt.%Bi.

- The high density of lead contributes to fuel dispersion instead of compaction in case of core destruction.
- The high heat of vaporization and high thermal capacity of lead provide significant thermal inertia in case of loss-of-heat-sink.
- Lead shields gamma-rays and retains iodine and caesium at temperatures up to 600°C, thereby reducing the source term in case of release of volatile fission products from the fuel.
- The low neutron moderation of lead allows greater spacing between fuel pins, leading to low core pressure drop and reduced risk of flow blockage, and higher natural circulation capability.

The most notable disadvantage of molten lead is that it is corrosive and erosive for structural materials at high-temperatures. This leads to a series of negative consequences, such as a limitation of the operating temperature range (400°C–480°C), a limited coolant flow velocity (max 2 m/s), and the need for coolant chemical (oxygen) control for prevention/limitation of lead erosion-corrosion effects on structural steels. The weight of lead introduces seismic/structural issues, too, while its opacity, in combination with its high melting temperature (327°C), presents challenges related to inspection and monitoring of reactor in-core components as well as fuel handling. In particular, in the case of a reactor system cooled by pure Pb, the high melting temperature of lead requires that the primary coolant system be maintained at temperatures adequately high to prevent the solidification of the lead coolant. Finally, an additional issue with lead-bismuth cooled reactors is related to the accumulation of volatile ^{210}Po which is a strong alpha emitter. In the Russian Federation, in which Pb-Bi reactors have been operated successfully in the submarine program, techniques to trap and remove ^{210}Po have been developed. Further R&D activity is however necessary to see how this experience could be adapted to the different characteristics of a power plant.

The scheme of Gen IV LFR with a pool-type primary system is shown in Fig. 9. The simple coolant flow path and low core pressure drop allow natural convection cooling of the primary system for shutdown heat removal. Because of chemical inertness of the coolant, the secondary side system (delivering high pressure superheated water) can be interfaced directly with the primary side using steam generators immersed in the pool. The expected secondary cycle efficiency of LFR systems is above 42%. The LFR system features a closed fuel cycle based on the U-Pu conversion, and envisions full MAs recycle. The first core is loaded with MOx fuel, while subsequently proliferation-resistant MA-bearing fuel are deployed.

The satisfaction of the four goals of GIF are fulfilled primary because of the coolant inertness (and corresponding simplified plant design, and improved economics) and the use of a closed fuel cycle. The safety goal

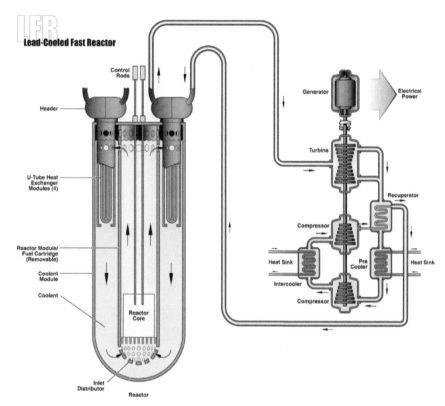

Fig. 9: Lead-Cooled Fast Reactor.

is intended to be achieved by taking advantage of inherent characteristics of the coolant such as chemical inertness as well as thermodynamic and neutron diffusion properties that permit the use of passive safety systems.

Strong efforts are being carried out on the important issue of material corrosion. The main strategy is presently based on oxygen control and/ or surface coating (aluminization or coating, especially for fuel cladding). Material studies have to address also the combined effect of the aggressive corrosion environment and the high radiation dose to which core internal materials will be subjected.

Besides material corrosion, during the next decade the main R&D efforts will be dedicated to the topics of core instrumentation, fuel handling technology and operational advanced modeling and simulation, fuel development (MOx and MA-bearing fuels) and actinide management (fuel reprocessing and fabrication), in-service inspection and repair techniques for opaque medium, and seismic impact.

The LFR systems identified by GIF include a wide range of plant ratings from the small to intermediate and large size. Detailed designs of several

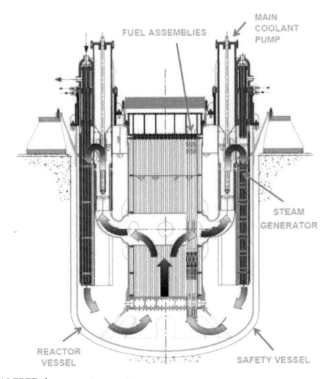

Fig. 10: ALFRED demonstrator reactor.

LFR-type systems will be completed in the near future. In the Russian Federation the construction of the lead-cooled BREST-OD-300 reactor and of the lead-bismuth-cooled SVBR-100 reactor are expected to be carried out, with first criticality close to 2020. In Europe R&D is focusing on the project ALFRED, a scaled demonstrator of the commercially viable, large scale electricity production European Lead Fast Reactor (ELFR) based on lead coolant technology. To present more details on the design of a loop-type Gen IV LFR, we provide an overview the ALFRED reactor (Frogheri et al. 2013), the schematic of which is shown in Fig. 10.

The heat source (the core), located below the riser, and the heat sink (the steam generators, eight in number) at the top of the downcomer, allow an efficient natural circulation of the coolant. The reactor vessel is cylindrical and is anchored to the reactor cavity from the top. A steel layer covering the reactor pit constitutes the safety vessel. The latter allows the primary coolant to always cover the steam generator inlet in case of reactor vessel leaks, and to maintain the lead flow path. The power is set to 300 MWt (125 MWe), with a primary loop inlet and outlet temperatures of 400–480°C, and a secondary cycle working at 180 bar and a temperature range of 335–450°C. The reactor core will contain 171 fuel assemblies, 4 safety rods, and 12 control rods, and 108 dummy elements shielding the inner vessel. ALFRED will investigate the feasibility of a core with a closed fuel cycle (adiabatic core, i.e., a core that burns its own MA), a crucial issue for the actual sustainability with respect to both natural resources and environmental impact.

The control/shutdown system consists of two diverse, independent and redundant shutdown systems. A first system, for both control and shutdown, is based on absorber rods passively inserted by buoyancy from the bottom of the core when the coupling electromagnet that pushes the rods down is switched off. The second system, for emergency shutdown only, is constituted by absorber rods which are inserted rapidly downwards from the top against buoyancy force by the actuation of a pneumatic system. In case of loss of this system, a tungsten ballast will force the absorber down by gravity in a slow insertion. The DHR system having the function of removing the decay heat power in case of unavailability of the normal path, consists of two passive (passive execution/active actuation), redundant and independent systems, DHR1 and DHR2, both composed of four heat exchangers (isolation condensers, ICs) immersed in a water pool, connected to the four steam generators secondary side. Valves equipped with redundant and diverse energy sources (batteries or locally stored energy) provide active actuation. Attention is posed not to provide excessive cooling of the primary coolant leading to fluid solidification. A scheme of the DHR system is shown in Fig. 11 (Frogheri et al. 2013).

MSR

Molten salt mixtures are much more efficient than compressed helium at removing heat from the core of a nuclear reactor, thereby reducing the need for pumping and piping and reducing the core size compared to HTRs. Compared to LWRs, a molten salt reactor (MSR) can run at higher temperatures and therefore achieve higher thermodynamic efficiency, while staying at low vapor pressure. Not surprising the research activity on molten salt reactors dates back as far as the late 1940s, focusing mainly on fluoride salts,[14] such as lithium and beryllium fluorides. A candidate compound is Li_2BeF_4, which has a melting point of 459°C, a boiling point of 1430°C, a density of 1.94 g/cm^3, and a volumetric heat capacity of 4540 kJ/m^{3}K, which is similar to that of water, more than four times that of sodium, and more than 200 times that of helium at typical reactor conditions (Ingersoll et al. 2005). When melted, fluoride salts generate a

---

[14] Fluoride salts are favored because fluorine has only one stable isotope (^{19}F), and does not easily become radioactive. Compared to chlorine, fluorine also absorbs fewer neutrons and moderates neutrons better. The less fast spectrum reduces irradiation damages (both displacement-per-atom and He production) by a factor 5–7.

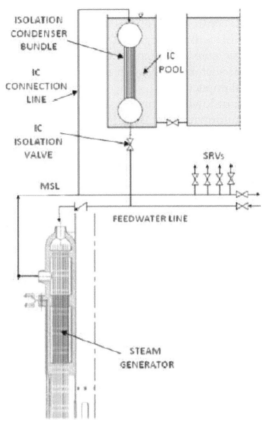

Fig. 11: ALFRED Decay Heat Removal system.

completely transparent fluid, therefore do not pose any problem from the point of view of core inspection, which is moreover exempt from chemical exothermal reactions with air and water. Positive neutronic aspects of a MSR are the limited amount of reserve reactivity, due to the mobility of its fuel, and a strong negative feedback coefficient.

Two different reactor concepts utilizing molten salt have been identified. In the first one, referred to simply as molten salt reactor, MSR, the fissile material is dissolved in the molten fluoride salt, usually in the form of uranium and thorium tetrafluoride, UF_4 and ThF_4 (generally 1% or 2% by mole is added). Since heat is produced directly in the heat transfer fluid, MSRs are characterized by no heat transfer delay and a very fast thermal feedback. The layout of the design of a generic Gen IV MSR is presented in Fig. 12.

Core heat is transported by the fuel salt to heat exchangers before returning to the core for re-heating. The cooling secondary circuit consists of another molten salt loop that is free of radioactive fuel and FPs. Heat from the secondary loop can be used to produce electricity in a tertiary circuit, and, thanks to the high temperature that molten salt fuel can reach, to produce in parallel high heat for industrial processes (water desalinization, production of cement and aluminum, or production of feedstock for synthetic, CO_2-free liquid fuels). An unconventional feature of the primary loop of a MSR is the presence of a chemical reprocessing plant, so that continuous reprocessing is possible. This plant is used to both remove undesired FPs and add more fuel to the reactor so that no reactivity reserve is necessary (fertile/fissile matter is adjusted during reactor operation). Note that purifying and reconstituting fluid fuel is in principle simpler than solid fuel.

Due to its recirculating liquid fuel, no core melt-down can occur in a MSR. Unlike conventional reactors, the avoidance of an abnormal increase in operating temperature does not rely on negative temperature coefficients:[15] in a MSR, should the fuel salt in the core become too hot, a freeze plug (made of salts kept solid by a cooling fan) below the

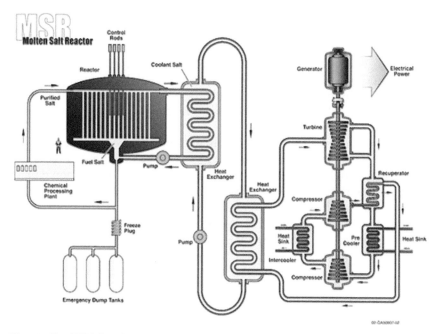

Fig. 12: Gen IV Molten Salt Reactor.

---

[15] A nuclear core has a negative temperature coefficient if, following a temperature increase due to an abnormal increase in fission rate, the density of the materials decreases, neutron leakage from the core increases, and the fission rate decreases.

reactor melts and the liquid content of the reactor core flows down into emergency dump tanks where it cannot continue to fission and can cool safely. For this novel safety feature, MSRs can be considered inherently stable. Another benign safety property of MSRs is that the fuel salt works at normal atmospheric pressure, so a breach of the reactor containment vessel would simply leak out the liquid fuel which would then solidify as it cools. Radioactive byproducts of fission like ^{131}I, ^{134}Cs and ^{137}Cs would remain physically bound to the hardened coolant.

The second reactor concept employing molten salts refers to reactors in which the molten fluoride salt serves as the coolant of a coated particle fueled core similar to that employed in VHTRs—it is custom to refer to this type as a fluoride salt-cooled high-temperature reactor, or FHR. FHR can be seen simply as a variant of VHTR having the capacity to potentially work at even higher temperatures (the boiling point of most molten salt candidates is > 1400°C), to reach higher electric conversion efficiencies, and to retain better FPs in the event of an accident (a molten salt has higher density than helium).

While in the beginning MSRs were mainly considered as thermal-neutron-spectrum graphite-moderator reactors, today the focus has shifted toward fast-spectrum molten salt reactors (MSFRs), which combine the generic assets of fast neutron reactors (extended resource utilization, waste minimization) with those related to molten salt fluorides as both fluid fuel and coolant (low pressure and high boiling temperature, optical transparency). Compared with thermal or epithermal spectrum configurations, MSFRs have a higher breeding ratio, and avoid the problem of graphite life-span.

Contrary to fast reactors with conventional solid ceramic fuel rods, the production of fission gases doesn't pose a problem to liquid fuel, since the gas simply bubbles up, typically to a gas unit in the coolant loop where it can be removed.[16] The liquid fuel can therefore be left in the reactor for a longer time than in conventional solid-fuel reactors, providing an unlimited burn-up capacity. If the MSR has a fast spectrum, the actinides remain in the reactor until they are fissioned, thereby reducing the burden of radioactive nuclear waste to a radio-toxicity that lasts for only few hundred years. For similar reason, MSRs are suitable to dispose current stockpiles of nuclear waste by using them as fuel.

Although MSFRs have been recognized as a potential long-term alternative to solid-fueled fast-neutron systems, mastering the challenging

---

[16] This gas removal capacity is favorable also in relation to the usual accumulation of xenon when a nuclear reactor reduces its power. Xenon is a fission product with high neutron capture cross section. When a reactor is shut down, xenon continues to accumulate, and several days must pass before the reactor is able to reach criticality again. In a MSR, xenon can be removed as it forms in the reprocessing plant connected with the primary loop.

technology will require a notably international research effort. Technology challenges are: neutronic irradiation damage to the structural materials (Ni-based alloys embrittlement) at high temperature; challenge to heat exchanger material mechanical performance and reflector/shield material temperatures due to high power density; proper chemistry control to reduce corrosion; necessity to operate near solubility limits for actinide trifluorides to maintain criticality. In addition, molten salt can generate substantial amounts of tritium which must be trapped at the primary to intermediate heat exchanger to preserve separation of nuclear and non-nuclear regions of plant.

While international collaboration on FHRs is relatively new and a common set of FHR focused projects has yet to be developed, six MSR projects have been proposed: material and components; liquid salt chemistry and properties; fuel and fuel cycle; system design and operation; safety and safety system; system integration and assessment. Within these projects, the following four R&D topics have been identified to assess the viability of MSR:

- Physical-chemical behavior of fuel salts and notably coupling between neutronic, thermal-hydraulic and chemistry.

- Compatibility of salts with structural materials for fuel and coolant circuits, as well as fuel processing material development. This topic is directly linked to instrumentation and control of liquid salt (next topic) because the corrosion of structural materials by molten salt is strongly dependent on the redox potential of the salt.

- Instrumentation and control of liquid salt: *in situ* measurements need to be developed in order to control the redox potential of the salt (and limit the corrosion) and to measure the composition of the salt containing fissile and fertile elements.

- On-site fuel processing: The electrochemical steps required for fuel processing have to be examined from both thermodynamic and kinetic point of view, as well as from the technological and engineering point of view, as the system moves from the lab-scale to the industrial scale.

In the long-term, due to the innovative technology underlying the MSR design, a step-by-step process is required. Initially, test pilot plants will aim to improve skills in handling large quantities of molten salt. At a later time larger demonstrators without induced fission will be dedicated to the study of corrosion under thermal gradient, control of the chemical potential of the molted salt fuel, and study the hydrodynamics of the fuel loop for safety demonstration. Finally, the last step will consider active demonstrators with induced fission. In particular, two reactors are presently foreseen. The first one, named MONO, is a full-scale unit with limited power (100 MWt) representative of a single loop of a larger reactor, a 1000 MWt reactor named DEMO, which would have 16 circulation loops. DEMO should establish the basis for obtaining the approval of safety authorities: demonstrate the control of the reactor; and test the managements of the active salt (drain-out, stop) with their volatile and FPs. It will allow testing of all the structural materials under real conditions.

Being characterized by a design that is simpler, in many respect, compared to conventional PWR reactors (no need of large pressurized containment vessel, reduced need of redundant safety systems, leading to a reduced number of parts), MSRs can be economically competitive. Their simplicity also allows MSRs to be small, which in turns make them ideal for factory-based mass production.

MSR are well suited for the Th-U fuel cycle due to the continuous reprocessing of fuel. The point is that ^{233}Pa has a high cross section for neutron capture, and it is thus a neutron poison: instead of rapidly decaying to the useful ^{233}U, a significant amount of ^{233}Pu converts to ^{234}U and consumes neutrons, degrading the reactor efficiency. To avoid this, ^{233}Pa can be extracted from the reactor active zone of a MSR during operation, so that it only decays to ^{233}U.

Currently, the conceptual design of two MSR systems are under consideration (Serp et al. 2014), the Russian Molten Salt Actinide Recycler and Transmuter (MOSART), and the European molten salt fast reactor (MSFR), both shown in Fig. 13. The first concept aims to be used as an efficient burner of TRU waste from spent LWR fuel with MA/TRU ratio up to 0.45 without any uranium or thorium support (Igniatiev 2012), while the second concept aims at high breeding when using the thorium fuel cycle, with high power densities to avoid excessive fissile inventories.

A promising single fluid configuration for the 2400 MWt MOSART is a cylindrical core with a graphite reflector filled with 100% of LiF-BeF$_2$ salt mixture. It is feasible to design a critical homogeneous core fueled only by TRU trifluorides from UOx or MOx LWR used fuel, while equilibrium concentration of trifluorides of actinides is truly below the solubility limit at minimal temperature in the primary circuit at 600–620°C. Maximum temperature in the fuel circuit does not exceed 720°C. The possibility of operation in self-sustainable mode using different loadings and make up have also been recently considered (Igniatiev 2012).

The reference MSFR is a 3000 MWt reactor with a total fuel salt volume of 18 m^3 (Mathieu et al. 2009), operated at a maximum fuel temperature of 750°C. The fuel salt is composed of lithium fluoride and thorium fluoride and the proportion of heavy nuclei is fixed at 22.5 mol.%. The preliminary design of the primary circuit of the MSFR consists of a single compact cylinder where the nuclear reactions occur within the liquid fluoride fuel salt acting also as coolant. The fuel salt flows in the central part of the core freely from bottom to top without any solid moderator. The return path

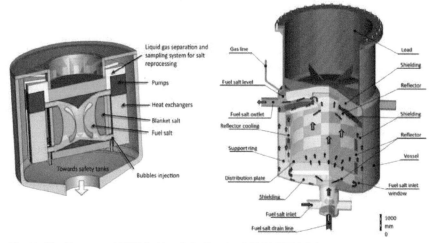

Fig. 13: The European MSFR (left) and the Russian MOSART (right).

is divided into 16 sets of pumps and heat exchangers located around the core. The breeding blanket is located annularly around the central core region.

To summarize, the common objective of MOSART and MSFR projects is to develop a conceptual design for intermediate/fast-spectrum molten salt reactors with an effective system configuration—resulting from physical, chemical and material studies—for the reactor core, the reprocessing unit and waste conditioning (Merle-Lucotte et al. 2013). The conceptual design activities are intended to increase the confidence that molten salt reactors can satisfy the goal of Gen IV reactors in terms of sustainability (Th breeder), non-proliferation (integrated fuel cycle, multi-recycling of actinides), resource savings (closed Th-U fuel cycle, no uranium enrichment), safety (no reactivity reserve, strong negative feedback coefficient) and waste management (actinides burner).

SCWR

A Gen II PWR employs water at p ~ 150 bar as working fluid in the primary circuit, with core inlet/outlet temperatures of 275/315°C. In comparison with Gen I PWR, built in the 1950–1960s, the efficiency of this type of nuclear reactor has increased only from ~ 34% to ~ 36%. Very different is the story of the evolution of coal fired power plants in the last 40 years, with a remarkable increase of net efficiency from around 37% in the 1970s to more than 46% today (Schulenberg et al. 2014). The last 20 years since 1990, in particular, were characterized by an increase of live steam temperature beyond 550°C, thanks to the availability of boiler steels. Along with the temperature increase, the live steam pressure went up to maximize the turbine power, finally exceeding the critical pressure of water. The next generation of coal fired power plants will even reach a net efficiency of ~ 50%, when live steam temperatures of 700°C or more can be realized.

Following this trend, Super-Critical Water-Cooled Reactors (SCWRs) are a class of thermal spectrum,[17] high temperature, high pressure, water-cooled reactors that operate above the thermodynamic critical point of water (374°C, 221 bar). These reactors can be seen as the evolution of Gen II PWR operating at higher pressure and temperature with a one-through direct coolant cycle, and thus can be developed step-by-step from current water-cooled reactors. Figure 14 shows the schematic of the Gen IV SCWR design.

The adoption of supercritical (SC) condition offers several advantages as compared to state-of-the-art water-cooled reactors. Thermal efficiency can be increased to 44%. No reactor coolant pumps are required, the only pumps driving the coolant under normal operational conditions

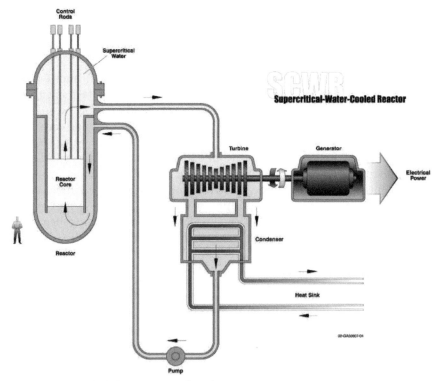

Fig. 14: Gen IV SCWR (pressure-vessel type).

---

[17] Some SCWR designs consider also mixed neutron spectra and fast spectrum cores.

being the feed water pumps and the condensate extraction pumps. The steam generators used in PWR and the steam separators and dryers used in BWRs can be omitted since the coolant is superheated in the core. The containment, designed with pressure suppression pools and with emergency cooling and residual heat removal systems (as in any LWR), can be significantly smaller than those of current water-cooled reactors. Finally, the higher steam enthalpy allows for a decrease in the size of turbine system, leading to a reduction in capital costs of the conventional island.

The technological challenges associated with the development of SCWRs derive mainly from the higher core outlet temperature and the higher enthalpy rise of coolant in the core, compared to current water-cooled reactors.[18] Because of higher fuel cladding temperatures, the zirconium alloys used for fuel cladding in current reactors must be replaced by steels or other high temperature materials.

Since SCWRs represent an evolution of current advanced LWR, the proven safety systems of the latter may be employed, but the strategy must be changed from control of coolant inventory to control of coolant mass flow rate due to the absence of recirculation inside the reactor. As in fossil-fired SC power plants, the problem of hot spots in the core must be addressed, too, most likely by introducing multiple heat-up steps plus intermediate coolant mixing which, however, adds more complexity to the core design. The large density variation within the core could lead to instabilities and subsequently large neutronic variation and high fuel cladding temperature. Finally, the unique water chemistry challenges related to water radiolysis and corrosion products transport must also be addressed.

A number of pre-conceptual core and plant design studies have been carried out in the last decade all around the world (Japan, Europe, Canada, Russia, and outside of the GIF framework, China). Europeans have developed a pre-conceptual plant design of a pressure-vessel-type reactor with a 500°C core outlet temperature and 1000 MWe. The core design is based on coolant heat-up in three steps so as to cope with the large enthalpy rise, as shown on the left of Fig. 15 (Schulenberg and Starflinger 2012). Additional moderator for thermal neutron spectrum is provided in water rods and in gaps between fuel assembly boxes. The fuel assembly is shown on the left of Fig. 16 (Schulenberg and Starflinger 2012). The design of the nuclear island and balance of the plant shows an efficiency improvement up to 43.5% and a cost reduction potential of 20 to 30% compared with the latest BWRs. Safety features, as defined by the stringent European Utility Requirements, are expected to be met.

---

[18] The enthalpy rise in the core would exceed the one of conventional LWRs by almost a factor of ten.

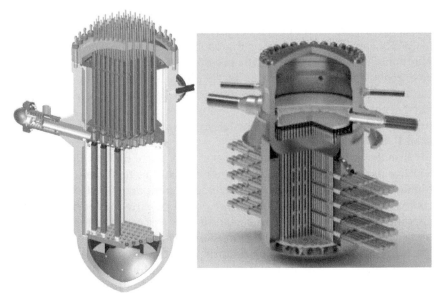

Fig. 15: European HPLWR (left) and Canadian SCWR (right) core concepts.

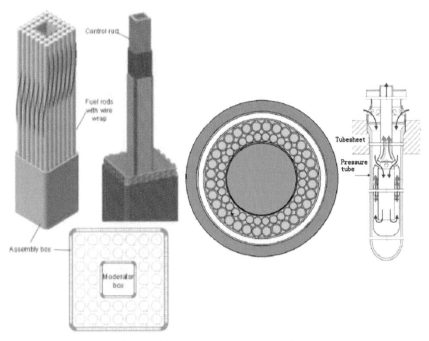

Fig. 16: European SCWR (left) and Canadian SCWR (right) fuel concepts.

A pressure-tube type SCWR concept with a 625°C core outlet temperature and a pressure of 250 bar is being developed in Canada (see right of Fig. 15 (Yetisir 2016)). It is designed to generate 1200 MWe, and has a modular fuel channel configuration with separate coolant and moderator (heavy-water), as shown on the right of Fig. 16 (Yetisir 2016). A high-efficiency, high-pressure fuel channel is incorporated to house the fuel assemblies. The moderator is in direct contact with the pressure tube containing the fuel, and is contained inside a low-pressure calandria vessel. In addition to providing moderation during normal operation, it is designed to remove decay heat from the high-efficiency fuel channel during long-term cooling, using a passive moderator cooling system. A mixture of thorium oxide and plutonium is introduced as reference fuel. With respect to safety, the introduction of the passive moderator cooling system coupled with a high-efficiency fuel channel should lead to a very small core damage frequency during postulated severe accidents such as large-break loss-of-coolant or station black-out events. The main parameters of the European and Canadian SCWR are summarized in the Table 3:

Table 3: Characteristics of the Canadian and European super-critical water-cooled reactors.

	Canadian SCWR	European SCWR
Type	Pressure Tubes	Pressure Vessel
Spectrum	Thermal	Thermal
Pressure (Mpa)	25	25
Inlet Temperature (°C)	350	280
Outlet Temperature (°C)	625	500
Thermal Power (MWt)	2540	2300
Efficiency (%)	48	43.5
Active Core Height (m)	5	4.2
Fuel	Pu-Th (UO_2)	UO_2
Moderator	D_2O	H_2O
Number of Flow Passes	1	3

Thanks to the high temperature of operation reached in SCWRs, the heat exiting the turbine and recovered at the condenser can be used for a number of applications, beside the primary application of electric power generation: hydrogen production, oil extraction (steam-assisted gravity drainage process), desalination, and process heat.

Having gained a better understanding of the expected design(s) of a SCWR, R&D within the next ten years will include more realistic testing of materials, thermal hydraulics and core components, first with out-of-core tests, and then in a reactor environment. Within ten years a decision about a prototype SCWR should be assumed.

5. Next Key Objectives for Gen IV Systems

The degree of technical progress of the different Gen IV systems over the last decade has not being uniform, having depended to a large extent on national priorities and efforts within GIF member Countries. Investments have been directed mainly on two systems, the SFR and the VHTR, in large part due to the considerable historical efforts associated with these technologies. More limited resources were invested on the remaining four concepts, even though each one of them still retains the four necessary characteristics (sustainability, safety and reliability, economic competitiveness, proliferation resistance and physical protection) to be considered a promising Gen IV system.

Time-lines and research needs were developed for each system, categorized in three successive phases (OECD Nuclear Energy Agency 2014):

- the viability phase, when basic concepts are tested under relevant conditions and all potential technical show-stoppers are identified and resolved;
- the performance phase, when engineering-scale processes, phenomena and materials capabilities are verified and optimized under prototypical conditions;
- the demonstration phase, when detailed design is completed and licensing, construction and operation of the system are carried out, with the aim of bringing it to a commercial deployment stage. The original (2002) and updated (2014) indicative time-lines for each GIF system are summarized in Fig. 17 (OECD Nuclear Energy Agency 2014).

The main milestones for each system for the next 10 years are summarized in the Table 4 (OECD Nuclear Energy Agency 2014). Additional R&D activities will be carried out to cope with lessons learned by the recent Fukushima-Daiichi accident. These lessons concern the capability of the nuclear power plant to respond to extreme natural events; consequential loss of safety systems, associated with long-term loss of electrical supplies and the ultimate heat sink; severe accident management systems; loss of core and spent fuel pool cooling; containment integrity.

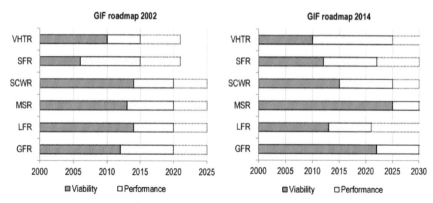

Fig. 17: System development time-lines as defined in the original 2002 GIF Road-map (left) and in the 2014 update (right).

6. Scenarios Employing Gen IV Systems and Accelerator Driven Systems

Evaluating any alternative nuclear fuel cycle strategy to augment or replace the current nuclear energy infrastructures in order to resolve the issues with present-day nuclear energy (poor fuel utilization, highly toxic waste, need for even higher safety standards, etc.) requires consideration of the entire integrated nuclear energy system, including natural resources, facilities and their operation, and the ultimate disposition of all waste materials. This approach allows to take advantage of synergistic features among the components of integrated systems (U.S. Department of Energy 2005).

The GIF technology road-map established an understanding of the ability of various reactors to be combined in so-called "symbiotic fuel cycles", for example, through combinations of thermal reactors and fast reactors to accommodate transition periods. This was one of the primary motivation of having a portfolio of Gen IV systems rather than a single system, since combinations of few systems in the portfolio would provide a symbiotic system worldwide.

An example of a nuclear system in equilibrium has already been shown in Fig. 2, in which the implementation of an advanced fuel cycle requires, besides reprocessing and fabricating facilities and the Gen II LWR operating with a once-through cycle (thus producing highly toxic spent fuels), a Gen IV FR having the role of TRU (Pu + MA) incinerators. Such a scenario, consisting of only one layer, is referred to as a single stratum scenario. Its implementation would allow for a drastic minimization of ultimate wastes in term of volume, radio-toxicity and heat load. It would

Table 4: Key objectives for the next 10 years for the six Gen IV systems.

SFR
• Three baseline concepts (pool, loop and modular configurations).
• Several sodium-cooled reactors operational or under construction (e.g., in China, India, Japan and Russia).
• Develop an advanced national SFR demonstrator for near-term deployment (France, Japan and Russia); proceed with respective national projects in China, Korea and India.

• In the coming years, the main R&D efforts will be concentrated on:
  - safety and operation (improving core inherent safety and I&C, prevention and mitigation of sodium fires, prevention and mitigation of severe accidents with large energy release, availability of ultimate heat sink, ISI&R);
  - consolidation of common safety design criteria;
  - advanced fuel development (advanced reactor fuels, MA-bearing fuels);
  - component design and balance of plant (advanced cycles for energy conversion, innovative component design);
  - used fuel handling schemes and technologies;
  - system integration and assessment;
  - implementation of innovative options;
  - economic evaluations, operation optimization.

VHTR
• In the near future, the main focus will be on VHTR with core outlet temperatures of 700–950°C.
• Further R&D on materials and fuels should enable higher temperatures up to above 1000°C and a fuel burnup of 150–200 GWd/t_{HM}.
• Development of further approaches to set up high-temperature process heat consortia for end-users interested in prototypical demonstrations.

• Development of the interface with industrial heat users—intermediate heat exchangers, ducts, valves and associated heat transfer fluids:
  - Advancing H_2 production methods in terms of feasibility and commercial viability to better determine process heat requirements for this application.
  - Regarding nuclear safety:
    ▪ Verify the effectiveness and reliability of passive heat removal systems.
    ▪ Confirm fuel resistance to extreme temperatures (~ 1800°C) through testing.
    ▪ Proceed with the safety analyses of coupled nuclear processes for industrial sites using process heat.

GFR
• Reference concept of 2400 MWt reactor capable of break-even breeding.
• Improving the design for the safe management of loss-of-coolant accidents including depressurization, and a robust removal of decay heat without external power supply.
• Advancing suitable nuclear fuel technologies with out-of-pile and irradiation experiments.
• Building experimental facilities for qualifying the main components and systems.
• Design studies for small experimental reactors (e.g., ALLEGRO).

Table 4 contd. ...

...Table 4 contd.

LFR

• Prototypes expected after 2020: Pb-Bi-cooled SVBR-100, BREST-300 in Russia.
• Proceeding with detailed design and licensing activities.
• Preliminary analyses of accidental transients including earthquakes and in-vessel steam generator pipe ruptures.
• Main R&D efforts will be concentrated on:
  - material corrosion and development of a lead chemistry management system;
  - core instrumentation;
  - fuel handling technology and operation;
  - advanced modeling and simulation;
  - fuel development (MOx for first core; then MA-bearing fuels); and possibly nitride fuel for lead-cooled reactors (BREST);
  - actinide management (fuel reprocessing and manufacturing);
  - ISI&R (techniques for opaque medium, seismic impact).

MSR
• A baseline concept: the molten salt fast reactor (MSFR). • Commonalities with other systems using molten salts (FHR, heat transfer systems). • Further R&D on liquid salt physical chemistry and technology, especially on corrosion, safety-related issues and treatment of used salts.

SCWR

• Two baseline concepts (pressure-vessel-based and pressure-tube-based).
• R&D over the next decade will include:
  - advancing conceptual designs of baseline concepts and associated safety analyses;
  - more realistic testing of materials to allow final selection and qualification of candidate alloys for all key components;
  - out-of-pile fuel assembly testing;
  - qualification of computational tools;
  - first integral component tests and start of design studies for a prototype;
  - in-pile tests of a small scale fuel assembly in a nuclear reactor.
• Definition of a SCWR prototype (size, design features) for decisions to be taken in coming years.

preserve resources (Pu is an essential resource) and provide enhanced resistance to proliferation (Pu and MA are kept together).

An alternative scenario is suggested by the desire to keep management of MA independent from the commercial fuel cycle. Its implementation would consists in two layers, and for this reason is termed a "double strata scenario": a first layer comprehending LWRs (and possibly FRs) which burn U and Pu, and a second layer consisting of a TRU (MA, and possibly Pu) transmuter. This scenario is represented in Fig. 18. The task of transmutation can be accomplished either by a critical FR, or alternatively by an accelerator driven system (ADS), i.e., a sub-critical system in which the fission reaction chain is sustained by an external neutron source. The scheme of an ADS is reported in Fig. 19.

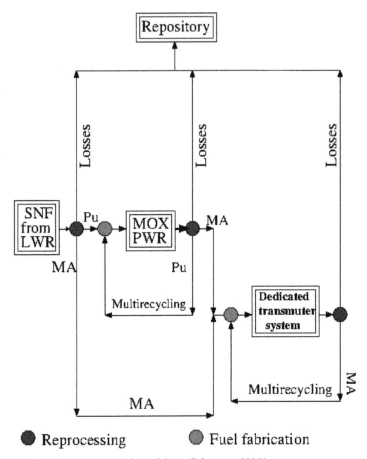

Fig. 18: Double strata scenario (adapted from (Salvatores 2006)).

Two main reasons suggest the employment of ADSs (Gandini 2000). First, achieving the assigned sub-criticality by reducing the fraction of fissile isotopes in the system core, the neutron flux level of an ADS must be accordingly enhanced, and this leads to higher burning rate of MA compared to a critical FR. Second, ADSs are characterized by two important safety features. One is the potential elimination of control elements for reactor operation and reactivity compensation during burn-up, with the consequence of excluding the reactivity accident following an inadvertent control rod extraction.[19] Another safety feature is related to the distance from a criticality condition, which is equivalent to an

---

[19] There are ADS proposal, however, in which control elements are introduced to allow an accelerator operation at constant beam current.

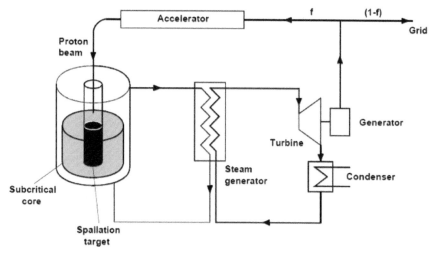

Fig. 19: Sub-critical, or accelerator-driven, system.

extra amount of delayed neutrons,[20] and therefore to a safe distance of the multiplication coefficient from prompt criticality condition. This latter property, in particular, makes the ADS the best candidate as MA incinerator, considering that these elements are characterized by a relative small fraction of delayed neutrons. The use of ADSs in place of critical reactors would of course introduce extra costs, in particular connected with the accelerator capital, operation, and maintenance, besides the plant efficiency penalty. Moreover, the accelerator introduces new safety concerns. In particular, the possibility of accidents following an inadvertent insertion of all the reserve current, relatively large at the beginning of life (cold condition), and the serious problem of a loss of flow rate, or a loss of heat sink accident without current interruption. In these latter cases, in fact, it can be shown (Gandini et al. 2000) that the coolant temperatures reach unacceptable levels (this problem could be however alleviated by coupling the accelerator current feed with the reactor power (Gandini et al. 1999)).

Additional scenarios can be envisaged for the transition period before the availability of Gen IV systems, or for systems having the only goal of reducing the TRU stockpiles (e.g., as a legacy from the past operation of power plants) in the case of phase-out of nuclear power plants (Salvatores 2006).

---

[20] While most fission neutron are emitted practically in concomitance with the fission event, a small but relevant fraction of them are emitted with some delay (up to almost a minute). It is the presence of these delayed neutrons that allows for an easy reactor control during normal operation (the flux dynamic is very much slowed down because of the presence of delayed neutrons).

7. Summary

This chapter provides an introduction to next generation nuclear reactors (Gen IV). International R&D efforts on Gen IV designs, under the coordination of the GIF (Gen IV International Forum), aim at several technological goals which are grouped into four broad areas: sustainability, economics, safety and reliability, proliferation resistance and physical protection. GIF has identified six innovative reactor systems, which can be differentiated according to the neutron spectra, the coolant medium, the core outlet temperature, and the adopted fuel cycle.

Due to their innovative characteristics, Gen IV nuclear systems are not expected to be operative before the mid of the century. The future of nuclear energy is then foreseen as the symbiotic operation of different types of Gen IV reactors with appropriate fuel fabrication and reprocessing facilities, in order to provide a nuclear energy system able to generate electrical energy as well as high-quality nuclear heat which can be employed as process heat by large industrial complexes (e.g., refineries, petro-chemistry, metallurgy), or utilized to desalinate seawater or produce hydrogen.

References

Allen, T., J. Busby, M. Meyer and D. Petti. 2010. Materials challenges for nuclear systems. Mater. Today 13: 14–23.

Aoto, K., P. Dufour, Y. Hongyi, J.P. Glatz, Y.-I. Kim, Y. Ashurko, R. Hill and N. Uto. 2014. A summary of sodium-cooled fast reactor development. Prog. Nucl. Energ. 77: 247–265.

Bayless, P.D. 2003. Prismatic core VHTR analysis using RELAP5-3D/ATHENA. Idaho National Engineering and Environmental Laboratory.

Chapin, D., S. Kiffer and J. Nestell. 2004. The very high temperature reactor: a technical summary. MPR Associates Inc.

Frogheri, M., A. Alamberti and L. Mansani. 2013. The lead fast reactor: demonstrator (ALFRED) and ELFR design. Proceedings of the International Conference. Fast Reactors and Related Fuel Cycles: Safe Technologies and Sustainable Scenarios. FR13, Vol. 1.

Gandini, A. 2000. New reactor concepts and scenarios. Workshop on Nuclear Reaction Data and Nuclear Reactors: Physics, Design and Safety.

Gandini, A., M. Salvatores and I. Slessarev. 1999. The power-current feed coupling in ADS systems. ADTTA '99 Conference, Prague.

Gandini, A., M. Salvatores and I. Slessarev. 2000. Balance of power in ADS operation and safety. Ann. Nucl. Energy 27: 71–84.

Grouiller, J.P., L. Buiron, G. Mignot and R. Palhier. 2013. Transmutation in ASTRID. FR13—Technical Session 6.4. Commissariat à l'énergie atomique et aux énergies alternatives.

IAEA. 2011. Status report 70 - Pebble Bed Modular Reactor (PBMR). https://aris.iaea.org/PDF/PBMR.pdf.

Ignatiev, V.V., O.S. Feynberg, A.V. Zagnitko, A.V. Merzlyakov, A.I. Surenkov, A.V. Panov, V.G. Subbotin, V.K. Afonichkin, V.A. Khokhlov and M.V. Kormilitsyn. 2012. Molten-salt reactors: new possibilities, problems and solutions. At Energy 112: 157–165.

Ingersoll, D.T., E.J. Parma, C.W. Forsberg and J.P. Reiner. 2005. Core physics characteristics and issues for the advanced high temperature reactor (AHTR) concept. Workshop on Advanced Reactors with Innovative Fuels, Oak Ridge National Laboratory.

Kelly, J.E. 2014. Generation IV International Forum: A decade of progress through international cooperation. Prog. Nucl. Energ. 77: 240–246.

Mathieu, L., D. Heuer, E. Merle-Lucotte, R. Brissot, C. Le Brun, E. Liatard, J.-M. Loiseaux, O. Meplan, A. Nuttin and D. Lecarpentier. 2009. Possible configuration for the thorium molten salt reactor and advantages of the fast non-moderated version. Nucl. Sci. Eng. 161: 78–89.

Merle-Lucotte, E., D. Heuer, M. Allibert, M. Brovchenko, V. Ghetta, P. Rubiolo and A. Laureau. 2013. Recommendations for a demonstrator of molten salt fast reactor. Proceedings of the International Conference on Fast Reactors and Related Fuel Cycles, Safe Technologies and Sustainable Scenarios (FR13).

OECD Nuclear Energy Agency for the Gen IV International Forum. 2002. Technology Roadmap Update for Gen IV Nuclear Energy Systems.

OECD Nuclear Energy Agency for the Gen IV International Forum. 2014. Technology Roadmap Update for Gen IV Nuclear Energy Systems.

Salvatores, M. 2006. Physics and Safety of Transmutation Systems. NEA Nuclear Science Report 6090.

Salvatores, M. 2016. Innovative reactor concepts and fuel cycle options. Joint IAEA-ICTP Workshop on Physics and Technology of Innovative Nuclear Energy Systems for Sustainable Development.

Schulenberg, T. and J. Starflinger. 2012. High Performance Light Water Reactor. Design and Analyses. KIT Scientific Publishing, Karlsruhe.

Schulenberg, T., K.H.L. Leung and Y. Oka. 2014. Review of R&D for supercritical water cooled reactors. Prog. Nucl. Energ. 77: 282–299.

Serp, J., M. Allibert, O. Benes, S. Delpech, O. Feynberg, V. Ghetta, D. Heuer, D. Holcomb, V. Ignatiev, J.L. Kloosterman, L. Luzzi, E. Merle-Lucotte, J. Uhlír, R. Yoshioka and D. Zhimin. 2014. The molten salt reactor (MSR) in generation IV: Overview and perspectives. Prog. Nucl. Energ. 77: 308–319.

Stainsby, R., K. Peers, C. Mitchell, C. Poette, K. Mikityuk and J. Somers. 2009. Gas cooled fast reactor research and development in the European union. Science and Technology of Nuclear Installations, 238624.

U.S. Department of Energy, Office of Nuclear Energy, Science, and Technology. 2005. Report to Congress—Advanced Fuel Cycle Initiative: Objectives, Approach, and Technology Summary.

Yetisir, M., M. Gaudet, J. Pencer, M. McDonald, D. Rhodes, H. Hamilton and L. Leung. 2016. Canadian supercritical water-cooled reactor core concept and safety features. CNL Nuclear Review 5: 189–202.

CHAPTER 6

Hydropower

K.A. Kavadias[1,*] and *D. Apostolou*[2]

1. Introduction

During the 20th century, the energy demand of contemporary societies significantly increased resulting in a profligate growth of fossil fuel based power generating capacity. Since the 1970s, many governments introduced measures in order to cope with this phenomenon that had a negative effect on both financial and environmental parameters. Promotion of Renewable Energy Sources (RES) technologies aimed at delivering clean energy to the final consumption without the drawbacks associated with the use of fossil fuels.

In 2015, the renewable energy sector contributed a share of 19.3% to the global final energy consumption. Traditional biomass, including heating and cooking activities in developing countries and rural areas, consisted 9.1% of total final consumption, followed by hydropower at 3.6% (REN21 2017). Figure 1 presents the share of energy sources in the global final consumption.

In the electrical power generating sector, RES comprised, at the end of 2016, 24.5% of the global electricity production. From this share, 16.6% was covered by hydropower, while the rest 7.9% included the rest of the RES technologies (see Fig. 2) (REN21 2017).

---

[1] Soft Energy Applications and Environmental Protection Laboratory, Piraeus University of Applied Sciences, P.O. Box 41046, Athens, 12201, Greece.

[2] Engineering & Technology Research Group, Center for Energy Technologies, Department of Business Development and Technology, Aarhus University, Birk Centerpark 15, 7400 Herning – Denmark.
Email: j.apostolou@puas.gr

* Corresponding author: kkav@puas.gr

Share of Power Generating Technologies in Global Energy Consumption in 2015

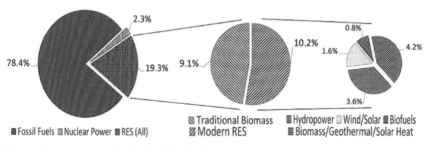

Fig. 1: Share of renewables in the global final energy consumption for 2015. Based on REN21 (2017).

Share of Power Generating Technologies in Global Electricity Production in 2016

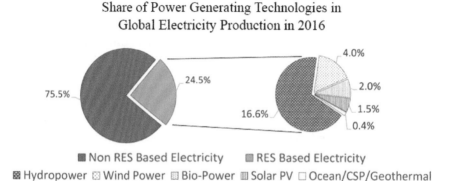

Fig. 2: Share of electrical power generating technologies in the global electricity production for 2016. Based on REN21 (2017).

According to Fig. 2, hydroelectricity is the most widely accepted and common renewable technology for electricity production. During the last 15 years, total installed capacity of hydropower has grown by almost 40%. Apart from being a mature technology supporting sustainability, this increase stems also from the side benefits that hydroelectric technology has to offer concerning water services and regional economic development (WEC 2017).

Hydropower is the technology where power is generated by harnessing the energy of flowing water. It is a versatile energy source where potential energy is transformed into electricity. For this reason, hydropower is highly dependent on local topology and hydrology. The hydrological cycle is driven by solar radiation which at a percentage close to 50% is used to evaporate water. Eventually, this amount of water returns to the surface as precipitation many kilometers away from the origin of evaporation. This phenomenon creates a water cycle from the oceans to the land and an equal flow of water back to the oceans through rivers and groundwater runoff (Kumar et al. 2012).

In this regard, hydroelectricity presents several intrinsic characteristics, compared to other RES and conventional electrical production methods, that enhance its establishment as one of the most mature electricity generating technologies (Liu et al. 2013, Kaldellis 2008):

- Resources are available worldwide.
- Energy conversion is adequately efficient (above 90%).
- Water compared to fossil fuels is not subjected to market fluctuations.
- Long life span and low operational cost.
- Flexible electricity production and supply.
- Improvement of living conditions in the surrounding areas.

The evolution of global hydropower capacity is illustrated in Fig. 3, where it may be noted that since the 1980s the cumulative capacity of hydroelectric stations (excluding pumped-hydro) has increased from 477 GW to around 1100 GW in 2016. The annual hydropower installations presented, during the last 35 years, an increasing trend with the most intense occurring after the beginning of the new millennium. However, it should be mentioned that at the end of the 1980s, new installations were halted for approximately two years.

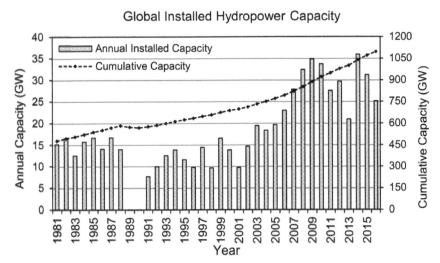

Fig. 3: Annual and cumulative global hydroelectric capacity excluding pumped-hydro, 1981–2016. Based on EPI (2015), IHA (2017a).

Hydropower can be categorized into three main typologies (see Fig. 4) based on the facility type and the size of the generating unit. The most mature category, which dates back many decades, comprises the reservoir storage hydropower, while the most recently developed technologies

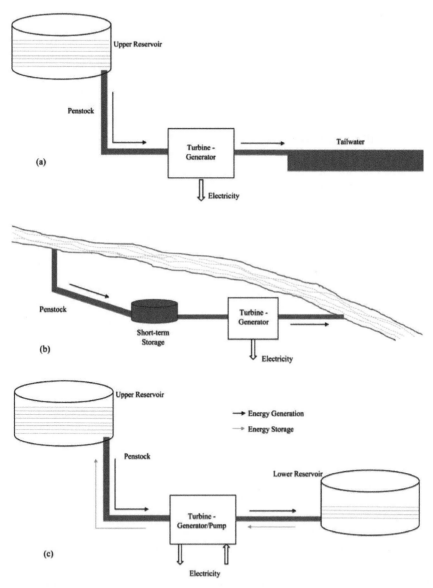

Fig. 4: Contemporary hydropower technologies.

include the run-of-river (RoR) and pumped hydro-storage plants (PHS) (IEA 2012).

- Reservoir Storage (Fig. 4a): This category of hydropower comprises a facility where a dam is used to impound water which, once released, is used to operate a turbine that in turn activates an electric generator. This kind of facility may provide base-load power or cover peak-load demand depending on the capacity of the station compared to the reservoir size. The short start up or shut down operation makes this layout very appealing for peak load regulation.

- Run-of-river (Fig. 4b): RoR installations comprise a facility where water from a river stream is used to drive a turbine through a canal or penstock. These plants include a short-term storage located higher than the turbine-generator configuration which is generally replenished from natural river water flow or released from an upstream reservoir. Due to its operational principle, these installations are highly dependent on weather conditions and therefore they present variable generation.

- Pumped hydro-storage plants (Fig. 4c): PHS plants are most appropriate for peak-load supply by using water flow in a cycle between a lower and upper reservoir. They are considered to be a zero sum electricity producing technology because, the generated operation energy is fed into the electrical grid, and almost the same amount of electricity is consumed in order to pump the water from the lower to the upper reservoir. Practically, PHS is an energy storage technology where excess energy during off-peak hours is used to store water in the upper reservoir. In turn, during peak-load hours the plant operates similarly as the reservoir storage technology. Introduction of PHS, especially in isolated electrical grids, is a promising solution able to encounter the increasing power demand and high electricity production cost (Kaldellis et al. 2010).

2. History

The use of water to perform several work tasks has been established thousands of years ago. Water wheels were used for grinding wheat since the Greek Classical Period (5th and 4th centuries BC) and the Han Dynasty in China (206 BC–220 AD), and subsequently as labor-serving devices in the 14th century. Other tasks included sawing mills, textile mills, and later manufacturing plants. Evolution of contemporary hydropower turbines began in the mid-1700s and at the end of the eighteenth century more than 10,000 water wheels were operating in New England alone (Gulliver and Arndt 2003, IHA 2017b, DOE 2017).

The power output of these early arrangements was in the order of 100 kW, just above the steam and internal combustion engines of that time. Just after the mid-1800s, hydropower was introduced into the generation of electrical energy. Although initial designs of hydropower turbines had been demonstrated by Bernard Forest de Bélidor, the first hydroelectric turbine dates back to 1833 and was developed by Benoit Fourneyron.

In 1849, James Francis developed one of the most widely used water turbines till date, the so-called "Francis" turbine, while in 1870, Lester Allan Pelton, an American inventor, developed the "Pelton" turbine (IHA 2017b).

In 1882, the first hydroelectric plant began its operation in Appleton USA where Direct Current (DC) technology was used. However, due to the drawbacks of DC, these early power plants had the capability to supply electricity only to the adjacent regions with. With the breakthrough of Alternative Current (AC), large hydroelectric plants were deployed in order to supply electrical energy to regions located at large distances. Multiple examples of this transition include the first commercial AC hydroelectric plant in California in 1883, the first three-phase hydroelectric system in 1891 located in Germany, and the hydropower plant located at Niagara Falls in New York State (Gulliver and Arndt 2003, IHA 2017b, DOE 2017).

The first half of the 20th century, saw the global spread of large hydropower. In 1913, an Austrian professor named Viktor Kaplan developed one of the most common water turbines, the "Kaplan" turbine which is based on an adjustable blades' propeller type configuration. The Hoover Dam on the Colorado River in 1936 was the largest hydroelectric plant with capacity of 1345 MW, and was surpassed in 1942 by the Grand Coulee Dam with an output power of 1974 MW. On the other hand, smaller hydroelectric plants that were constructed at the beginning of the 1900s, were retired due to the high operating and maintenance (O&M) costs which could not be covered from their income. In contrast, between 1960 and 1980 large hydropower installations were implemented in many regions worldwide and more specifically in Canada, USSR, and Latin America. In fact, the biggest hydroelectric power station until 2008 was the Itaipu Dam being in service since 1984, straddling Brazil and Paraguay, with a power capacity of 14 GW (Gulliver and Arndt 2003, IHA 2017b).

Nowadays, the largest hydroelectric power plant is the Three Gorges Dam in China with 22.5 GW of power output and energy production between 80 and 100 TWh/year (Kumar et al. 2012). In order to comprehend the magnitude of this power plant it could be compared to the total installed power capacity of Greece in 2017. Power capacity in Greece, including thermal units, hydroelectric units and RES, reached in April of 2017 was 18.9 GW (LAGIE 2017, HEDNO 2017). Additionally, in

2015, gross electricity production in Greece reached 51.8 TWh (Eurostat 2017). Hence, one may notice that the size of the Three Gorges Dam power plant is capable of covering the entire power demand of Greece while maintaining a power reserve for emergency cases.

3. Hydro Energy Potential—Hydropower Market

3.1 Global and Regional Hydro Electricity Generation

Since the previous century, hydropower was established as one of the most mature electricity generating technologies. In fact, at the beginning of the 1970s it was covering almost the 1/3 of global electricity demand (see Fig. 5) reaching approximately 1800 TWh. In 2016, hydroelectric energy production reached 4100 TWh which indicates an almost three-fold increase during the last 40 years (IEA 2017a, IHA 2017b).

According to Fig. 5, hydropower nowadays provides more than 16.5% of the world's electricity which is by far higher than the other RES (7.9%) and significantly higher than the nuclear based electricity production share which reaches 10.6%. On the other hand, fossil fuel based electricity production remains the main technology for covering the global electricity demand reaching a share of more than 65% (IEA 2017).

In the European continent hydroelectric energy covers approximately 15% of total electricity demand (BP 2017). Figure 6 presents the share of hydroelectricity to the gross electricity generation in the major EU countries. In most countries, the contribution share of hydropower to electricity demand in 2016 was below or reached 10%. On the other hand, Scandinavian countries utilizes hydropower to a higher degree for covering their electricity needs. Norway, by taking into advantage its physical terrain and climatic conditions, is presently covering above 90% of its electricity demand with hydroelectric energy. Similarly, Sweden's hydro-based electricity contribution exceeds 40%, while Finland's share to gross electricity production reaches 20%. The country which ranks second in the utilization of hydropower is Austria with a contribution share to its gross electricity generation above 53% for 2016. Apart from Scandinavian countries and Austria, Romania is also one of the European countries that has invested in hydroelectric energy. In 2016, hydroelectric generation reached 18.5 TWh out of 64.8 TWh of total electricity generation, contributing more than 28% (BP 2017).

Electricity production via hydropower technologies in comparison with other RES, is not considered variable since the operator of a plant is able to control the inlet stream according to the storage capabilities of the installation in conjunction with predictions concerning precipitation. However, in long-term operation even hydropower can be considered

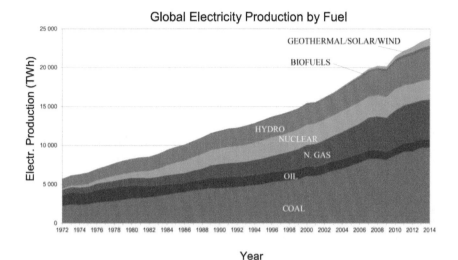

Fig. 5: World electricity generation by fuel type, 1972–2014. Adapted from (IEA 2017).

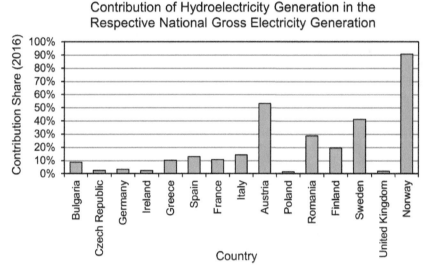

Fig. 6: Contribution of hydro-based electricity to total gross electricity demand in selected European countries. Based on (BP 2017).

variable, as it depends on the annual precipitation and water run off (IEA 2012). Since the mid-1960s, hydro-based annual generated electricity has been increased from around 900 TWh to above 4000 TWh. According to Fig. 7, although Europe and N. America were the leading markets half a century ago covering more than 77% of total hydroelectricity, nowadays

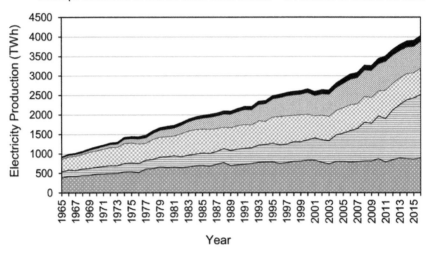

Fig. 7: Hydropower based electricity production by region, 1965–2016. Based on (BP 2017).

the leading market is Asia with a share of approximately 40% of the total hydro-based electrical production in 2016. Significant growth is observed also in the L. America region, particularly after the 1980s, which today ranks third with a production in 2016 of 690 TWh. On the other hand, Europe presented a steady increase and compared to the mid-1960s hydroelectricity has been doubled reaching to 890 TWh in 2016. Electricity production from hydropower plants in N. America reached 680 TWh in 2016. 50 years ago, N. America was the second highest producer of hydroelectricity, while nowadays it ranks fourth just above the African region.

3.2 *Hydropower Potential*

As already mentioned above, hydropower depends on local topology and hydrology. In this context, hydro resource potential is higher in mountainous areas or regions where rivers flow. The potential can be derived from the total available water flow multiplied by the elevation differences and a conversion factor. According to Kumar et al. (2012) the annual surface runoff is estimated to 28,000 m^3 of water and the corresponding theoretical hydropower potential is estimated to be around 40,000 TWh/yr.

However, it is obvious that this amount of potential cannot be technically exploited due to restrictions set by topology and technology limitations. Hence, the definition of hydropower's technical potential can be introduced, indicating the exploitable energy that can be harnessed to produce hydroelectricity. The estimated total technical potential reaches 14,500 TWh/yr, almost 4 times higher than the 2016 hydroelectric energy production, implying that today there is an unexploited potential of around 10,000 TWh. Figure 8 indicates the technical potential of hydropower by region where it is obvious that Asia has the highest which may reach 7680 TWh/yr, a fourfold greater amount than the 2016 hydroelectricity production. The region with the second higher potential is Latin America with an annual value of 2850 TWh, three times higher than hydro-based electrical production of 2016. It is worth mentioning that Africa with current hydroelectric production of around 100 TWh/yr and a potential of 1170 TWh/yr, constitutes the region with the least exploited hydropower resource (IHA 2017a).

The countries exhibiting the highest technical hydropower potential are China and Russian Federation with 2140 and 1670 TWh/yr respectively. However, in terms of current utilization, China exploits 41% of its total potential while Russia only 10%, suggesting that there is a substantial ground for large investments. Figure 9 presents the technical potential at the top 20 countries along with the amount already exploited based on their hydroelectric plants in operation. According to the figure, most of the countries present significant hydropower potential that remains unexploited. There are three countries however that currently utilize above 45% of their entire technical potential. Among these countries, USA uses 52% of its hydropower technical potential (i.e., ≈ 265 TWh in 2016), followed by Brazil at 48% (≈ 410 TWh in 2016), and Norway at 45% (≈ 144 TWh in 2016).

Nevertheless, it should be noted that the estimations of hydropower potential are based on historical and current climatic data, and thus may be subject to alterations caused by a future climatic change. For example, river flows are greatly related to the local climate and particularly with precipitation and temperature. A change in these weather phenomena may substantially affect runoff volume, and seasonal water flows. Additionally, extreme conditions in many regions such as floods and droughts may be prohibitive to hydroelectric investments due to increased cost and risk. Another contingent effect that may also arise from extreme weather conditions is alterations in sediment loads. An increase of sediment could reduce storage capacity of reservoirs and may cause abrasions on turbines and subsequently decrease of the overall efficiency of a hydroelectric plant.

Consequently, it is obvious that climate change is a challenge for hydropower technologies, and therefore mitigation of these effects should

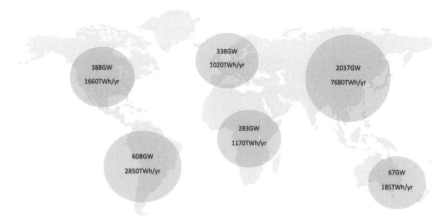

Fig. 8: Hydropower technical potential (power capacity and electrical production) by region. Values based on Kumar et al. (2012).

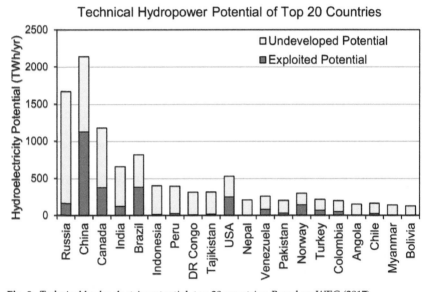

Fig. 9: Technical hydroelectric potential, top 20 countries. Based on WEC (2017).

be compulsory. In general, according to Hamududu and Killingtveit (2010), computations and model simulations indicate that hydropower production in some regions will be negatively affected from climate change (i.e., S., E. and W. Europe, S. and N. Africa, Middle East, Central America, Australasia), while the rest are expected to present a slightly higher hydropower generation compared to current values.

3.3 Major Markets in Hydro-Based Electricity Production

During the last decade, hydropower capacity has significantly increased by approximately 250 GW, reaching in 2016 1100 GW. The majority of these new installations are found in the Asian and Pacific region where more than 330 GW have been installed increasing the hydropower capacity from 218 GW in 2005 to 553 GW in 2016 (EIA 2017, IHA 2017a). In this context, several markets presented significant increase of the hydro-based electricity production contributing to the gross electricity generation. Figure 10 illustrates the distribution of global hydro-based electricity production by region for 2005 and 2016.

According to Fig. 10, although the Asian and Pacific region along with N. America, C. and S. America and Europe shared almost the same percentage in global distribution of hydroelectricity production, a tremendous growth of the hydroelectric energy generation in the Asian and Pacific region increased its share up to 41.3% in 2016. On the other

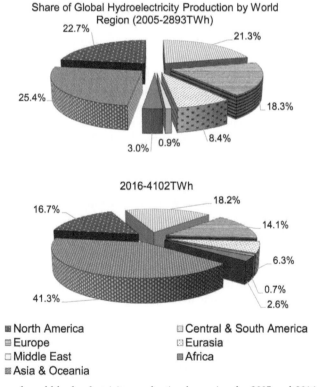

Fig. 10: Share of world hydroelectricity production by region for 2005 and 2016. Based on EIA (2017), IHA (2017a).

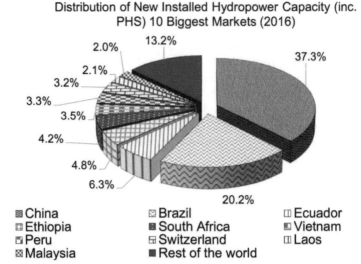

Fig. 11: Market share of new hydropower capacity in 2016. Based on IHA (2017a).

hand, although the hydro-based electrical energy increased in the other regions too, their share has been reduced compared to the 2005 values.

By considering the above, it would be also beneficial to analyze the 2016 hydropower market by country in order to evaluate the future potential of new hydroelectric investments. Hence, Fig. 11 indicates the share of annual hydropower installations for the top 10 markets. Based on the pie chart, China was by far the biggest market in 2016 by adding 11.74 GW of new hydro-capacity reached almost 40% of total new installations. The second largest market for 2016 was Brazil with new capacity of 6365 MW, reaching 20% of total added capacity, while the third was Ecuador with 6.3%. By a closer look to the data of Fig. 11, one may also notice that most of the new hydropower investments were developed in countries with either intense precipitation or where the terrain is characterized by mountainous areas. Four out of the ten biggest markets belong to the Asian region, and three belong to S. America indicating that the regions with the higher hydropower potential (see Fig. 8) present already an increasing trend concerning new hydroelectric investments.

4. Operational Principles

4.1 Capture of Hydro-Based Energy

As already mentioned above, hydropower is greatly related to the hydrological cycle initiated by solar energy which evaporates water from the oceans, lakes, etc., and subsequently form clouds of condensed water that finally falls back to the surface as precipitation. In order to exploit the available streams of water returning to the seas, a dam is placed in specific sites to form a reservoir at a given elevation head between the water storage and the downstream water flow (tailrace).

A typical hydroelectric power station (see also Fig. 12) comprises the following major components (Kaldellis and Kavadias 2001):

- The dam, which is used to store water in a reservoir.
- The hydraulic auxiliary components which include the penstock, gates, valves, and the tailrace.
- The turbine/generator unit including the turbine, guide vanes, speed regulator, the draft tube and the generator.
- The electrical equipment, a transformer and the transmission lines.
- The control room and safety systems.

The power that is generated via this hydraulic head can be calculated by the following relation (Kaldellis and Kavadias 2005):

$$P = \eta_{ov} \cdot \gamma \cdot Q \cdot H \tag{1}$$

where, P is the generated power in W, η_{ov} is the overall efficiency of the hydroelectric plant, γ is the specific weight of water in N/m^3, Q is the water flow (i.e., discharge) in m^3/s, and H is the hydraulic head in m.

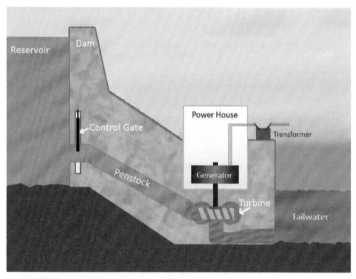

Fig. 12: Typical hydropower installation. Adapted from (Daware 2017).

The most variable parameter is the water flow (Q) which is highly dependent on precipitation, weather data, and the type of the plant (see Section 1). On the other hand, the other parameters of Eq. 1, are specified by the characteristics of the hydroelectric facility.

The hydraulic head (H) is defined as the geometric head (h_g) reduced by head losses due to friction of the water against the pipe walls (h_f) and the local head losses (h_k) due to geometric changes at entrances, bends, elbows, joints, racks, valves and at sudden contractions or enlargements of the pipe section. More precisely, friction losses can be calculated as (Kaldellis and Kavadias 2005):

$$h_f = f \cdot \frac{L}{D} \cdot \frac{U^2}{2 \cdot g} \tag{2}$$

where, f is the friction factor which can be found graphically from Moody chart or diagram based on the type of flow determined by the Reynolds number and the relative surface roughness of the pipe, L is the length of the pipeline in m, D is the inner diameter of the pipe in m, U is the average velocity of the water and g is the gravitational acceleration which can be sufficiently considered to be constant and equal to 9.81 m/s². In case of a pipeline which includes pipes with different characteristics, the calculation should be repeated separately for each group of pipes.

The local (or minor) head losses are referred to all losses excluding those in straight pipelines, i.e., bends, valves, fittings, etc. These losses are expressed as the product of an experimental coefficient k multiplied by the kinetic energy of the water for each of the N parts associated with local head losses in the pipeline:

$$h_k = \sum_{i=1}^{i=N} k_i \cdot \frac{U_i^2}{2 \cdot g} \tag{3}$$

where, k_i is the resistance coefficient of each cause that results in resistance to water flow.

The amount of electric energy that is available from the water at the inlet of the hydro turbine is subject to several losses occurred in the turbine and the electric generator. Hence, one may point out that the overall efficiency of a hydropower plant is defined as (Kaldellis and Kavadias 2001):

$$\eta_{ov} = \eta_Q \cdot \eta_h \cdot \eta_m \cdot \eta_e \tag{4}$$

where, η_Q is the volumetric efficiency which refers to possible losses in water volume (e.g., via leakage), η_h is the hydraulic efficiency of the turbine, η_m is the mechanical efficiency taking into consideration the friction losses in the mechanical parts of the hydro turbine and η_e is the efficiency of the electrical generator.

Acknowledging the above, one can calculate the amount of electricity that is injected in the grid for a given time period as:

$$E = \int_{t=t0}^{t=t0+\Delta t} P(t) \cdot dt \tag{5}$$

where, E is the energy in Wh, and Δt is the time period in hr.

4.2 Turbine Classification and Selection

Capturing the available potential energy from the stored water of an elevated reservoir can be accomplished with the utilisation of water turbines that can be classified into two main categories, the impulse turbines and the reaction turbines. A typical impulse turbine is the Pelton turbine, while reaction turbines consist of Francis turbines and axial flow turbines which are of the Kaplan type or propeller type. The reversible pump turbines that are based on the Francis turbine concept are often considered as a third or special category of water turbines.

During the design stage of a hydro turbine, model analysis is carried out to test the actual performance of the prototype. A properly designed model based on hydraulic similitude can simulate the prototype with low cost. In order to classify hydro turbines based on hydraulic similitude, the "Specific Speed" (K_n) is commonly used. The specific speed combines the power and the head coefficient and is proportional to the operating speed (Papantonis 1995). It can be calculated as:

$$K_n = n \cdot \frac{\sqrt{P/\rho}}{(g \cdot H)^{5/4}} \tag{6}$$

where, n is the turbine speed in rpm, P is the output power in kW, g is the gravitational acceleration in m/s², ρ is the water's density in kg/m³, and H is the net head in m.

However, specific speed has been adopted by engineers in different forms. For example, a short form, taking into account that density and acceleration of gravity are constant, is given as:

$$n_s = n \cdot \frac{\sqrt{P}}{H^{5/4}} \tag{7}$$

where n_s is the specific speed.

Depending on the power output and the available head of a hydropower installation, the type of the turbine should be recognized during the techno economic evaluation of a project. Figure 13 distinguishes the types of turbines used in hydropower installations, based on the head, the power output, and the ratio of turbine rotational speed n to the specific speed n_s.

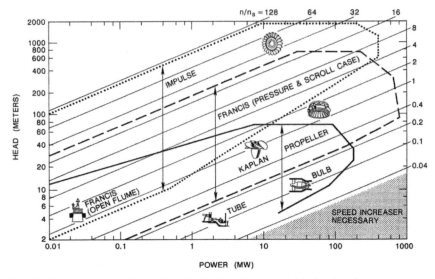

Fig. 13: Hydropower turbines' classification based on available head and power output. Taken from Gulliver and Arndt (2003).

According to the Fig. 13, classification of hydropower turbines comprises of two main categories. The impulse turbines (i.e., Pelton and Turgo) and the reaction turbines (i.e., Francis, and the Kaplan and propeller based turbines). Each of them is appropriate and more efficient depending on the application. The turbines which are most commonly used include:

- Pelton turbines: Impulse turbines are used mostly for high heads (above 300 m), low flow rates, and generally in a wide range of loads. Their main characteristic is the use of curved buckets where one or more nozzles launch water jets upon them, thereby producing a force and consequently a torque that enables rotation (see Fig. 14a). Specific rotational speed is relatively low and proportionate with the square root of the number of the nozzles and therefore can be increased by adding extra nozzles, with the number of nozzles being limited to a maximum of six due to constraints on orderly outflow.

- Francis turbines: Compared to other turbine types, Francis turbines (see Fig. 14b) are considered more flexible due to their efficacy in a wide range of hydropower applications. They are suitable for low/ medium head resources (between 15 and 500 m), large water flows and power loads that may reach 1000 MW. Their operational principle comprises the utilization of the kinetic and potential energy of the water to produce work. Initially, the water enters an annular channel and flows between the guide vanes. The guide vanes are arranged

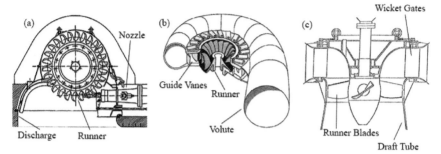

Fig. 14: (a) Pelton turbine, (b) Francis turbine, (c) Kaplan turbine. Adapted from Gulliver and Arndt (2003), Boyle (2004).

on the casing in such a way where the pressure energy of the water is imparted generating impulse that results in rotation of the runner. One of the main components of a Francis turbine is also the volute, located prior to the runner, which is a closed passage whose cross sectional area decreases resulting to an increased velocity of the water (Sawhney 2015).

- Kaplan turbines: Similar to Francis turbines, Kaplan or propeller-type turbines (see Fig. 14c) are mostly vertical axis machineries with a main characteristic of axial-flow. The main difference between a propeller and a Kaplan turbine is that in a Kaplan the runner blades are adjustable in order to achieve higher efficiencies whilst a propeller turbine has fixed runner blades in the airfoil section. Kaplan turbines were developed to utilize large water flows at a low head between 2.5 and 50 m. Their specific speed is usually very high, and are suitable for lower loads compared to a Francis or a Pelton turbine (Venkannna 2009).

In terms of operational efficiencies, the above described turbines present values above 90%. More specifically, Table 1 comprises indicative efficiency values for the most common hydro turbines.

Moreover, Fig. 15, indicates how efficiency changes with the ratio of water flow to the nominal flow. The Pelton turbine, compared to the rest turbine types, is able to sustain its optimum efficiency for a wide range of water flow. In contrast, propeller turbines are more sensitive to changes of the flow, and consequently, maximum efficiency is obtained in a short range of flow rates.

There are also some inherent differences between the two main categories of the hydropower turbines (i.e., impulse and reaction) that are quoted in Table 2.

Table 1: Typical efficiency values of hydropower turbines. Based on Kaunda et al. (2012), Papantonis (1995).

Pelton	Francis	Kaplan	Propeller
90–93%	93–95%	92–94%	87–91%

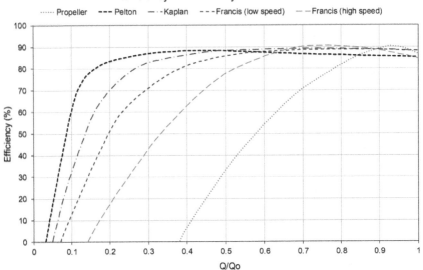

Fig. 15: Efficiency curves of most widely used hydropower turbines. Based on Kaltschmitt et al. (2007).

One of the main factors affecting the efficiency of hydropower turbines and should be considered prior a project implementation, is the effect of cavitation. Reaction turbines and more specifically Francis, operate under variable pressure and are more prone to cavitation. It has been observed that the performance of such turbines declines after some years of operation due to the erosive wear caused by cavitation. The phenomenon of cavitation occurs whenever the pressure of water drops at any part of the turbine below the evaporation pressure resulting in the creation of small bubbles of vapor. Thereafter, the bubbles are curried through the stream to areas with higher pressure, the vapor condenses and the bubbles collapse formatting a cavity. The surrounding water rushes to fill the cavity and therefore pressure may increase up to 7000 atm. This continuously repeating process results in the pitting of the metallic surface causing corrosion (Kumar and Saini 2010).

Table 2: Differences in operation and overall arrangement of hydroelectric turbines. Based on Venkannna (2009).

Impulse Turbines	Reaction Turbines
Turbine can be installed above tail race	Turbine submerged below tail race
Draft tube is not necessary	Draft tube is necessary
Runner is not submerged	Runner is completely submerged
Impinging of water jet causes rotation	Water pressure reaction causes rotation
Pressure of the flowing water remains constant (equal to atmospheric)	Pressure of water reduces during the flow over the vanes
Potential energy is fully converted to Kinetic energy prior entering the turbine	Potential energy is partly converted to Kinetic energy prior the turbine
Power output is based on the jet's Kinetic energy	Power output is based on the Kinetic energy and the Pressure energy
Not subjected to cavitation	Prone to cavitation
Water jet strikes directly the runner	Water enters initially a row of fixed blades and then enters the runner

5. Technology Status and Future Trends

The development of hydropower during the last 150 years and its establishment as one of the major electricity generating technologies, testifies that it is a well-advanced and mature technology covering nowadays more than 16% of global electrical demand. Hydropower is a very flexible power technology with high efficiency. Notwithstanding its maturity, technological development in the hydropower sector continues to expand in order to cope with the increasing electricity demand in most regions of the world.

5.1 Key Recent Innovations

Improvements in turbine technologies during the 20th century resulted in increased hydraulic efficiencies. The introduction of Computational Fluid Dynamics (CFD) facilitated examination of fluid flows inside the turbines and contributed significantly to optimization of the respective turbine components. In this regard, the power output of hydropower stations is increasing and higher values of efficiency are being achieved. The maintenance costs have been reduced due to the minimization of cracking and cavitation, while environmental impacts associated with hydropower have been mitigated with the introduction of low pressure aerating turbines that increase the dissolved oxygen in the water streams and thus protecting the aquatic habitat (IEA 2012).

On top of that, advances in materials induced improvements of their properties and thus resistance to cavitation and corrosion has been increased resulting in a higher life expectancy of a turbine. In addition, the reduction of runner's weight along with increased strength of materials have improved machinability and increased output power and consequently efficiency (IEA 2012). The utilization of such materials also enables operation under harsher environments where for example sediment transport is more intense and has enabled the manufacturing of larger machines. Additionally, the mitigation of sedimentation problems has been accomplished by better management of the land in the upstream watersheds through the avoidance of erosion-induced civil works, and by mechanical removal of the sediment loading (Kumar et al. 2012). Improvements also in civil works during the last decades, including new technologies and materials for designing and constructing dams, have allowed lower cost and environmental impacts, and minimized the required period of construction activities (IEA 2012).

Another significant innovation that has occurred since the development of the electronic industry, was the resource management optimization. Unlike fossil fuels, resources (i.e., stored water) of a hydropower station are highly dependent on climatic conditions. Hence, optimization of the reservoir management was crucial in order to maximize revenues and maintain continuous power generating. Compared to the past, advances in software optimization and decision support tools, and utilization of contemporary Automatic Voltage Regulators (AVR), have improved resource management and response speed of a hydro-based power station respectively (WEC 2017).

Recent advances in the hydropower sector applied also to PHS facilities. More specifically, although PHS is considered, similarly with conventional hydroelectric plants, a mature technology, a number of recent innovations improved its efficacy. A new pump technology employed in existing PHS plants in Japan comprises the use of an asynchronous motor/generator, which allows the rotation speed of the pump/turbine component to be fully adjustable. In this way, energy storage (i.e., pumping mode) can be accomplished even when the available energy in the electric network is low. This technology also allows the operation of the turbines closer to their optimum efficiency. More advances in PHS systems include the utilization of a multi-blade turbine/pump runner in the Kannagawa PHS plant in Japan. This "splitter runner", as it is called, improved generation and pump efficiencies up to 4% compared to the previous installation (Deane et al. 2010).

In the area of site and resource management, the first marine pumped system began its operational service in Okinawa Japan in 1999. Its development started during early 1980s and the main drawback that had to be overcome was the prevention of corrosion caused by seawater. The head of this plant is 136 m and the output power reaches 30 MW. Although this technology presents several advantages compared to conventional PHS, such as lower construction cost and substantially greater site availability, development is ongoing and new projects are still in the designing process (Deane et al. 2010).

5.2 Future Trends in Hydropower Technology

As already mentioned, although hydropower is a mature technology, future trends in technological advancements indicate that there is potential for further growth and development. In this context, many countries have initiated programs to renovate existing hydroelectric plants that have been in operation for more than 40 years. Renovations and upgrading of old hydropower stations cost less than constructing a new facility (see also Fig. 16) and often present less environmental impacts. For this reason, upgrade of existing facilities is an appealing investment in order to improve performance, efficiency, and lifespan of a hydropower plant, and therefore achieve higher revenues.

For example, generating equipment such as the turbine and the generator can be replaced with more technologically advanced equipment more than two times during the lifespan of a project, resulting in increased efficiency and output power at the same water flow (UNESCO 2006). In

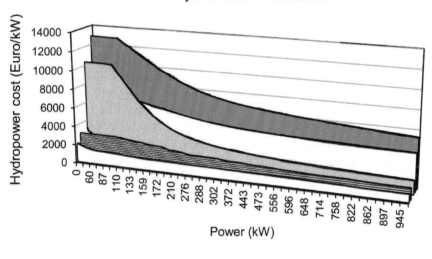

Fig. 16: Cost of hydroelectric power stations per output power based on different construction works. Based on Kaltschmitt et al. (2007).

fact, the US Department of Energy reported that due to these proposed renovations of old hydropower plants, generation may increase by 6.3%.

Presently, ongoing research focusses on aims to extend the operational range of hydropower technology concerning head and efficiency while improving the environmental performance of new or under renovation projects. In this regard, new under development technologies aim to utilize low head infrastructures, below five meters. Apart from this, new contemporary techniques such as fuzzy logic, neural networks, and genetic algorithms are used to improve operation, reliability, and therefore reduce maintenance costs (Kumar et al. 2012).

New R&D projects aiming at improving hydropower technology include (Schneeberger and Schmid 2004, ERPI 2011):

- Introduction of matrix technology where a number of small turbine-generator units are inserted in a matrix shape frame, is under research. This technology is based on the operation of a number of those units under optimal flow conditions according to the available resources. It is appropriate for low head installations and existing irrigation dams.

- In terms of environmental performance, an emerging technology consists of the fish-friendly hydro-turbines. This technology focusses on minimizing the risk of fish deaths during operation without compromising the output power. In fact, a study published from Alden Laboratory in cooperation with the US Department of Energy, suggests that fish-friendly turbines will present efficiencies up to 94% while increasing fish survival to approximately 98%.

- Hydrokinetic turbines are expected to be utilized in sites with very low head, under two meters, by taking into advantage the kinetic energy of a water stream (e.g., river) without requiring a dam or stream diversion. These turbines are underwater systems where the blades capture the kinetic energy of the moving water similarly to wind turbines capturing the wind. Although they were developed for ocean tide power systems it is suggested that in the near term this technology will be more beneficial in river and stream schemes.

- Nowadays, research on new materials in the hydropower sector aims to improve resistance of a plant's components to corrosion. Fiberglass and material combination have been already studied, revealing a positive effect on the lifespan of a hydroelectric facility while increasing its efficiency. In addition, due to the problems arising from large sediment amounts, especially in developing countries where sediment concentration is generally high, new solutions are under development including ceramic coating on steel surfaces to delay erosion and new turbine designs aiming to minimize cavitation damages.

- Concerning civil works, new developments comprise innovations in tunneling and dam technologies. Directional drilling of penstocks, based on oil-drilling processes, has been tested for small hydropower projects with a distance between intake and power station up to one kilometer. This advancement is expected to reduce cost and visual impact associated with above ground penstocks. On the other hand, new dam manufacturing processes such as the one used in roller-compacted concrete dams (i.e., using much drier concrete during construction works), are expected to be used extensively in the future due to their reduced cost compared to conventional construction methods.

- Another important factor that is going to enhance efficiency and optimize operation of a hydropower station, is the development of improved hydrological forecasts and new optimization models. The upgrade of currently used models, is going to optimize water management and therefore contribute to increased energy output of existing and new hydroelectric plants.

By taking into account all the aforementioned innovations it would be beneficial to quote the growth projections of hydropower capacity concerning 2030 in order to get a picture of how hydroelectricity is going

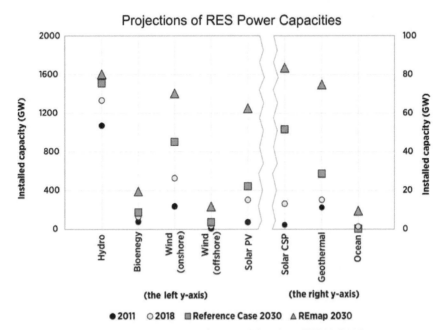

Fig. 17: RES power capacities projections for 2030. Taken from IRENA (2014).

to contribute in the future fuel mix among the other RES technologies. Figure 17 indicates the power capacity of RES technologies in 2011 with the one that is expected in 2018 and in 2030 based on the roadmap for 2030 (REmap 2030) published by IRENA (2014).

According to Fig. 17, it is obvious that hydropower constitutes the major RES technology with significantly more installed capacity compared to other RES, reaching around 1100 GW while the total cumulative capacity of the remaining RES technologies does not exceed 920 GW. The Remap 2030 scenario presents an increase of hydropower capacity to around 1600 GW, suggesting that hydropower will continue to be the major RES contributor in the future electricity fuel mix. However, it should be mentioned that compared to other RES 2030 installations, the capacity share is expected to be more equally distributed between the most mature renewable technologies including onshore wind and solar PV (IRENA 2014, REN21 2017).

6. Conclusions

This chapter provides an analysis of the hydropower, including the major markets and different hydro-based power plant categories. The evolution of hydropower capacity has been investigated quoting also the contribution of hydroelectric energy in the global final electricity demand. The undertaken study provided also a historical review of hydropower since the invention of modern hydro-turbines to the implementation of the large hydroelectric projects of the 20th century. Hydropower potential and major markets were analyzed revealing that Asia and South America present an enormous hydro potential that may reach 4900 TWh/yr for both. In this context, it is demonstrated that the three biggest markets for 2016 reaching 64% of total new installations belong to the above mentioned regions and comprise China, Brazil, and Ecuador.

Besides the above, this chapter also examined the operational principles of hydropower technologies by giving detailed information concerning the main components of a hydroelectric facility and describing the major turbine technologies available today. Further, current technology status and future trends of hydropower are quoted suggesting the prospects of optimizing operation and increasing efficiency of existing and under design hydroelectric power projects.

In general, hydropower is a mature renewable technology that is going to contribute significantly to the carbon emissions reduction efforts. It is a price-competitive technology mostly due to its low O&M costs and compared to other RES, electrical generation can be predicted more accurately, providing a reliable way of covering the electricity demand. To this end, research on technological innovations and new materials is ongoing and is expected to further improve the environmental and operational performance of hydroelectric facilities.

References

Boyle, G. 2004. Renewable Energy: Power for a Sustainable Future. The Open University Oxford, UK.

BP. 2017. Hydroelectricity Consumption—Electricity Generation. Energy Charting Tool. http://tools.bp.com/energy-charting-tool.

Daware, K. 2017. Hydroelectric Power Plant: Layout, Working and Types I Electricaleasy. com. Accessed July 25. http://www.electricaleasy.com/2015/09/hydroelectric-power-plant-layout.html.

Deane, J.P., B.P. Ó Gallachóir and E.J. McKeogh. 2010. Techno-economic review of existing and new pumped hydro energy storage plant. Renewable and Sustainable Energy Reviews 14(4): 1293–1302.

DOE, Department of Energy. 2017. History of Hydropower. https://energy.gov/eere/water/history-hydropower.

EIA, Energy Information Administration. 2017. Hydroelectric Capacity. International Energy Statistics. http://www.eia.gov.

EPI, Earth Policy Institute. 2015. World Installed Hydroelectric Generating Capacity, 1980–2013. Data Center—Climate, Energy, and Transportation. http://www.earth-policy.org/data_center/C23.

ERPI, Electric Power Research Institute. 2011. "Fish Friendly" Hydropower Turbine Development and Deployment: Alden Turbine Preliminary Engineering and Model Testing. U.S. Department of Energy.

Eurostat. 2017. Energy—Total Gross Electricity Generation. Main Tables. http://ec.europa.eu/ eurostat/web/energy/data/main-tables.

Gulliver, John S. and R.E.A. Arndt. 2003. Hydroelectric power stations. In Encyclopedia of Physical Science and Technology. 489–504. Elsevier.

Hamududu, B. and Å. Killingtveit. 2010. Estimating effects of climate change on global hydropower production. In 6th International Conference on Hydropower. Tromsø, Norway.

HEDNO, Hellenic Electricity Distribution Network Operator. 2017. Monthly Electricity Production Report for the Non-Interconnected Islands. https://www.deddie.gr/en/themata-tou-diaxeiristi-mi-diasundedemenwn-nisiwn/stoixeia-ekkathariseon-kai-minaion-deltion-mdn/miniaia-deltia-ape-kai-thermikis-paragwgis-sta-mi/2017.

IEA, International Energy Agency. 2012. Technology Roadmap-Hydropower. Paris.

IEA, International Energy Agency. 2017. Electricity and Heat Statistics. https://www.iea.org/statistics/statisticssearch/report/?country=WORLD&product=electricityandheat&year=1990.

IHA, International Hydropower Association. 2017a. Hydropower Status Report 2015, 2016, 2017.

IHA, International Hydropower Association. 2017b. A Brief History of Hydropower. https://www.hydropower.org/a-brief-history-of-hydropower.

IRENA, International Renewable Energy Agency. 2014. REmap 2030. A Renewable Energy Roadmap—Summary of Findings. Abu Dhabi, UAE.

Kaldellis, J.K. and K.A. Kavadias. 2001. Laboratory Applications of Soft Energy Resources. Athens, Greece: Stamoulis (in Greek).

Kaldellis, J.K. and K.A. Kavadias. 2005. Computational Applications of Soft Energy Resources: Wind Energy, Hydro Power. Athens, Greece: Stamoulis (in Greek).

Kaldellis, J.K. 2008. Critical evaluation of the hydropower applications in Greece. Renewable and Sustainable Energy Reviews 12(1): 218–34.

Kaldellis, J.K., M. Kapsali and K.A. Kavadias. 2010. Energy balance analysis of wind-based pumped hydro storage systems in remote island electrical networks. Applied Energy 87(8): 2427–37.

Kaltschmitt, Martin, W. Streicher and A. Wiese (eds.). 2007. Renewable Energy: Technology, Economics, and Environment. Springer, New York.

Kaunda, C.S., C.Z. Kimambo and T.K. Nielsen. 2012. Potential of small-scale hydropower for electricity generation in Sub-Saharan Africa. ISRN Renewable Energy 2012: 1–15.

Kumar, A., T. Schei, A. Ahenkorah, R. Caceres Rodriguez, J.-M. Devernay, M. Freitas, D. Hall, A. Killingtveit and Z. Liu. 2012. Hydropower. In Renewable Energy Sources and Climate Change Mitigation: Special Report of the Intergovernmental Panel on Climate Change, edited by Ottmar Edenhofer, Ramón Pichs Madruga, Y. Sokona, United Nations Environment Programme, World Meteorological Organization, Intergovernmental Panel on Climate Change, and Potsdam-Institut für Klimafolgenforschung. Cambridge University Press, New York.

Kumar, A. and R.P. Saini. 2010. Study of cavitation in hydro turbines—A review. Renewable and Sustainable Energy Reviews 14(1): 374–83.

LAGIE, Operator of Electricity Market. 2017. DAS Montly Reports—April 2017. http://www.lagie.gr/en/market/market-analysis/das-monthly-reports.

Liu, Jian, J. Zuo, Z. Sun, G. Zillante and X. Chen. 2013. Sustainability in hydropower development—A case study. Renewable and Sustainable Energy Reviews 19(March): 230–37.

Papantonis, D. 1995. Hydrodynamic Machines: Pumps—Hydro Turbines. Athens, Greece: Symeon (in Greek).

REN21. 2017. Renewables 2017—Global Status Report.

Sawhney, G.S. 2015. Fundamentals of Mechanical Engineering: Thermodynamics, Mechanics, Theory of Machines, Strength of Materials and Fluid Dynamics. Third edition. Eastern Economy Edition. PHI Learning Private Limited, Delhi.

Schneeberger, M. and H. Schmid. 2004. StrafloMatrix™, 2004—Further Refinement to the HYDROMATRIX® Technology. In Proceedings of the Hydro 2004 Conference, Porto, Portugal.

UNESCO, United Nations Educational, Scientific and Cultural Organization. 2006. Water: A Shared Responsibility. United Nations World Water Development Report 2.

Venkannna, B.K. 2009. Fundamentals of Turbomachinery. PHI Learning Private Limited, New Delhi.

WEC, World Energy Council. 2017. World Energy Resources-Hydropower 2016.

CHAPTER 7

Geothermal Power

F. Donatini

1. Introduction

Geothermal energy consists of the heat present inside the earth, which, at considerable depths, can reach temperatures of several thousand degrees. Geothermal means "Earth Heat" and it refers to the heat contained in the ground that is continuously transferred toward the earth's surface. This endogenous energy comes from two main sources: from the heat initially contained in the planet, and from the one generated by the decay of radioactive isotopes of elements, such as uranium, thorium and potassium in the earth's crust.

The heat diffuses towards the earth's surface with an average, rather low, thermal flow, of the order of 0.05 W per square meter. It is very much lower than the flow of solar energy that hits the ground with a maximum specific power of 1.35 kW per square meter. For this reason, the energetic use of the geothermal resource requires far greater spaces than those required by other renewable sources. Moreover, unlike other sources, geothermal energy is not equally distributed over the planet, since the earth's crust does not present homogeneous characteristics as regards its thickness and its composition. In fact, there are areas of the earth characterized by a discontinuity in the surface rock structure, located in correspondence of the plates that form the earth's crust, favoring the emerging of magmatic masses at depths ranging between 5 and 10 km. These areas are essentially; the west coast of the American continent including the US, Mexico and

University of Pisa, Largo Lucio Lazzarino, 56122 Pisa, Italy.
Email: franco.donatini@yahoo.it

Latin American countries, the Atlantic ocean zones that include Iceland, Azores and Canary Islands, the Pacific ocean including Hawaii and Far East area countries such as China, Indonesia, Philippines, Japan, New Zealand (Fig. 1).

Fig. 1: The boundaries of tectonic plates that form the earth's crust.

The outcrop of magmatic masses determines a significant increase in these areas of the temperature, which may reach values of 200 or 300°C already at some thousands of meters deep, making possible an adequately efficient exploitation of the geothermal source.

Usually in the crust, the thermal gradient is equal to 3°C for every hundred meters, while in those particular areas, the gradient may be more than five times, allowing to reach temperatures of 300°C at depths of 2000–3000 m which can be easily reached by the current drilling techniques. The mean terrestrial heat flow is 65 mW/m² in continental areas and 100 mW/m² in oceanic areas, with an overall weighted average value of 87 mW/m² (Pollack et al. 1993). In geothermal areas, the heat flow can reach much higher values as 300 to 500 mW/m². The heat flow is based on conductive and convective heat transfer mechanisms. Considering the low thermal conductivity of rocks, such high heat flows can be explained assuming a considerable contribution of convective mechanisms. In fact, the geothermal fields are usually characterized by the presence of fluids, which provide the double function of energy storage and of heat transfer enhancing.

2. Geothermal Resource

A geothermal system essentially consists of three main elements: a heat source, a reservoir and a fluid, which is the carrier of heat transfer. The heat source is a very high temperature (> 200°C) magmatic intrusion that has reached relatively shallow depths (5–10 km). The reservoir is a volume of hot permeable rocks from which the circulating fluids extract heat. The reservoir is generally overlain by a cover of impermeable rocks and connected to a superficial recharge area through which the meteoric waters can replace or partly replace the fluids that escape from the reservoir through springs or are extracted by boreholes. The geothermal fluid is water, in the majority of cases meteoric water, in the liquid or vapour phase, depending on its temperature and pressure (Fig. 2).

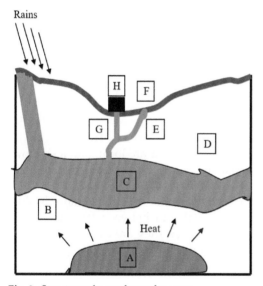

A - Magma
B – Igneous Rocks
C - Porous and Permeable Rocks
D - Compact an Impermeable Rocks
E - Cracks
F - Fumarole
G - Geothermal Well
H - Geothermal Power Station

Fig. 2: Structure of a geothermal system.

With reference to Fig. 2, the formation of a geothermal system takes place through the following phases:

- The hot Magma (A), near the earth's surface, solidifies forming igneous or volcanic rocks (B).
- The heat of magma is transferred by conduction through the igneous rocks or by convection of gas through the cracks, heating the water present in an overlying layer of porous and permeable rocks (C).

- A further layer of solid rock and waterproof (D) covers the layer (C) by trapping the hot water or steam as in a tank ("geothermal tank"). This water often carries chemicals and gases such as CO_2, H_2S and others compounds.
- The solid rock (D), however, has slits (E), which act as vents of a giant boiler, giving rise to fumaroles or hot water basins (F).
- So it is possible to construct wells (G), that take the steam from the cracks to transform it into electric energy using a thermal machine (H).
- The water of meteoric precipitation can in some cases restore the underground tank.

The geothermal sources are classified into three groups:

2.1 Hydrothermal Systems

Hydrothermal systems, the only ones currently industrially explored, consist of geothermal reservoirs that contain water in both liquid and vapor phase, covered by impermeable rocks. They can be distinguished in "dominant water systems" and in "vapor-dominated systems", depending on the status of fluid.

- Dominant water systems: They consist of storages of water in the liquid phase or water-steam mixtures and are of two types. The first ones do not have the cover of impermeable rock and are located at a shallow depth. As the temperature is usually below 100°C, they are exploited for direct thermal uses. On the contrary, when the geothermal field is enclosed by a waterproof cover, the water at high pressure is able to reach temperatures higher than 100°C or between 180 and 350°C, remaining still in the liquid phase. When they are drilled, the water begins to boil due to the associated decrease of pressure, producing a mixture of water and steam, which can be exploited for electricity generation.
- Vapor-dominated systems: They are geothermal reservoirs, geologically similar to the above ones, where the geothermal fluid is present in the form of dry steam, including other gases or soluble substances, and can be directly used in a turbine for power generation, without the separation of water. Vapor-dominant systems are the most suitable and efficient for electricity generation, but, unfortunately, very few are available in the world. The largest reservoir of this type, used since the '60s, is "The Geysers", located in California (USA), with a potential of 1400 MW. The first field of this type in the world has been exploited in Italy (Larderello Tuscany), where in 1904 the count Pietro Ginori produced electricity through a dynamo driven by a motor fed by geothermal steam. The Larderello field has reached

today a power of about 700 MW, achieving nearly 1000 MW with the close field of Monte Amiata. Other fields of this type are present in Japan, New Mexico and Indonesia.

2.2 Geo-pressured Systems

They are systems containing water at temperatures higher than hydrothermal ones and with a higher pressure than the hydrostatic value corresponding to their depth. They are therefore at high temperature and pressure, but, due to their very great depth, their energetic exploitation has not yet developed.

2.3 Hot Dry Rocks

Hot Dry Rocks (HDR) are geothermal fields, in which the circulation of fluids as waterproof is not present. As they have high temperatures, between 200°C and 350°C, their energetic exploitation could be carried out with an adequate efficiency, but it requires the employment of a fluid, which acts as thermal vector. The applied technic is the hydro-fracturing using pressurized water. Through a specially drilled well, high pressure water is pumped into a hot, compact rock formation, causing its hydraulic fracturing, making it permeable and forming an artificial tank. The cold water, circulating in the hot rocks, increases its temperature and, through a second drilled well, reaches the surface as liquid water or steam. Hot fluid is finally fed into a heat exchanger or directly in the turbine, to produce electricity. Another potential way to exploit hot dry rock, is to create an artificial heat exchanger inside the rock, avoiding the technic of hydro-fracturing, which can produce an increase of local seismicity. This technic, known as "Enhanced (Engineered) Geothermal Systems" or "EGS", has not yet been applied, as, due to the very low geothermal flux, it requires enormous heat transfer surfaces.

3. Geothermal Energy Potential

The energy potential of the three geothermal sources are reported in Fig. 3. The geothermal energy contained inside the earth is very high. The heat contained in all hydrothermal systems is equal to primary world energy consumption for only up to two years.

Referring to the heat balance developed by Stacey and Loper (1988), the total heat power from the Earth is estimated 42×10^{12} W, which includes conduction, convection and radiation. A heat power of 8×10^{12} W comes from the crust, which, although represents only 2% of the total volume of the Earth, is rich in radioactive isotopes. A power of $32, 3 \times 10^{12}$ W comes from the mantle, which represents 82% of the total volume of the Earth,

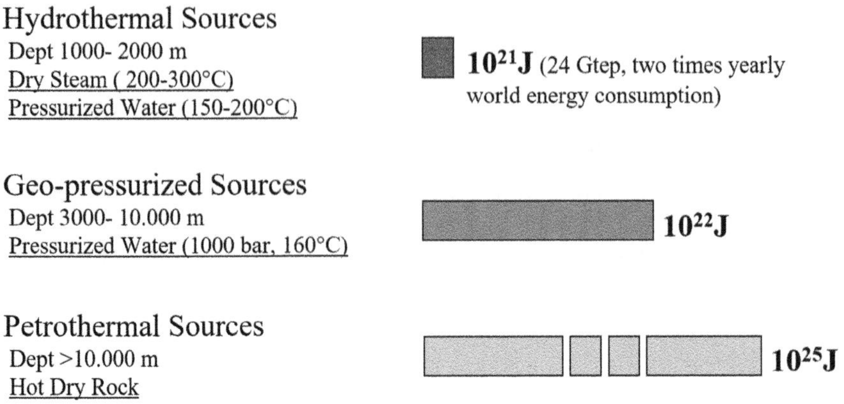

Hydrothermal Sources
Dept 1000- 2000 m
Dry Steam (200-300°C)
Pressurized Water (150-200°C)

10^{21}J (24 Gtep, two times yearly world energy consumption)

Geo-pressurized Sources
Dept 3000- 10.000 m
Pressurized Water (1000 bar, 160°C)

10^{22}J

Petrothermal Sources
Dept >10.000 m
Hot Dry Rock

10^{25}J

Fig. 3: Energy potential of geothermal sources.

and 1.7×10^{12} W comes from the core, which accounts for 16% of total volume and contains no radioactive isotopes. Since the radiogenic heat of the mantle is estimated in 22×10^{12} W, the cooling of this part of the Earth provides 10.3×10^{12} W (Dickson and Fanelli 2005).

The available geothermal energy is enormous, but only a small part can be effectively exploited, due to different reasons, as, too elevate depth drilling to reach adequate level of temperature and the fact that most rocks are not permeable and do not have fluid inside. So, the actual present production of electricity from geothermal resources is low, as is its exploitation for direct uses.

Temperature is a very important parameter to evaluate the opportunity of exploiting a geothermal resource, especially for electricity generation. The most common criterion for classifying geothermal resources is based on the enthalpy of the geothermal fluids that quantifies the thermal level of the resource. The resources are divided into low, medium and high enthalpy, according to the criteria of different researchers (Table 1).

The energetic potentiality of a geothermal resource depends, not only on the energy content, but especially on its temperature. For temperatures

Table 1: Classification of geothermal resources.

Resource enthalpy	Muffler and Cataldi (1978)	Hochstein (1990)	Benderitter and Cormy (1990)	Nicholson (1993)	Axelsson and Gunnlaugsson (2000)
Low	< 90°C	< 125°C	< 100°C	≤ 150°C	≤ 190°C
Medium	90–150°C	125–225°C	100–200°C	-	-
High	> 150°C	> 125°C	> 200°C	> 150°C	> 150°C

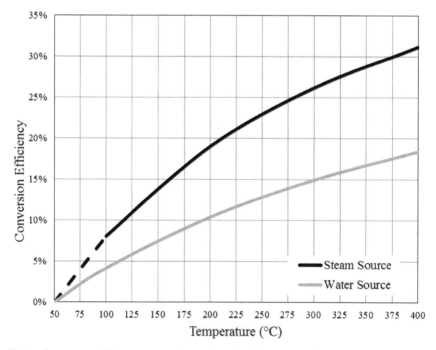

Fig. 4: Conversion efficiency versus temperature for steam and water source.

below 150°C, the production of electricity is not at present economically practicable, due to the low conversion efficiency; in such cases, the resource is used for direct uses of heat. Medium and high enthalpy resources are exploited for power generation, and the conversion efficiency depends both on the status of fluid (steam or water) and on temperature, as showed in Fig. 4.

The trends of Fig. 4 outline's the efficiency in geothermal plants, which is usually much lower (less than one third) than in fossil power stations. Considering that the largest fraction of existing geothermal fields is represented by medium enthalpy liquid sources, the mean value of electrical efficiency of about 10% can be assumed.

The global production of electricity from geothermal resources today is about 14 GW and it could reach nearly 18 GW with the announced capacity addition by 2020 (Fig. 5). If it were possible to exploit all the available energy in earth's crust, it could produce a power of nearly two orders of magnitude, of the current production. This estimate is only true in theory, due to the limitations related with the depth and the low enthalpy of the majority of sources. Moreover, the exploitation

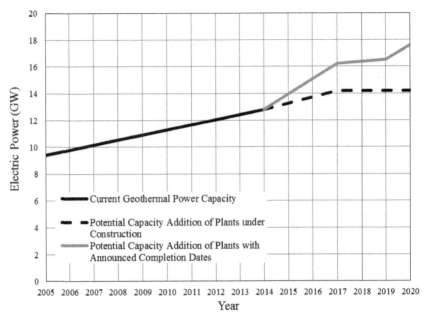

Fig. 5: Worldly Installation of Geothermoelectric Power (GEA 2015).

of geothermal resources today, must allow the resource to maintain itself over a period of time, instead of, frequently building "as large as possible" geothermal power plants, as done in the past, in order to meet economic convenience in a reduced time. In recent times the concept of renewability has been improved with the introduction of the idea of sustainability: this concept includes the idea of expending geothermal resources by using some compensative actions, like cultivation strategies and reinjection. The proper matching between the reservoir capability and the plants parameters (power size, extraction/reinjection rate) is a critical key point. In fact the main task of potential assessment and sustainable design of a plant is the evaluation of resource durability (Franco and Donatini 2017).

According Doe-US estimation, geothermal power could realistically increase in the future to above 100 GW, which is nearly ten times the current installed power.

Today the countries with the highest installed geothermal power are, in order, United States, Philippines, Indonesia, New Zealand, Mexico Italy, Iceland, Kenya, Turkey, Japan (Fig. 6), but the situation is rapidly changing, due to the intense development plans in some countries as Iceland, Kenya and China.

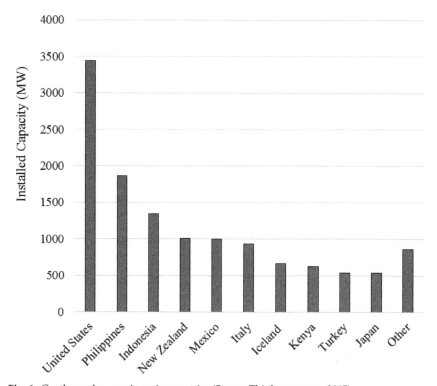

Fig. 6: Geothermal power in main countries (Source Thinkgeoenergy 2015).

4. Geothermal Plants for Power Generation

The configurations of plants employed to produce electricity are different and are according to the typology of available geothermal source, but they are usually based on Rankine cycle. The geothermal plants can be classified in the following typologies: dry steam, flash steam, binary steam and enhanced geothermal systems for the exploitation of hot dry rocks.

4.1 Dry Steam Power Plant

Dry steam power plants (Fig. 7) draw from underground resources of steam. The steam is piped directly from underground wells to the power plant where it is directed into a turbine/generator unit. The incondensable gas contained inside the steam (usually with 5% fraction) is extracted from the condenser by a compressor or by an ejector. The mixer condenser is cooled by a wet cooling tower, which consumes the water flow of about 70% of steam, allowing the reinjection of only 30% water fraction. It is the

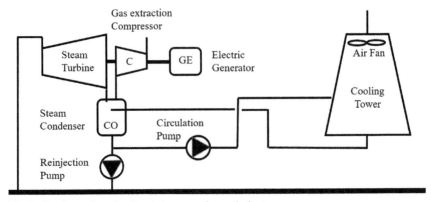

Fig. 7: Configuration of a direct steam geothermal plant.

most efficient solution of geothermal power conversion but there are few available underground steam resources in the world.

The conversion efficiency in terms of steam flow rate per electric power depends on the thermodynamic condition of geothermal fluid, as shown in Fig. 8, while Fig. 9 depicts the trends of parasitic losses of power due to gas extraction and to air ventilation in the cooling tower.

The figures show that on reducing steam temperature, steam consumption as parasitic losses increase, resulting in a strong penalization when temperature falls below 200°C, making the resource exploitation not economically practicable, with this technology.

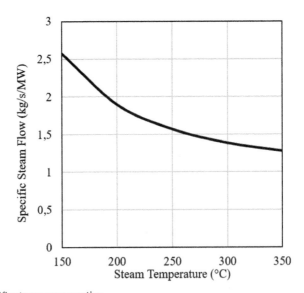

Fig. 8: Specific steam consumption.

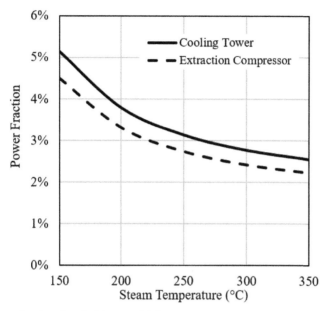

Fig. 9: Power fraction absorbed by parasitic losses.

4.2 Flash Steam Power Plants

Flash steam power plants are the most common and used for geothermal reservoirs of water with temperatures greater than 180°C. This very hot water flows up through wells in the ground under its own pressure. As it flows upward, the pressure is decreased and some of the hot water boils into steam. The steam is then separated from the water and used to power a turbine/generator. Separated water and condensed steam are re-injected into the reservoir. The configuration of the plant is similar to the direct steam one, with the addition of the water/steam separator (Fig. 10).

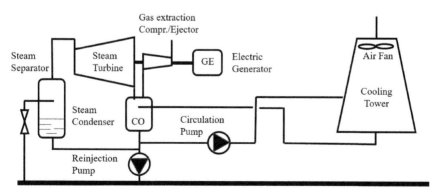

Fig. 10: Configuration of a flash steam geothermal plant.

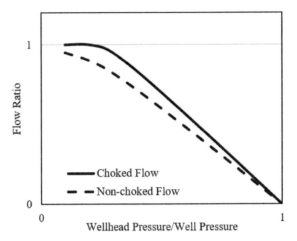

Fig. 11: Geo-fluid mass flow versus pressure ratio.

In a geothermal resource that is dominated by liquid, the geo-fluid mass flow rate is a function of wellhead pressure, as seen in Fig. 11, and is related, respectively, with choked and non-choked well flow conditions.

The maximization of the geo-fluid flow rate is obtained through adopting low wellhead pressure, but the objective of electricity generation consists in maximising the extracted power. The optimization procedure results, as demonstrated below, in an optimised value for pressure and temperature, that is intermediate between well pressure and atmospheric one (Di Pippo 2005).

The goal of the demonstration is to evaluate the value of flash temperature T_1 that maximises the specific work w, expressed according to eq. (1).

$$w = x(h_1 - h_c) \tag{1}$$

where x is the fraction of flashed steam, h_1 and h_C the enthalpy of steam respectively at turbine inlet and at condenser discharge.

To a first approximation, eq. (2) gives the expression of h_1.

$$h_1 - h_c = c(T_1 - T_c) \tag{2}$$

where, T_C is the temperature at the condenser and c is the specific heat of steam, considered, for simplicity of demonstration, as a perfect gas. The fraction of flashed steam can be expressed by eq. (3).

$$x = \frac{h_0 - h_{1l}}{h_{1g} - h_{1f}} = \frac{c(T_0 - T_1)}{h_{gf}} \tag{3}$$

where, T_0 is the geo-fluid temperature inside the well and h_{fg} is the latent evaporation enthalpy at flash pressure.

The specific work, therefore, can be expressed by eq. (4).

$$w = \frac{c^2(T_0 - T_1)(T_1 - T_c)}{h_{gf}} \qquad (4)$$

In order to maximise w, it is necessary to differentiate w with respect to T_1 and to equal it to 0, as expressed by eq. (5).

$$\frac{dw}{dt_1} = \frac{c^2}{h_{fg}}(T_0 - 2T_1 + T_c) = 0 \qquad (5)$$

obtaining the temperature (see eq. (6)) corresponding to the pressure of flash that optimise the specific work of the cycle:

$$T_{1opt} = \frac{T_0 + T_c}{2} \qquad (6)$$

A possible way of improving the efficiency is to pass from a single flash cycle to multiple flash one, according the plant configuration of Fig. 12.

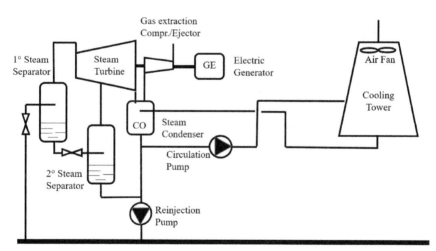

Fig. 12: Double flash plant configuration.

The optimized values of flash temperatures for a double flash can be evaluated in the same way as in the case of single flash obtaining the expressions of eq. (7–8).

$$T_{1opt} = \frac{T_0 + T_c}{2} \qquad (7)$$

$$T_{2opt} = \frac{T_{1opt} + T_c}{2} \qquad (8)$$

4.3 Binary Cycle Power Plants

Binary cycle power plants usually operate on water dominant resources at lower temperatures between 100°–180°C. Binary cycle plants use the heat from the hot geothermal water to boil a working fluid, usually an organic compound with a low boiling point. The working fluid is vaporized in a heat exchanger and used to feed a turbine. The organic fluid operates in a Rankine cycle, and for this reason the plant is conventionally called ORC (Organic Rankine Cycle). The water is then injected back into the ground to be reheated. The water and the working fluid are kept separated during the whole process, avoiding the emissions of incondensable gases, which are usually contained in the geothermal water (Fig. 13).

ORC plants combine the advantages of the intrinsic zero emissions characteristics and of the complete possibility of geo-fluid reinjection to their extreme modularity and flexibility in off design operation, making

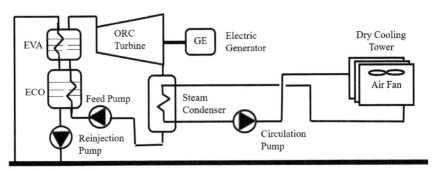

Fig. 13: Binary ORC plant configuration.

them particularly suitable for resources with not well assessed and/or variable capacity. The modularity allows reducing the time of installation and accelerates the phase of resource exploitation and electricity production.

The only ORC penalization is represented by the increase of parasitic losses related with the use of dry cooling. Considering the low efficiency due to the exploitation of low enthalpy resources, the fan consumption due to the very high air flow in dry cooling towers, is high, as shown in Fig. 14.

As in the case of flash plants, the ORC efficiency can be improved, through increasing the number of pressure/temperature levels (n). The optimization of temperature levels can be developed in theory by considering a series of Carnot cycles that operate between the water dominant resource (Q_G) at temperature T_G and the environment at temperature T_0 (Fig. 15).

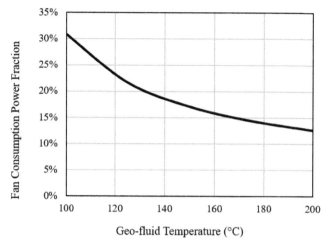

Fig. 14: Fraction of power absorbed by air fan of dry cooling towers.

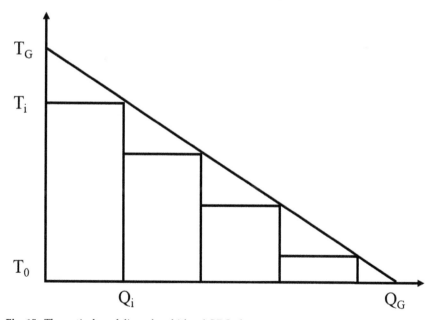

Fig. 15: Theoretical modeling of multi-level ORC plant.

The overall power produced by the system can be evaluated by eq. (9).

$$P=\sum_i P=\sum_i Q_i\frac{T_i-T_0}{T_i}=Q\sum_i\left(\frac{T_i-T_{i-1}}{T_c-T_0}\right)\left(\frac{T_i-T_0}{T_i}\right)=\frac{Q}{T_c-T_0}\sum_i\left(T_{i-1}+T_0-\frac{T_{i-1}T_0}{T_i}-T_i\right) \quad (9)$$

The optimization of T_i in order to maximize the overall power can be obtained deriving the above expression and equaling to 0. The eq. (10) gives the optimized value for each temperature level (i).

$$Ti_{opt} = T_0 \left(\frac{T_0}{T_G}\right)^{\frac{i}{n+1}} \tag{10}$$

The results of the optimization are reported in Figs. 16–17 for one and two pressure/temperature levels versus the geothermal resource temperature. Efficiency improves with the number of levels, but the improvements are more and more reduced as the number of levels increases; thus in real plants, for reasons of simplicity and economy, there is a tendency not to introduce more than two or three levels.

Another application of ORC technology is represented by the bottoming of conventional steam and flash plant, in order to avoid the problems of the absence of water and of the emission of gas. This solution consists in interrupting the expansion of steam at a temperature higher than the one of the condenser and in using the exhausted steam to power a low temperature ORC cycle, transforming the original cycle into a zero-emission one.

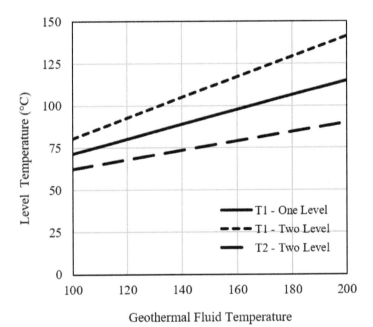

Fig. 16: Optimized level temperature.

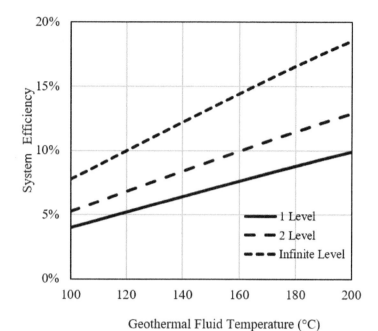

Fig. 17: System efficiency in optimized condition.

4.4 Kalina Binary Cycles

Binary cycles are penalized in term of energy efficiency by the irreversibility related with the heat exchange of the working fluid with the brine and with the cooling water, which is not isothermal in both cases. In order to reduce the irreversibility, binary cycles should increase the number of pressure level or operate with supercritical fluids. Another efficient solution is represented by Kalina binary cycles (Di Pippo 2005).

The features that distinguish the Kalina cycles from other binary cycles are the following:

- The working fluid is a binary mixture of H_2O and NH_3.
- Evaporation and condensation occur at variable temperature in order to match the variable temperatures of brine and cooling water at condenser.
- Cycle incorporates heat recuperation from turbine exhaust.
- Composition of the mixture can be varied during cycle, in order to perform evaporation and condensation at variable temperature.

The simplest configuration of Kalina cycle with variable working fluid composition is shown in Fig. 18.

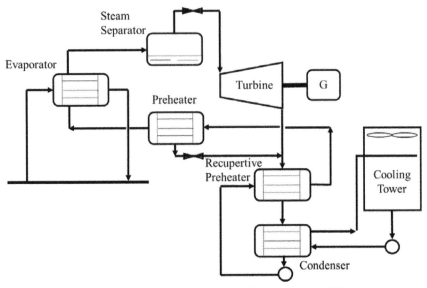

Fig. 18: Basic Configuration of Kalina cycle with variable composition of the water-ammonia working fluid.

The separator allows a saturated vapor, rich of ammonia, to flow to the turbine. The weak solution, a liquid rich in water, is used in the preheater and then throttled down to the turbine exhaust pressure before mixing with the strong solution to restore the primary composition. The mixture is then used in a recuperative preheater, before being fully condensed.

Though Kalina cycles are, in theory, interesting and promising, their commercial exploitation has been limited due to some intrinsic complexities and to the difficulty of maintaining very tight pinch point temperature differences in the heat exchangers.

4.5 Enhanced Geothermal Systems

Enhanced Geothermal Systems (EGS) aim to exploit hot dry rocks, characterized by elevated temperatures, but, unfortunately, by absence of water and low permeability of rock. The exploitation of these fields is made possible through the hydro-fracturing. The first Hot Dry Rock project has been launched in United States in the early 70s: by means of a specially drilled well, highly pressurized water is pumped into a hot, compact rock formation, causing its hydraulic fracturing. The water permeates these artificial fractures, extracting heat from the rocks that act as a natural reservoir. This tank is later penetrated by a second well, used to extract the water, which has received the heat from the rock. So the system is constituted by the well-used for hydraulic fracturing, through which cold water is injected into the artificial reservoir and by the well for hot water extraction. The overall system, comprising also the utilization plant for power generation, forms a closed circuit, avoiding any contact between the fluid and the external environment. The drawn hot water exploitation takes place through heat exchange with an ORC cycle (Fig. 19).

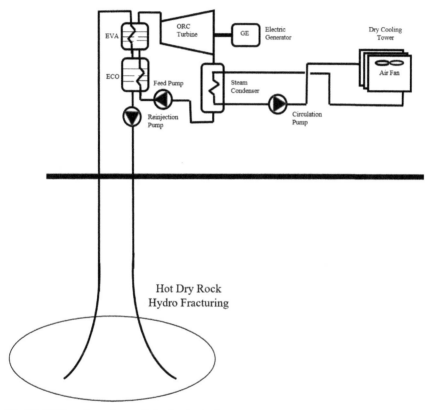

Fig. 19: EGS configuration with hydro fracturing.

The exploitation of HDR sites for energy purposes presents today a number of critical issues, summarized as follows:

- Difficulties in the drilling of wells with significant deviation from the vertical, which is necessary for the considerable extension of the underground reservoir.

- Need to carry out drilling in harder and more compact rocks compared to oil drilling.

- Need of perforations at a considerable depth, about 5000 m, in order to have temperatures higher than 200°C; which as yet are not yet economically feasible.

- Difficulty in correct interception of the field in drilling the extraction well.
- Risk of seismicity induced by hydro-fracturing and water passage.
- Water consumption linked to the losses in the underground reservoir; in the first projects in the US losses of almost 30% were reached.

In order to overcome some of the above problems, as risk of seismicity and water losses, a new plant configuration has been conceived. It consists of designing and constructing an engineered heat exchanger underground at elevate depth (Fig. 20). Due to the low geothermal heat flux, the required heat exchanger surface results enormous and it is not economically sustainable.

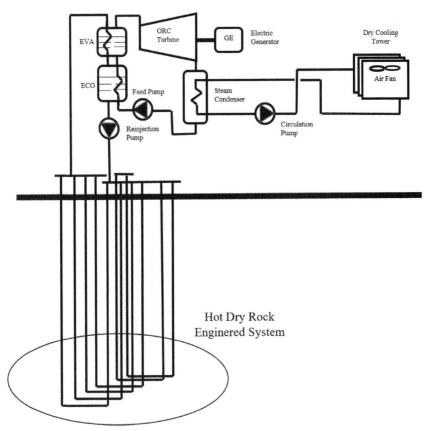

Fig. 20: Engineered EGS configuration.

5. Emissions in Geothermal Plants

Geothermal plants emit in air a high variety of non-condensable compounds as Carbon Dioxide (CO_2), Hydrogen Sulphide (H_2S), Ammonia (NH_3), and Methane (CH_4) which have a critical local and global impact on environment. The above compounds are contained in the geothermal fluid and they escape in to the environment, through the gas extractors at condenser and through the cooling towers. In fact, the geothermal steam contains gases and impurities that have an impact both on the power generation system and on the environment; a typical composition, referred to the Italian geothermal field of Larderello, is shown below:

- Pure steam   95%
- Non-condensable gas   5%
    of which:
    - Carbon Dioxide   98,4%
    - Hydrogen Sulphide   1%
    - Methane   0,4%
    - Others   0,2%
    - Mercury   traces

The most critical emissions impacting local pollution, are represented by H_2S and Mercury. ENEL developed the AMIS process (Abatement of Mercury and Hydrogen Sulphide) in order to reduce the impact of geothermal power plants. The process almost entirely treats the mercury present in the geothermal fluid and the part of the H_2S present in the non-condensable gas (which represents the highest fraction). The fraction of H_2S treated by AMIS represents about 80% of the total, the remaining 20% is dissolved in the water leaving the condenser, and part of it is released in the atmosphere through the cooling tower.

A comprehensive analysis of emissions from geothermal plants was developed by Bravi and Basosi (2014), with reference to the plants of Bagnore and Piancastagnaio, in the Mount Amiata area (Tuscany, Italy). Carbon dioxide is the main emission from the geothermal field, the actual range being from 245 to 779 kg/MWh with the weighted average being 497 kg/MWh. Ammonia emissions range between 0.086 and 28.94 kg/MWh with a weighted average of 6.54 kg/MWh. Hydrogen sulfide has a mean range of 3.24 kg/MWh, with values varying between 0.4 and 11.4 kg/MWh. In this case, the average values of Piancastagnaio are 4 times higher than those of the geothermal field of Bagnore. These values are related to the characteristics of the geothermal fluid and to the fact that only since the end of 2008, the plant has been equipped with AMIS. Geothermal gases emitted from the power plants also contain traces of mercury (Hg), arsenic (As), antimony (Sb), selenium (Se) and chromium (Cr). Mercury emissions range between 0.063 and 3.42 g/MWh with a weighted average of 0.72 g/MWh.

The emission of pollutants from geothermal plants are a critical concern that requires the installation of abatement systems, in order to mitigate the associated phenomena. The introduction of AMIS has essentially solved the problems of local environmental impact and of plants acceptability by population, but it has not solved the problem of Greenhouse Gas emission (GHGs), such as Carbon Dioxide, which represents the main component of non-condensable gas.

Based on the emission data reported above, the GHGs emission is not negligible, even though it is less than fossil fuel plants. GHGs emission depends not only by the composition of the geothermal fluid, i.e., the fraction of incondensable contained in the fluid, but also by the enthalpy level of the specific geothermal resource, which affects the electrical conversion efficiency.

The emission of CO_2 can be evaluated by means of eq. (11), as function of gas fraction (f), of geothermal fluid enthalpy (h_g) and of electrical efficiency (η).

$$\epsilon \left(\frac{gCO_2}{kWh} \right) = \frac{3600\,F}{h_g\,\eta} \tag{11}$$

The results are reported in Fig. 21, which shows the emission trends of carbon dioxide per unit of produced electricity (e), as a function of both the incondensable content in geothermal fluid and the conversion efficiency. In most practical cases, the emission of CO_2 is lower than that of natural gas fired plants, but, under certain conditions, it can also be higher.

According to some studies on the subject, the real impact of greenhouse gas emission, resulting from exploitation of a geothermal resource, should be calculated by subtracting the current emission from the natural one present in the territory (Holm et al. 2012). In fact, because geothermal systems naturally contain these gases, they also naturally vent them to the atmosphere through diffusive gas discharges from areas of natural leakage, including hot springs, fumaroles, geysers, hot pools, and mud pots. These natural discharges have taken place throughout the history of the Earth and continue today independently of geothermal power production. Carbon dioxide is the most widely emitted gas because geothermal systems tend to be found in areas with large fluxes of carbon dioxide. Methane is the second most common greenhouse gas emitted naturally from geothermal systems, but those emissions are minimal. Because geothermal systems are natural sources of greenhouse gases,

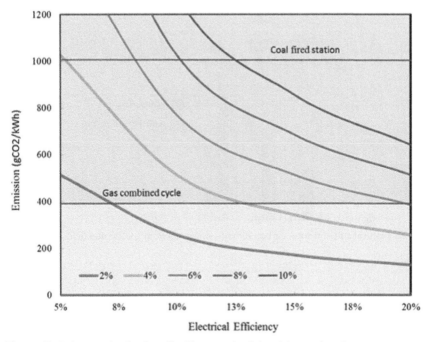

Fig. 21: Emission trends of carbon dioxide per unit of electricity produced.

isolating the emissions attributable to human activities requires more than a simple measurement of emissions from a particular site. Understanding how natural emissions are altered by industrial utilization would require a baseline determination prior to power development since GHGs are present in both producing and non-producing geothermal systems. If power production has already begun, baseline information would have to be collected both before and after new capacity is brought into service, but this data can be difficult to obtain.

One estimate by Ólafur G. Flóvenz, General Director of Iceland GeoSurvey, was that about half of the carbon dioxide emissions from geothermal power plants would be emitted anyway through natural processes (Holm et al. 2012). Another estimate from Iceland suggests that natural carbon dioxide emissions from geothermal fields significantly exceed those from power plants, as shown in Fig. 22 (Flóvenz 2006).

In any case, the future exploitation of geothermal resources must be made ensuring the environmental sustainability and, therefore, the complete greenhouse gases pollution control.

One solution to this problem consists in the application of ORC cycles that also allow a full reinjection of geothermal fluid, which represents a key objective for the sustainability of geothermal energy (Fig. 23).

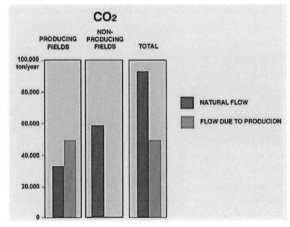

Fig. 22: Natural and anthropic carbon dioxide emissions from geothermal fields.

Fig. 23: Greeneco Saraykoy I 12 MW ORC plant, Denizli, Turkey (courtesy of Exergy S.p.A).

6. GHGs Capture and Sequestration

The most difficult challenge to make geothermal energy environmentally sustainable is represented by the capture and sequestration of Carbon Dioxide that is a prevalent contributor to global warming. Even if the modest worldwide installed geothermal power plants, are far lower emitters of CO_2 than the coal fired power station, the potential global expansion of geothermal power requires the solution of GHG emission.

Two ways have been explored: the natural sequestration inside the geothermal system by mineralization of carbon dioxide inside the rocks and its chemical conversion into methanol using hydrogen produced by geothermal electricity.

In 2007, a project was launched that aimed at storing CO_2 in Iceland's lavas by injecting greenhouse gases into basaltic bedrock where it literally turns to stone. CO_2 turning into calcite is a well-known natural process in volcanic areas. Now scientists of the University of Iceland, Columbia University N.Y. and the CNRS in Toulouse, France are developing methods to imitate and speed up this transformation of CO_2 into calcite. Injecting CO_2 at carefully selected geological sites with large potential storage capacity can be a long lasting and environmentally benign storage solution. Today, CO_2 is stored as gas in association with major gas production facilities. The uniqueness of the Icelandic project is that whereas other projects store CO_2 mainly in gas form, where it could potentially leak back into the atmosphere, the current project seeks to store CO_2 by creating calcite in the subsurface. Calcite, a major component of limestone, is a common and stable mineral in the Earth that is known to persist for tens of millions of years (Armansson 2003).

Iceland has also developed the chemical conversion of CO_2 from geothermal steam to produce methanol. The processing of carbon dioxide from emissions of geothermal power plants and the electrolyzing of water provide necessary carbon sources and hydrogen gas to produce clean fuel. Carbon Recycling International (CRI) and HS Orka (HS) cooperate in demonstrating the use of geothermal energy and capture carbon dioxide to produce Renewable Methanol (RM). RM can be upgraded to dimethyl ether or gasoline using available standard conversion technology.

The process shown in Fig. 24 uses carbon dioxide, which is the primary component of an effluent stream of Non Condensable Gas (NCG) discharged from the geothermal power plant as a raw material for methanol production. The electrical power produced from the geothermal power is used for the electrolysis of water to produce hydrogen, which is also used as a raw material. Hydrogen sulphide is contained in the NCG

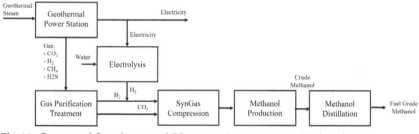

Fig. 24: Conceptual flow diagram of CO_2 conversion process into Methanol.

effluent and it must be removed prior to use in the methanol conversion reactor to ensure longevity of the downstream catalyst's lifetime. The purification of the carbon dioxide is achieved through an amine scrubbing system, which will discharge the H2S removed from NCG. Following this purification step, the carbon dioxide is combined with the hydrogen and compressed from the atmospheric pressure to the pressure required in the methanol conversion process. After the raw syngas is compressed, it is passed through a heat exchanger for pre-heating and then sent to a reactor where it is partially converted to methanol. The partially converted syngas is then cooled and sent to a gas liquid separator where the gas from the separator must be recompressed and recycled back to the methanol converter, in order to improve the overall conversion of the process. The liquid from the separator contains crude methanol, which can be further processed for fuel grade methanol or to other fuels such as dimethyl ether or gasoline (Tran and Albertson 2010).

An approximate energy balance of the process can be derived from Harp et al. (2015), with reference to CRI methanol production site at Grindavik near the Svartsengi geothermal power station in Iceland. The annual production capacity is 4000 t of methanol, which corresponds to the use of 5500 t of CO_2. Taking into account that the Svartsengi plant produces 75 MWe power and its CO_2 emission is 0.18 t/MWh (Harp et al. 2015), those 5500 t CO_2 corresponds to almost 10% of the total CO_2 emission of the power station (Table 2).

Table 2: Methanol production data at Svartsengi geothermal power station.

Electric power (MW)	75
Specific CO_2 emission (kg/MWh)	180
Annual CO_2 emission (t/y)	55.000
Fraction of CO_2 conversion	10%
Annual CO_2 conversion (t/y)	5.500
Annual Methanol production (t/y)	4.000
Process energetic efficiency	65%
Annual electricity consumption (MWh/y)	35.000

Table 2 outlines the consumption levels of a relevant portions of the plant's electricity for converting merely 10% of emitted carbon dioxide, essentially for the production of hydrogen, necessary for the synthesis of methanol. So the process is not very effective in capturing a large amount of CO_2 but it has an industrial value for countries. In fact the methanol, produced in the process can be used in other energy sectors as in transport and distributed energy uses.

7. Power and Typology of Geothermal Plants, Today Installed in the World

In 2015, the total geothermal power globally was 12,6 GW, consisting of different plant categories, back pressure, dry steam, binary, single flash, double flash, triple flash, and hybrid that is integrated with other renewable sources. Table 3 reports the installed power per country and per category; today the lowest geothermal power is in Africa, but geo resources in this continent are very large and unexplored, so it is forecasted there will be a big trend of power installation increase in the following years to come.

Figure 25 reports the pie charts of produced energy and number of plants per category.

Table 3: Plant category per country (2015) (Bertani 2015).

Country	Back Pressure (MW)	Binary (MW)	Double Flash (MW)	Dry Steam (MW)	Hybrid (MW)	Single Flash (MW)	Triple Flash (MW)	Total (MW)
Africa	48	11				543		602
Asia		236	525	484		2514		3758
Europe		268	273	795		796		2133
Latin America	90	135	510			908		1642
North America		873	881	1584	2	60	50	3450
Oceania	44	266	356			259	132	1056
TOTAL	181	1790	2544	2863	2	5079	182	12641

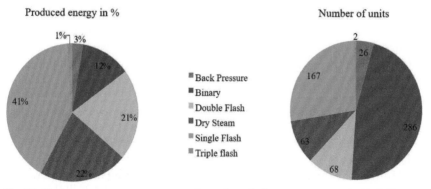

Fig. 25: Produced energy percentage and number of plant per category in 2015 (Bertani 2015).

The most of power production comes from single flash followed by dry steam plants, while the most installed plants are represented by a binary cycle (usually ORC), due to the relatively low size of these units. With the increase of size, which toady has overcome 10 MW per unit, these plants will, in the future strongly increase their contribution to the world wide power generation.

Table 4 reports the list of plants installed in the world during 2010–2015. The most plants installed in this period were binary units, by Ormat that is today the dominant supplier in the world of this category of plants.

8. Perspectives of ORC for a Sustainable Geothermal Power Generation

As briefly described in Section 4.2, binary geothermal power plants, and in particular ORC, which represent almost the entirety of the binary plants, are the available technology for improving the sustainability of geothermal power generation. In the recent years, the geothermal community has started looking into ORC as an alternative to flash power plants not only for residual applications, in terms of quality and quantity of the geothermal resource, but also as the reference technology for projects which were traditionally considered as monopoly of the direct expansion technologies (Spadacini 2017).

There are many reasons behind this gradual technology shift: technical, environmental and economic.

First of all, after 2010 there has been a massive improvement in turbine technology in the ORC geothermal plants: in a market having one incumbent player, Ormat, with a spot competition from solution providers using technology taken from the oil and gas sector, such as Rotoflow (GE) and Mafi Trench (Atlas Copco) radial inflow turbines, two Italian ORC producers entered the geothermal market with innovative solutions. Turboden (Mitsubishi), coming from a long experience in biomass and heat recovery small modules manufacturing, provided a total of approximately 20 MW, in modules rated 5 to 8 MW, of geothermal plants in Central Europe, while Exergy supplied a first 1 MW unit in Italy in 2012 to Enel Green Power, followed by approximately 200 MW of geothermal plants in Turkey, rated between 3 to 28 MW each, in the following 4 years.

The success of the new players and the massive growth of the binary market were based on technological improvements: the new radial outflow technology developed by Exergy increased the reference turbine efficiency over 90%, pushing the competitors to new solutions, with new axial turbines featuring more than 3 stages that are going to be in operation between year 2016 and 2017 (Spadacini 2015).

Table 4: List of geothermal plant installed in the period 2010–2015 (Bertani 2015).

Country	Plant	Unit	COD	Type	Manufacturer	Capacity (MWe)	Operator
Japan	Oguni Matsuya	1	2015	binary		0.06	
Indonesia	Ulu Belu	1	2014	Single Flash	Toshiba	55	PLN
Indonesia	Ulu Belu	2	2014	Single Flash	Toshiba	55	PLN
Italy	Bagnore 4	2	2014	Single Flash	Ansaldo_Tosi	20	Enel Green Power
Italy	Bagnore 4	1	2014	Single Flash	Ansaldo_Tosi	20	Enel Green Power
Japan	Yumura Spring	1	2014	binary		0.03	
Japan	Hagenoyu	3	2014	binary		2	
Japan	Shichimi Spring	1	2014	binary		0.02	
Japan	Goto-en	2	2014	binary		0.09	
Japan	Beppu Spring	4	2014	binary		0.5	
Kenya	Olkaria I	4	2014	Single Flash	Toyota	70	KenGen
Kenya	Olkaria I	5	2014	Single Flash	Toyota	70	KenGen
Kenya	Olkaria III	3a	2014	Single Flash	ORMAT	13	ORMAT
Kenya	Olkaria III	3b	2014	Single Flash	ORMAT	13	ORMAT
Kenya	Olkaria IV	1	2014	Single Flash	Toyota	70	KenGen
Kenya	Olkaria IV	2	2014	Single Flash	Toyota	70	KenGen
Kenya	WellHEad OW43	1	2014	Back Pressure	Elliot	12.8	Oserian Development

Table 4 contd. ....

...Table 4 contd.

Country	Plant	Unit	COD	Type	Manufacturer	Capacity (MWe)	Operator
Kenya	WellHEad	1	2014	Back Pressure	Elliot	30	Oserian Development
New Zealand	Te Mihi	1	2014	Double Flash	Toshiba	83	Contact Energy
New Zealand	Te Mihi	2	2014	Double Flash	Toshiba	83	Contact Energy
Philippines	Maibarara	1	2014	Single Flash	Fuji	20	Maibarara Geothermal
Turkey	Dora	3b	2014	Binary	ORMAT	9	MENDERES
Turkey	Degirmenci	1	2014	Binary		2.5	Turcas Güney Elektrik
Turkey	Alasheir	1	2014	Binary	ORMAT	24	Turkeler
USA	Lightening Dock	2	2014	Binary		4	Raser Technologies
USA	Don Campbell	1	2014	Binary	ORMAT	16	ORMAT
Australia	Habanero	1	2013	Binary	Peter Brotherhood	1	Geodynamics
Germany	Dürrnhaar	1	2013	Binary	UTC_Turboden	5.6	Municipality Germany
Germany	Kirchstockach	1	2013	Binary	UTC_Turboden	5.6	Municipality Germany
Germany	Insheim	1	2013	Binary	UTC_Turboden	4.3	Municipality Germany
Indonesia	Mataloko	1	2013	Single Flash	Fuji	2.5	PLN

Country	Name	No.	Year	Type	Turbine	Capacity	Developer
Japan	Abo-tunnel	1	2013	binary	Elliot	0.003	
Kenya	WellHEad OW37	1	2013	Back Pressure		5	Oserian Development
Mexico	Los Humeros II	B	2013	Single Flash	Alstom	27	Comisión Federal de
New Zealand	Ngatamariki	1	2013	Binary	Ormat	20.5	Mighty River Power
New Zealand	Ngatamariki	2	2013	Binary	Ormat	20.5	Mighty River Power
New Zealand	Ngatamariki	3	2013	Binary	Ormat	20.5	Mighty River Power
New Zealand	Ngatamariki	4	2013	Binary	Ormat	20.5	Mighty River Power
New Zealand	Te Huka	1	2013	Binary	Ormat	12	Contact Energy
New Zealand	Te Huka	2	2013	Binary	Ormat	12	Contact Energy
Nicaragua	San Jacinto-Tizate	4	2013	Single Flash	Fuji	36	Ram Power
Turkey	Germencik	2	2013	Double Flash	Mitsubishi	51	GURMAT
Turkey	Gümüsköy	1	2013	Binary	TAS	6.6	BM
Turkey	Pamukören	1	2013	Binary	Atlas-Copco	24	Çelikler Jeotermal Elektrik
Turkey	Pamukören	2	2013	Binary	Atlas-Copco	24	Çelikler Jeotermal Elektrik
Turkey	Kizildere	2	2013	Double Flash	Fuji	80	ZORLU
USA	EGS	1	2013	Binary	ORMAT	1.7	ORMAT
USA	Cove Fort	B1	2013	Binary	ORMAT	13	Enel Green Power
USA	Cove Fort	B2	2013	Binary	ORMAT	13	Enel Green Power

Table 4 contd. ...

...Table 4 contd.

Country	Plant	Unit	COD	Type	Manufacturer	Capacity (MWe)	Operator
Germany	Sauerlach	1	2012	Binary	UTC_Turboden	5	Municipality Germany
Indonesia	Ulumbu	1	2012	Single Flash	Fuji	2.5	PLN
Indonesia	Ulumbu	2	2012	Single Flash	Fuji	2.5	PLN
Indonesia	Lahendong	B1	2012	Binary		7.5	BPPT
Indonesia	Lahendong	4	2012	Single Flash	Fuji	20	BPPT
Italy	Bagnore Binary	1	2012	Binary	Exergy	1	Enel Green Power
Japan	Niigata	1	2012	Binary	EcoGen	2	Wasabi
Mexico	Los Humeros II	A	2012	Single Flash	Alstom	27	Comisión Federal de
New Zealand	Kawerau-opp	1	2012	Binary	ORMAT	23	Norske Skog Tasman
Nicaragua	San Jacinto-Tizate	3	2012	Single Flash	Fuji	36	Ram Power
Romania	Oradea	1	2012	Binary	UTC_Turboden	0.05	null
Taiwan	Qingshui	2	2012	Binary		0.05	null
Turkey	DENIZ	1	2012	Binary	ORMAT	24	MAREN
Turkey	SINEM	1	2012	Binary	ORMAT	24	MAREN
Turkey	Dora	3a	2012	Binary	ORMAT	9	MENDERES
USA	Hudson Ranch I	1	2012	Triple Flash	Fuji	49.5	EnergySource

Country	Plant	No.	Year	Type	Manufacturer	MW	Developer
USA	Florida Canyon Mine	1	2012	Binary		0.1	Electratherm
USA	McGinness Hill	1	2012	Binary	ORMAT	52	ORMAT
USA	San Emidio	1	2012	Binary	Turbine Air System	12.75	US Geothermal
USA	Tuscarora	1	2012	Binary	ORMAT	32	ORMAT
USA	Neal	1	2012	Binary		23	US Geothermal
China	Yangyi	1	2011	Single Flash		0.9	Jiangxi Huadian Electric
Costa Rica	Las Pailas	1	2011	Binary	ORMAT	21	Instituto Costarricense de
Costa Rica	Las Pailas	2	2011	Binary	ORMAT	21	Instituto Costarricense de
Iceland	Hellisheidi IV	1	2011	Single Flash	Mitsubishi	45	Orkuveita Reykjavikur
Iceland	Hellisheidi IV	2	2011	Single Flash	Mitsubishi	45	Orkuveita Reykjavikur
Italy	Nuova Radicondoli	2	2011	Dry Steam	Ansaldo_Tosi	20	Enel Green Power
Kenya	Eburro	1	2011	Single Flash	Elliot	2.5	KenGen
Turkey	IREM	1	2011	Binary	ORMAT	20	MAREN
USA	Puna	11	2011	Binary	ORMAT	8	ORMAT
USA	Beowawe	2	2011	Binary	Turbine Air System	1.9	Beowawe Power

Table 4 contd. ...

...Table 4 contd.

Country	Plant	Unit	COD	Type	Manufacturer	Capacity (MWe)	Operator
USA	Dixie Valley	2	2011	Binary	Turbine Air System	6.2	Terra Gen
China	North Oil Field	1	2010	Binary		0.4	
China	Longyuan	2	2010	Binary		1	Longyuan Co
Italy	Chiusdino 1	1	2010	Dry Steam	Ansaldo_Tosi	20	Enel Green Power
Kenya	Olkaria II	3	2010	Single Flash	Mitsubishi	35	KenGen
New Zealand	Nga Awa Purua	1	2010	Triple flash	Fuji	132	Mighty River Power
Turkey	Dora	2	2010	Binary	ORMAT	9.5	MENDERES
Turkey	Tuzla	1	2010	Binary	ORMAT	7.5	Dardanel
USA	Jersey Valley	1	2010	Binary	ORMAT	19.4	ORMAT

With a new reference efficiency of the ORC, improved by turbine design and cycle improvements (e.g., 3-pressure-level system by Ormat), the distance between flash solutions and ORC solutions in terms of power output and specific cost of the complete plant is greatly reduced, and more projects are becoming feasible.

Two projects, completely different each other, give a good example supporting this statement.

In 2014 Akca Enerji Tosunlar, in the Denizli area in Turkey, started up a 3.5 MW geothermal plant (Fig. 26) fed by a resource at 100–105°C, flowing from existing wells and used at the time for greenhouse warming only. The very low temperature of the resource combined with the low quantity of resource, made the classical 1-pressure-level system too inefficient and the 2-pressure-level (and, thus, 2 turbines) system too expensive to be feasible. Exergy solution of the very first single-disk 2-pressure-level turbine granted the highest efficiency with a competitive cost, and the project was successful and is in operation since then.

Fig. 26: Akca Tosunlar 3.5 MW 2-pressure-level single-turbine plant (courtesy of Exergy S.p.A).

Geotérmica del Norte (GDN), a joint venture of Enel Green Power and ENAP, developed the Cerro Pabellon project in Chile with the intention to apply flash technology, as the resource is extremely good (approximately 240°C bottom-hole temperature). The project encountered many challenges due to the special environmental conditions of the site: high altitude (4600 m a.s.l.), extreme climate (minimum temperature reaching 40°C below zero), and remote location impacting on both transportation and construction activities. After a deep analysis of technical and economical features of the flash and binary scenarios, GDN finally decided to proceed with a series of 20 MW net modules, two already installed in 2017 and others to come in a second phase.

As already underlined, the environmental impact of ORC plants is lower, especially because they do not need to consume clean water, which is a scarce resource in many geothermal areas, for cooling, and to reinject most of the geothermal fluid, granting zero emissions.

Additionally, many areas where flash plants are installed they are starting the conversion into combined flash and bottoming cycles, applying an ORC unit downstream the existing plant (on exhaust steam or, more frequently, on the brine flow after separator) and providing additional 15–25% power with no additional mining investment and zero risk on the resource.

Other prospects for binary geothermal are rising due to the increasing integration of district heating, greenhouse farming and solar concentrated power (Astolfi 2011).

Market observers agree that many projects in the future will proceed with binary units, representing at least half of the new geothermal installations.

Acknowledgments

Eng. Luca Xodo of Exergy S.p.A is acknowledged for the preparation of paragraph 8.

References

Armannsson, H. 2003. CO_2 emissions from Geothermal Plants. International Geothermal Conference, Reykajavik, Sept.

Astolfi, M., L.G. Xodo, M.C. Romano and E. Macchi. 2011. Technical and economical analysis of a solar–geothermal hybrid plant based on an Organic Rankine Cycle. Geothermics.

Axelsson, G. and Gunnlaugsson, E. 2000. Long-term Monitoring of High- and Low-enthalpy Fields under Exploitation. International Geothermal Association, World Geothermal Congress.

Benderitter, Y. and G. Cormy. 1990. Possible approach to geothermal research and relative cost estimate. pp. 61–71. *In*: Dickson, M.H. and M. Fanelli (eds.). Small Geothermal Resources. UNITARRJNDP Centre for Small Energy Resources. Rome, Italy.

Bertani, R. 2015. Geothermal Power Generation in the World 2010–2014 Update Report. Proceedings World Geothermal Congress. Melbourne, Australia.

Bravi, M. and R. Basosi. 2014. Environmental Impact of Electricity from Selected Geothermal Power Plant in Italy. Elsevier.

Dickson, M.H. and Mario Fanelli. 2005. Geothermal Energy: Utilization and Technology. Routledge.

DiPippo, R. 2005. Geothermal Power Plant: Principles, Application and Case Studies. Elsevier.

Flóvenz, Ó. 2006. The Power of Geothermal Energy. Iceland GeoSurvey. July 2006.

Franco, A. and F. Donatini. 2017. Methods for the estimation of the energy stored in geothermal reservoirs. IOP Conf. Series: Journal of Physics: Conf. Series 796(2017): 012025.

GEA Geothermal Energy Association. Annual Report 2015. March 10, 2016.

Harp, G., K.C. Tran, C. Bergins, T. Buddenberg, I. Drach, E.I. Kovtsoumpa and O. Sigurbjornsson. 2015. Application of Power to Methanol Technology to Integrated Steelworks for Profitability. Conversion Efficiency, and CO_2 Reduction: 2nd European Steel Technology and Application Days, June 2015, Duesseldorf, Germany.

Hochstein, M.P. 1990. Classification and assessment of geothermal resources. pp. 31–59. *In*: Dickson, M.H. and M. Fanelli (eds.). Small Geothermal Resources. UNITAEWNDP Centre for Small Energy Resources. Rome, Italy.

Holm, A., D. Jennejohn and L. Blodgett. 2012. Geothermal Energy and Greenhouse Gas Emissions. GEA, Geothermal Energy Association. November 2012.

Hofmann, M. and M.A. Morales Maqueda. 2009. Geothermal heat flux and its influence on the oceanic abyssal circulation and radiocarbon distribution. Elsevier.

http://www.thinkgeoenergy.com/newest-list-of-the-top-10-countries-in-geothermal-power/.

Muffler, L.J.P. and R. Cataldi. 1978. Methods for regional assessment of geothermal resources. Geothermics 7: 53–89.

Nicholson, K.N. 1993. Geothermal Fluids. Chemistry and. Exploration Techniques.

Pollack, H.N., S.J. Hurter and J.R. Johnson. 1993. Heat flow from the Earth's interior: analysis of the global data set - Reviews of Geophysics. Wiley Online Library.

Spadacini, C., L. Centemeri, M. Danieli, D. Rizzi and L.G. Xodo. 2015. Geothermal Energy Exploitation with the Organic Radial Outflow Turbine. Proceedings World Geothermal Congress 2015.

Spadacini, C., L.G. Xodo and M. Quaia. 2017. Geothermal energy exploitation with organic rankine cycle technologies—Chapter 14 of Organic Rankine Cycle (ORC) Power Systems, by E. Macchi, M. Astolfi, Woodhead Publishing.

Stacey, F.D. and D.E. Loper. 1988. Thermal history of the Earth. A corollary concerning non-linear mantle rheology. Physics of the Earth & Planetary Interiors 53: 167–174.

Tran, K.C. and A. Albertson. 2010. Utilization of Geothermal Energy and Emission for Production of Renewable Methanol. World Geothermal Congress 2010. Bali, Indonesia.

II

Advances in Renewable Energy-based Thermal Technologies

CHAPTER 8

Power-to-Gas Conversion Technologies and Related Systems

S. Pérez,* and M. Belsué Echevarria

1. Introduction

Wind and solar energy must play a major role to achieve 20% energy consumption from renewable energy in the EU by 2020, as proposed by the European Commission Renewable Energy Roadmap target. However, as renewable energies are variable and intermittent, they cannot be expected to provide a safe and steady supply by themselves. As a result, to achieve the transformation to a renewable energy based system, large-scale energy storage is required to make use of renewable energy surplus which cannot be supplied to the grid and to offset variations.

This electric power surplus can be used to produce hydrogen, which may be injected back to the existing natural gas energy grid. This conversion is carried out through water electrolysis. Hydrogen production based on renewable energy sources may help to improve management of these types of plants as they may be viewed as administrable plants by grid operators. Thanks to this technique, hydrogen is generated through water electrolysers, making the most of renewable energy surplus and becoming a sustainable hydrogen generation source.

Tecnalia R&I. Mikeletegi, 2. 20009 San Sebastián – Guipúzcoa – SPAIN.
Email: mikel.belsue@tecnalia.com
* Corresponding author: susana.perez@tecnalia.com

Nevertheless, the use of hydrogen entails some issues such as: lack of distribution and consumption infrastructures, low volumetric density, and high costs to guarantee safety and storage. Moreover, the amount of hydrogen which may be admitted to the gas grid is limited by specific standards and regulation of each country. In general, this amount is typically up to 12% of volume (Götz et al. 2016).

To solve these issues, hydrogen may be turned into methane through a methanation process and then injected and distributed through the existing natural gas grid without limit, as long as the methane to be injected meets the necessary quality requirements. Thus, hydrogen and methane produced from renewable energy surplus may be used to connect electricity and gas grids. The process to convert electric power into hydrogen is known as Power-to-Gas.

Methane production through methanation is based on a process which produces combustible methane from CO_2 hydrogenation, called the Sabatier reaction. This process recovers CO_2, generated in thermal processes or biogas generations plants, to obtain Synthetic Natural Gas (SNG) without the technical or safety constraints associated with hydrogen distribution.

Therefore, the Power-to-Gas (PtG) system transforms electric power into fuel gas (hydrogen or methane) for storage and further use. To increase the system profitability, it is recommended to recover the excess electricity generated by Renewable Energies (RE) to produce H_2 at valley hours of energy demand; while electricity generation is more convenient at peak hours when the sale price is higher. Moreover, this system represents a major environmental benefit as CO_2 is used as raw material.

This bi-directional electricity and gas conversion facilitates energy storage and management of electrical grid stability, so that renewable energy surpluses, which may not be supplied to the grid due to grid stability or lack of demand problems, may be stored as natural gas and used when required.

Methane production in PtG plants is based on a two-stage process (Fig. 1):

- the first stage involves hydrogen electrolysis generation using excess renewable energy electricity with the following reaction:

$$H_2O \rightarrow H_2 + \tfrac{1}{2} O_2$$

- the second CO_2 hydrogenation stage uses hydrogen generated in the previous stage for the Sabatier reaction to take place:

$$CO_2 + 4H_2 \rightarrow CH_4 + 2H_2O$$

The main components of a PtG plant for methane production are: an electrolyser, a methanation reactor, a generated water condenser and a SNG compressor. The global process efficiency is usually below 65% as Fig. 2 shows.

A description of the different technological processes to carry out these two stages to obtain SNG directly injectable into the grid is included as follows.

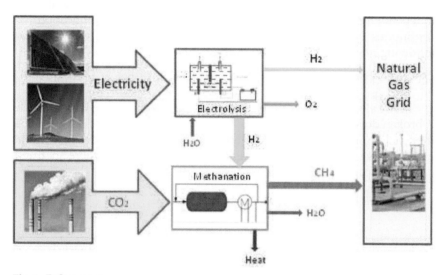

Fig. 1: PtG concept.

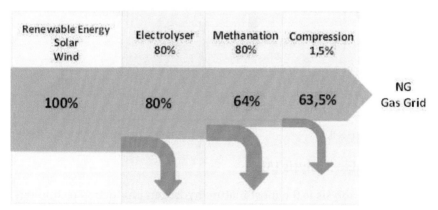

Fig. 2: Sankey diagram of the PtG process.

2. Water Electrolysis

Water electrolysis is the main process in a PtG system and the electrolytic cell (Fig. 3) is the equipment required to convert electricity into chemical energy (hydrogen and oxygen). The electrolytic cell must have suitable efficiency, flexibility and life cycle to maximize its performance.

In general, electrolysis is the process where water is broken down into two components: hydrogen and oxygen. Both components are independently generated in gas form through the application of continuous electric power through metal electrodes according to the following reaction:

$$H_2O \ (l) \rightarrow H_2 \ (g) + \tfrac{1}{2} O_2 \ (g) \qquad \Delta H_R \ (25°C, 1 \ atm) = 286{,}43 \ kJ/mol$$

Metal electrodes are submerged in a solution containing water, which is broken down, and an electrolyte. The electrolyte is essential to produce the reaction, as it increases the medium conductivity, and may be made up of salt, acid or bases. Depending on the electrolyte used there are three different types of electrolysers.

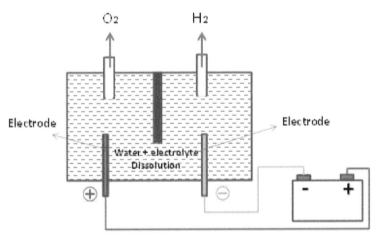

Fig. 3: Electrolytic cell.

3. Alkaline Electrolysis (AEL)

Alkaline electrolysis is the most mature hydrogen production technology and the most developed to date. An electrolytic cell comprises two electrodes, an alkaline solution and a separation membrane. Electrodes mainly consist of stainless steel with nickel, cobalt or iron coating and are submerged in an alkaline solution made by the electrolyte which normally has a concentration of 20–40% (weight) potassium hydroxide (KOH) or sodium hydroxide (NaOH) in water. A porous membrane is used to separate the cathode and anode chambers. The operating temperature of the alkaline electrolyte system ranges between 70–90°C and can work either at atmospheric pressure or under high pressure.

In these systems, electrodes work with relatively low current density around 0.4 A/cm², which translates into less compact electrolysis systems. System efficiencies depend on the quality of hydrogen to be achieved and the pressure level desired. Efficiencies are typically close to 70% in relation to the High Heating Value (HHV) of hydrogen, to meet energy demands of around 5 kWh/Nm³ of hydrogen added (SGC Rapport 2013:284). Hydrogen purity in this type of system is relatively lower than in other technologies, around 99.5%. To achieve purer hydrogen a subsequent purification stage has to be followed.

As already mentioned, alkaline electrolysis is the most mature technology currently used at industrial scale, and there are commercially available modules of 2.5 MWe (Schiebahn et al. 2015), offering hydrogen production capacities of up to 760 Nm³/h of hydrogen. The main advantages of this technology are durability, maturity, availability and low specific costs, as these systems are not based on components containing noble metals unlike other technologies. According to the manufacturers, alkaline cells can work from 20 to 100% of their design capacity, which makes them highly versatile and suitable for application to PtG type systems, where electric current is fluctuating or intermittent. The disadvantages of this technology include low current density and the use of a highly corrosive electrolyte solution entailing high maintenance costs (SGC Rapport 2013:284).

Regarding durability, commercial cells offer 10,000 hours of operation; however, this may vary when it works with variable loads. Frequent on/off operation affects cell durability, reducing the amount of gas obtained and efficiency achieved is also lower (Lehner et al. 2014).

R&D efforts, currently being made in this technology, are mainly focused on developing new membranes to act as separators between electrolytic cells and increase operation times as well as developing new materials preventing electrode corrosion (Götz et al. 2016).

4. Proton Exchange Membrane Electrolysis (PEM)

In PEM cells a solid polymeric membrane acts as an electrolyte and separation system. This membrane is directly connected to the electrodes allowing the passage of protons from one electrode to another. These systems are less developed than alkaline systems and are only commercially available for small scale applications. Their major advantages are: fast start-up; high flexibility; good operation in fluctuating systems; and production of highly pure hydrogen (Götz et al. 2016). Figure 4 shows a PEM electrolysis cell block diagram.

As PEM electrolysis systems may operate at current densities around 1.5 A/cm², i.e., four times the current density of alkaline electrolysers, this is one of its strengths.

The maximum operation limit temperature of 80°C is due to the polymeric membrane material limitation. The efficiency of these systems is in the 70% range compared to hydrogen HHV. As the purity of hydrogen produced in this type of systems is typically over 99.99%, no auxiliary purifying equipment is required. On the other hand, as no highly corrosive electrolyte liquids are required (as in the case of solutions used in alkaline electrolysis) this technology is safer and less harmful to the environment.

However, the main weakness of this method is the cost of cell materials, mainly due to the need for noble materials, such as platinum, to be used as the catalyst. All in all, this makes the cost of these systems treble that of the alkaline electrolysis systems. Moreover, as this technology is not very mature, there are durability problems with shorter life cycle times compared to alkaline electrolysis systems (Lehner et al. 2014, Schiebahn et al. 2015).

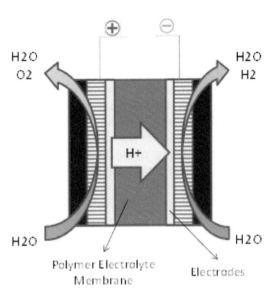

Fig. 4: PEM cell.

5. High-temperature Water Electrolysis (SOEC)

The Solid Oxide Electrolysis Cell is the least mature and less developed technology in this field. This method, currently being developed at laboratory level, offers greater efficiency than low-temperature electrolysis as some of the energy is supplied as heat. A solid oxide electrolyte (Y_2O_3 doped ZrO_2) provides high conductivity at high temperatures and good thermal and chemical stability (Schiebahn et al. 2015, Götz et al. 2016).

Solid oxide cell systems have operating temperature ranges of 700–1000°C and use steam instead of liquid water. From a thermodynamic viewpoint these temperatures are an advantage as electrolysis voltages are reduced; as well as from a kinetic point of view, as catalysts based on precious metals are not required.

Figure 5 shows a SOEC diagram. Electrodes are based on porous materials with Ni/YSZ composite cathode and Perovskite-type lanthanum, strontium, manganite (LSM)/YSZ composite anode.

Current densities, which could be achieved in these systems, may be within the same ranges as PEM electrolysis systems; however, due to high degradation at process temperatures, they are usually maintained at the same level as current densities used in alkaline electrolysers, 0.4 A/cm². The corresponding cell voltages are around 1.3 V and generate significantly reduced power consumption compared to the other technologies, around 3 kWh/H_2. Efficiencies may be over 90% taking into account electricity as well as heat recovery. Although these systems operate at atmospheric pressure, tests are being conducted with pressures of up to 25 bars (SGC Rapport 2013:284).

The main critical point of these systems is the high degradation rates achieved due to high temperatures; therefore, a major part of R&D efforts focuses on reducing operating temperatures to 500–600°C. Furthermore,

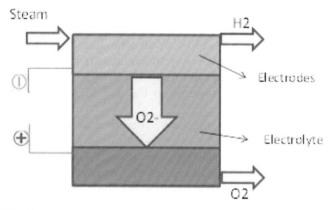

Fig. 5: SOEC cell.

due to the temperatures reached, the product obtained is a hydrogen and water steam mixture which needs to be subsequently purified. Finally, it is worth highlighting that this technology is not stable against power supply variations (Götz et al. 2016).

To summarize the state-of-the-art technology described in this section, Table 1 shows the main features and parameters of the three technologies.

Table 1: Features and parameters of electrolysis technologies.

Feature	AEL	PEM	SOEC
Type of electrolyte	KOH	Polymeric membrane	Y_2O_3 doped ZrO_2
Working temperature, °C	40–90	20–100	800–1000
Current density, A/cm^2	0.2–0.5	0.5–2	0.3–0.5
Voltage efficiency, as HHV, %	60–80	65–80	> 80
Specific energy consumption, kWh/Nm^3	> 4.6	> 4.8	< 3.2
Stage of maturity	Commercial	Commercial at small scale	Laboratory
Advantages	Available for large plants	High energy density	High electricity efficiency
Disadvantages	Corrosive electrolyte	High level of degradation Expensive technology	Low stability with time Expensive technology
Operation with intermittent load	Possible	Good	Poor

6. Methanation

As we have already mentioned, the use and distribution of hydrogen as an energy vector entails problems related to the lack of infrastructure for distribution and consumption, low volumetric density, high storage costs and lack of guaranteed supply safety. In this context, hydrogen conversion into methane and direct injection into the already existing natural gas infrastructure may play a key role in the development of a hydrogen economy and optimal harness of RE surplus.

The methanation process obtains Synthetic Natural Gas (SNG) from Hydrogen/Carbon Dioxide reaction, yielding a major environmental advantage as CO_2 may come from the atmosphere or from industrial emissions as raw material. Therefore, this process may take place in biological reactors and catalytic reactors. For the purpose of this chapter, only catalytic methanation will be referred to.

6.1 Catalytic Methanation

Over the last 50 years many breakthroughs in CO_2 methanation process have been made to encompass a reduction in greenhouse gases with SNG energy storage.

Methanation is a hydrogenation reaction implemented at commercial level to obtain methane from CO or CO_2 gases. This reaction, also known as selective methanation of carbon oxide, takes place at temperatures of 250–450°C and pressures of 1–30 bars. Usually alumina-supported nickel and ruthenium catalysts are used to achieve conversion efficiency of 80 to 90%. This reaction is used in industrial processes when obtaining a carbon oxide-free hydrogen current is required, and oxides are in low concentration (approx. 2%). One application example is the elimination of CO in hydrogen feeding for ammonia synthesis.

CO_2 hydrogenation combines CO hydrogenation and reverse reaction of the water gas shift reaction.

$$CO + 3H_2 \leftrightarrow CH_4 + H_2O \text{ (g)} \qquad -206.3 \text{ kJ/mol (at 298 K)}$$

$$CO_2 + H_2 \leftrightarrow CO + H_2O \text{ (g)} \qquad 41.2 \text{ kJ/mol (at 298 K)}$$

$$CO_2 + 4H_2 \leftrightarrow CH_4 + 2H_2O \text{ (g)} \qquad -165.1 \text{ kJ/mol (at 298 K)}$$

In thermodynamic balance, high pressures favor the production of methane. On the contrary, high temperatures limit methane generation. As for main characteristics, this reaction is exothermic and, as a result, for every 1 m³ of synthetic methane produced per hour, 1.8 kW of heat is emitted. Moreover, this reaction yields a volume reduction of approximately 40% of reaction gases due to water generation (Rönsch et al. 2016). Finally, it is worth mentioning the downside of this reaction entails major kinetic limitations in the process of CO_2 to methane reduction and requires the use of a suitable catalyst to contribute to the achievement of acceptable reaction selectivity and velocity.

Therefore, the main characteristics of Sabatier reaction are as follows:

- is an equilibrium
- is exothermic
- requires a catalyst.

7. Sources of H_2 and CO_2

To carry out the methanation reaction, the following supply sources of H_2 and CO_2 have to be available in the process.

- **Sources of H_2:** There are several commercial technologies to obtain high purity hydrogen continuously, and mainly steam reforming from natural gas and water electrolysis. As reforming uses a fossil source to obtain energy, electrolysis of water seems to be the best technology available at the moment to produce H_2 from non-fossil sources. Nevertheless, electrolysis consumes a large amount of energy which significantly raises the cost of the process. For this reason, it is interesting to use renewable energy electricity surplus to be harnessed to obtain H_2.

- **Sources of CO_2:** CO_2 may come from biogas plants; biomass gasification plants; CO_2 by-product from industrial processes (mainly from the production of ammonia and derivatives); or CO_2 from fossil fuels converted into electricity or heat. Those sources of CO_2 associated with biomass plants and biomass gasification plants are more interesting in combination with PtG processes, due to their renewable nature and added value yield.

8. Methanation Catalyst

CO_2 methanation is catalyzed by Group VIII metals, Mo and Ag as active phase deposited on Al_2O_3, SiO_2 or TiO_2 type porous oxides (Brooks et al. 2007). Methanation activity of metals varies according to the following sequence: Ru > Fe > Ni > Co > Rh > Pd > Pt > Ir (Rönsch et al. 2016). Ni-based catalysts are still the most used as they show good catalytic activity and stability while having lower cost than other alternatives. These types of catalysts are active in methanation within the temperature range 250–400°C. Ru catalysts usually have greater intrinsic and higher activity at operating temperatures below 200°C, but with the downside of their high cost.

The purpose of catalysts supports is to enable the maximum exposure of active metal particles (Ni or Ru) to reaction gas while preventing sintering under reaction conditions. The activity of methanation catalysts is related to the amount of metal present on the support surface, so a catalyst with higher metal dispersion entails greater activity during methanation. Due to the significant influence of the support and the mode of particle deposition on the morphology and dispersion of metal active phases, the preparation and type of support to be used are two key aspects which have attracted much interest in the research of the most active catalytic systems. The most widely used support is aluminum oxide (Al_2O_3) as it offers the possibility of obtaining materials with different surface and porosity (150–250 m^2/g).

The selectivity and concentration of active elements (Ni or Ru) in commercial catalysts are determined by the concentration of carbon oxides to be passed to methane and by the degree of conversion required, so the greater concentration and conversion of carbon oxides to be transformed, the higher concentration of active components required in the catalyst. Nickel concentration in commercial systems ranges from 25 to 77% by weight, while in Ru-based catalysts Ru concentration is around 0.5% by weight.

Due to its high activity at moderate temperatures and lower cost, Ni/Al_2O_3 catalyst is the most used. The main problem with the Ni-based catalyst is that it is easy to deactivate due to the generation of carbon on its surface (Bartholomew 2001). For that reason, this catalyst is doped with another element (Mg, Zr, Mo, Mn, Co, Fe and Cu) which prevents deactivation of active places and offers improved selectivity. Mg, Ce, Cu and Fe offer high conversion in methanation reaction along with the Ni catalyst. MgO is used to increase both resistances to carbon deposition as well as Ni/Al_2O_3 catalyst thermal stability. Co modifies the catalyst surface structure to improve its resistance to sulfur. In addition to that, if Ni/Al_2O_3 catalyst is doped with CeO_2, a high degree of reduction and higher duration are achieved (Rönsch et al. 2016).

9. Reactor Concept

Methanation is generally achieved in fixed bed or fluidized bed reactors, although in recent years other types of reactors such as bubble column or micro-structured reactors have emerged, as can be seen in Fig. 6. This process was commercially developed by Lurgi, Tremp™, Conoco/BGC, HICOM, Linde, RMP and ICI/Koppers, but it still presents difficulties related to the process efficiency and heat removal. The design of new reactors focuses above all on temperature control because unless temperature is appropriately controlled, a temperature increase may lead to catalyst sintering and methane decomposition into carbon.

The main differences between fixed-bed and fluidized bed methanation reactors are: the temperature level, at which generated heat is transferred, and operating pressure. Normally, fixed-bed reactors have adiabatic operation and fluidized reactors isothermal operation. In adiabatic

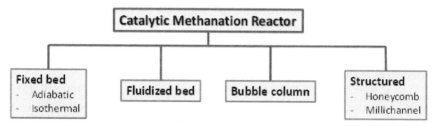

Fig. 6: Classification of catalytic methanation reactors.

operation, temperatures up to 650°C are reached, while isothermal operation is carried out at around 300°C. Lower reaction temperatures may be favorable from a thermodynamic point of view; however, they limit the catalyst activation and may generate carbon deposition problems. Taking into account that high temperatures facilitate methane performance, with stronger impact at high temperatures, adiabatic fixed-bed reactors need to operate at high pressure unlike isothermal reactors which reach high conversions even at atmospheric pressure.

To prevent problems detected in fixed-bed or fluidized bed reactors, other types of reactors are currently under research to maximize gas-solid contact and enable good energy transfer to prevent hot points on the catalytic bed, which deactivate them.

9.1 Fixed Bed Reactors

The main characteristics of fixed-bed reactors in methanation reaction are:

- catalyst high load per volume unit;
- high conversion;
- no catalyst attrition problems;
- temperature profiles are generated in the reactor due to an increase in bed temperature which may lead to the generation of hot spots;
- a concentration profile is created throughout the reactor;
- there may be load loss problems.

The main aim of fixed-bed reactors is the continuous removal of heat generated in the reaction. Therefore, fixed-bed reactors may be adiabatic, where gas is cooled down externally, or isothermal, where the reactor is continuously refrigerated (SGC Rapport 2013:284). In adiabatic fixed-bed reactors there are 2 to 5 reactors in series with internal heat exchange and occasionally with gas recirculation.

In the first reactor, reactive concentration is the highest of the entire process, dropping in each stage and increasing the methane content. Before gas is fed to the next reactor, the reactor is cooled down and, as a result, water formed during the reaction is condensed. The purpose of gas cooling is also to reduce internal temperature of the second reactor to improve thermodynamically the formation of methane. This process is repeated until gas contained in the final reactor has a high methane content. The heat removed from the gas produced in the final reactor is used to heat the gas supplied to the first reactor.

In the isothermal fixed-bed, the reactor has a constant temperature profile during the reaction as cooling takes place on the catalytic bed itself. As a result, the Sabatier reaction favors methane generation. There are different types of isothermal reactors and the most widely used are packed-bed reactors (PBR) where cooling fluid flows through tubes located inside the reactor.

The main advantage of an adiabatic reactor is its construction and maintenance but on the downside, more phases are required to effectively remove heat. In turn, the main advantage of an isothermal reactor is better mass and heat transfer, although the reactor complexity is a disadvantage.

Established Methanation Concepts for Fixed Bed Reactors

- In 1970, Lurgi developed a methanation unit with two adiabatic fixed-bed reactors in series with internal recirculation and intermediate cooling. The results obtained by Lurgi and Sasol led to the construction of the first and only commercial plant to obtain Synthetic Natural Gas (SNG) from carbon, Great Plains Synfuels Plant located in North Dakota (USA) with a capacity of 4,800,000 m³/day of SNG (Rönsch et al. 2016).

- The Tremp™ (Topsoe's Recycle Energy-efficient Methanation Process) was developed by Haldor Topsøe. This process is similar to Lurgi process, but using 3 to 4 reactors in series instead of two. It may operate at high temperatures (up to 700°C) facilitating heat recovery in the form of high-pressure steam. Recently, the TREMP technology started operation at the GoBiGas plant in Gothenburg (Sweden) (Rönsch et al. 2016, Götz et al. 2016).

- British Gas Corporation developed the HICOM process, which combines shift and methanation reactions. The process consists of three fixed-bed reactors in series with internal cooling and gas recirculation. Temperature is controlled through cooled gas recirculation and water steam is passed through the reactor to prevent the formation of coal. This technology is marketed by Johnson Matthey (Davy Technologies) (Rönsch et al. 2016).

- Imperial Chemical Industries (ICI) and Koppers have also developed several fixed-bed reactors in series for the methanation of synthesis gas obtained in a gasifier (Rönsch et al. 2016, Götz et al. 2016).

- Vesta methanation. Developed by Clariant and Foster Wheeler. The system comprises three fixed-bed reactors in series with steam injection to control temperature (Rönsch et al. 2016).

- In the 70's, Linde AG (Germany) developed an isothermal fixed-bed reactor with indirect heat exchange. In this reactor, as the bundles of cooling tubes are embedded in the catalyst bed, the reactor is able to produce steam from exothermic reactions, and part of that steam can be added to feed gas to minimize the risk of carbon deposition.

No information regarding the commercial availability of this type of reactor has been found (Rönsch et al. 2016, Götz et al. 2016).

9.2 Fluidized-bed Reactors

In the fluidized bed reactor, the catalyst is set in the bed. The reaction gases are inserted below the reactor at a determined velocity, achieving the suspension of the catalyst and its expansion along the reactor. This allows a high energy transfer, but a poor mass transfer. So, the main advantage of fluidized-bed reactors is the high heat transfer achieving an isothermal process, which facilitates process temperature control. In addition, the following characteristics can be highlighted:

- easy to renew catalyst;
- uniform particle/fluid mixing;
- lower conversion and selectivity than fixed-bed reactors due to reduced gas-solid contact; and
- there may be catalyst attrition which reduces efficiency.

Established Methanation Concepts for Fluidized Bed Reactors

- Bituminous Coal Research Inc. developed a fluidized-bed reactor for methanation of gas obtained from coal gasification. The reactor is fitted with two heat exchangers on the reactor bed.
- The Comflux process consists of a fluidized-bed reactor with internal heat transfer for methanation reaction, developed by Thyssengas GmbH and the University of Karlsruhe (Germany). A pilot pre-commercial plant with a production capacity of 2,000 m^3/h of SNG was built (Rönsch et al. 2016, Götz et al. 2016).

9.3 Three-phase Reactors

In three-phase reactors, the reaction takes place when gas is fed into a liquid and catalyst mixture which fills the reactor; this is why this process is said to have 3 phases (solid, liquid and gas). Catalyst particles are suspended on the fluid. The fluid where the catalyst is suspended is usually a DBT-type (dibenzyltoluene) cooling liquid, which facilitates heat transfer (Götz et al. 2016). This design is similar to Air Products layout for methanol synthesis process LPMEOH™ and the one used by SASOL for the Fischer-Tropsch synthesis process.

The main advantages of these types of reactors are:

- Only one reactor is necessary;
- The fluid has high thermal capacity, improving heat transfer;

- Heat removal is easy;
- Possible operation in isothermal conditions; and
- The effect of operating with fluctuating feeds is eliminated.

The main disadvantages of this process are:

- Partial evaporation and decomposition of the catalyst suspension fluid; and
- Limitations of mass transfer in the fluid.

9.4 Structured Reactors

Structured reactors (monoliths or microchannel) have been developed to reduce hot spots, which may be generated due to the reaction exothermicity, and bed load loss. Due to its design, structured reactors achieve higher transfer of heat and mass compared to fixed-bed and fluidised reactors. The main disadvantage of this type of reactor is higher complexity in catalyst deposition (Götz et al. 2016).

Microstructured reactors are miniature reaction systems used in process intensification systems where the catalyst usually refills the channels which conduct reaction gases. As these reactors have large surface areas in relation to their volume, mass and heat transfer rise, they achieve a 10–20% increase in reaction performance in relation to conventional reactors. Other important characteristics of this configuration are as follows:

- low volume, yielding savings in production materials, space and energy;
- safer than conventional reactors, as they limit flame spread in the event of a fire, so they are also suitable for reactions where explosion risk is high;
- modular nature, which facilitates upgrading by multiplying the number of reactors according to the process needs and not by increasing reactor size; and
- a suitable reactor design offers the possibility of managing heat efficiently and with flexibility, as coolant may be introduced through reaction gaps but without coming into contact with the gases which are going to react. In this way, heat removal is continuous through the entire reactor.

The most important factor for the correct operation of a micro-structured reactor is its suitable design and technique used for its manufacture. Normally, each millichannel reactor is made up of individual plates which are welded or joined together through another bonding process. This entails difficulties in correct assembly and operation, mainly due to leaks in the reactor.

Developed Methanation Concepts for Structured Reactors

According to the literature, one of the first microstructured reactors dates back to 1989 (Delsman et al. 2003). Since then, many research teams have investigated these types of reactors for different applications, due to their advantages compared to conventional reaction systems. Research teams testing these types of reactors for CO_2 reaction to obtain methane are carried out at laboratory scale and develop both the catalyst and specific reactor for the reaction, proposing and testing different designs and configurations.

The following examples of research studies on different thermochemical processes can be highlighted:

- Pacific Northwest National Laboratory USA. Ru/TiO_2 catalyst and oil is used as cooling system. At 400°C conversions close to thermodynamic balance are achieved, as well as high selectivity to the product desired (Brooks et al. 2007).

- Institut fur Mikrotechnik Mainz. Germany Reaction at 250°C with Ni or Ru catalyst on alumina (Men et al. 2007).

- Tecnalia R&I has developed a new microstructure reactor based on a bundle of tubes with millimetric inner diameter (Fig. 7). This reactor may contain from 16 to 400 channels and catalyst deposition takes place inside each millichannel. The total length of the reactor is 10 cm. The reactor is submerged in a thermal fluid flowing between channels and achieving high heat transfer.

A reactor with 388 millichannels was tested and demonstrated in the methanation reaction. 97% conversion of 2 m^3/h of hydrogen and 2 m^3/h of biogas (with an approximate CO_2 content of 25%), during 100 hours of isothermal operation, has been achieved.

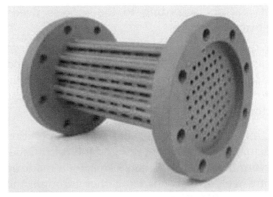

Fig. 7: Millichannel reactor developed by Tecnalia R&I.

10. Conclusions

The PtG system to obtain and inject methane into the existing natural gas grid is bound to become an autonomous and decentralized means of energy production and storage. This system has great benefits: environmental, as CO_2 is recovered and therefore its release into the atmosphere is reduced; and economic as energy surplus generated by RE is used. Nevertheless the most important outcome is that this system connects electricity and gas grids ensuring energy supply stability, as can see in Fig. 8.

Synthetic Natural Gas obtained through PtG may be fully renewable if CO_2 content in biogas is used and has the right quality for direct injected into the natural gas infrastructure for use in different applications: transport, electricity and/or heat.

Some of the breakthroughs achieved in this field worth highlighting include research under way on new more efficient reactors, as well as innovative catalysts for methanation process.

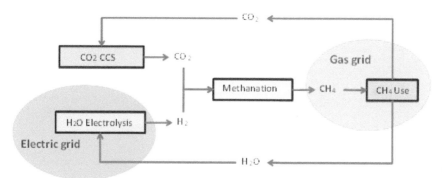

Fig. 8: Electric and Gas grid connection.

References

Bartholomew, C.H. 2001. Mechanisms of catalyst deactivation. Appl. Catal. A 212: 17–60.

Brooks, K.P., J. Hu, H. Zhu and R.J. Kee. 2007. Methanation of carbon dioxide by hydrogen reduction using the Sabatier process in microchannel reactors. Chemical Engineering Science 62: 1161–1170.

Delsman, E.R., M.J.H.M. De Croon, A. Pierik, G.J. Kramer, P.D. Cobden, Ch. Hofmann, V. Cominos and J.C. Schouten. 2003. Design of an integrated microstructured reactor—heat exchanger: a selective co oxidation device for a portable fuel processor. Proceedings of the 7th International Conference on Microreaction Technology (IMRET 7) pp. 74–76.

Götz, M., J. Lefebvre, F. Mörs, A. McDaniel Koch, F. Grafa, S. Bajohr, R. Reimert and T. Kolb. 2016. Renewable power-to-gas: A technological and economic review. Renewable Energy 85: 1371–1390.

Lehner, M., R. Tichler, H. Steinmüller and M. Koppe. 2014. Power-to-Gas: Technology and Business Models. ISSN 2191–5520.

Men, Y., G. Kolb, R. Zapf, V. Hessel and H. Löwe. 2007. Ethanol steam reforming in a microchannel reactor. Process Safety and Environmental Protection 85(5): 413–418.

Rönsch, S., J. JensSchneider, S. Matthischke, M. Schlüter, M. Götz, J. Lefebvre, P. Prabhakaran and S. Bajohr 2016. Review on methanation—From fundamentals to current projects. Fuel 166: 276–296.

Schiebahn, S., T. Grube, M. Robinius, V. Tietze, B. Kumar and D. Stolten. 2015. Power to gas: Technological overview, systems analysis and economic assessment for a case study in Germany. Int. Journal of Hydrogen Energy 40: 4285–4294.

SGC Rapport 2013: 284.

CHAPTER 9

Stationary Fuel Cells and Hybrid Systems

A. Moreno,[1] *V. Cigolotti,*[1,*] *M. Minutillo*[2] *and A. Perna*[3]

1. Introduction

The electricity system of the future must produce and distribute reliable, affordable and clean electricity. Currently, most of the electricity, in the world, is still produced in centralized large fossil fuels power plants and transported through the transmission and distribution grids to the end consumers. These plants require high management costs of large infrastructures and are susceptible to unreliability and instability under unforeseeable events (Momoh 2014). Moreover, issues linked to the climate change are converging to drive fundamental change in the way energy is produced, delivered and utilized (Hidayatullah et al. 2001).

In fact, the aging of energy transmission networks, the grid congestion, the increase of remote areas and the need of energy in developing countries are pushing utilities to find alternatives to traditional forms of supplying power. The answer lies with distributed generation (DG) that is defined

[1] ENEA, Italian National Agency for New Technologies, Energy and Sustainable Economic Development, Via Anguillarese 301, 00123 Roma, Italy; Piazzale Enrico Fermi 1, 80055 Portici (Napoli), Italy.
[2] University of Naples Parthenope, Centro Direzionale di Napoli, Isola C4, 80143 Napoli, Italy.
[3] University of Cassino and Southern Lazio, Viale dell'Università, 03043 Cassino (FR), Italy.
Emails: angelo.moreno@enea.it; mariagiovanna.minutillo@uniparthenope.it; perna@unicas.it
* Corresponding author: vivana.cigolotti@enea.it

as small scale electricity generation, where electricity is produced next to its point of use. Rather than build a costly transmission system to provide power to remote or congested areas, utility companies can use distributed generation to supply customers in those areas with a constant supply of power (McPhail 2015).

Therefore, in the last decade, the number of small generators (usually in kW to MW range) that use renewable energy sources (such as wind and PV) or fossil fuels is significantly growing. Most of these generators are connected to the distribution network level and referred to as DG. The major factors that contribute to the rise of DG are the developments in distributed generation technologies, the constraints on the construction of new transmission lines, the increased customer demand for highly reliable electricity, the electricity market liberalization and the concerns about the climate change (ElMubarak and Ali 2016).

DG can satisfy the specific heat and power demands of the consumer on site by using several small scale technologies such as internal combustion engines, fuel cells or micro-turbines. Thus, DG allows to support the development and deployment of combined heat and power systems that permit to reach a higher primary energy saving.

Fuel cell power plants are an ideal solution for DG applications; several prior studies have examined the potential of fuel cells in supporting a future grid with high renewable penetrations (Shaffer et al. 2015, Owens and Mcguinness 2015). In particular, stationary fuel cells as distributed generation technology, produce power and heat at the site of the consumers, providing an immediate supply of energy, and have the potential to be one of the technologies for the future energy transition. Fuel cell for stationary applications promises significant benefits: high energy efficiencies (electrical efficiency of up to 60%, combined efficiency in cogeneration—combined heat and power generation CHP—of more than 90%), low acoustic emissions, the ability to change electricity output levels to meet rapid fluctuations in electricity demand, low noise level and fast load following. These benefits make fuel cell technology ideally suited to enable integrated power networks and because of the low noise level it is easy to place fuel cells in urban areas. Moreover, from the environmental point of view, fuel cells can substantially reduce the CO_2 emissions when compared to conventional fossil-based power plants. Depending on the fuel used and its source, fuel cells can virtually eliminate CO_2 and other emissions like NOx or SOx.

The general principle of fuel cells is the conversion of primary chemical energy from hydrogen directly into electrical and thermal energy by means of an electrochemical processes. Fuel cells consist of an anode and a cathode, where the electro-oxidation of hydrogen and the electro-reduction of oxygen occur, respectively, and an electrolyte which allows the ions flowing. The electrons, delivered by the hydrogen electro-oxidation, pass through an external circuit providing electrical energy. Fuel cells are generally classified according to the nature of the electrolyte (except for direct methanol fuel cells which are named for their ability to use methanol as a fuel) that determines the flowing ions, the operating temperature, the materials and the feeding fuel. In Fig. 1 and Table 1 the

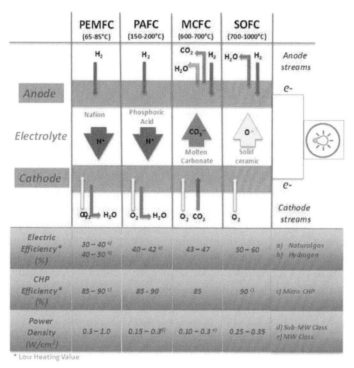

Fig. 1: Fuel cells classification and characteristics (Ellamla et al. 2015, Ryan et al. 2009).

Table 1: Fuel cell operating details.

Fuel Cell Type	Fuel Type	Internal Reforming	Primary Contaminants	Anodic and Cathodic Semi-Reactions
PEMFC	Hydrogen Syngas	No	$CO < 10$ ppm	$H_2 \rightarrow 2H^+ + 2e^-$ $\frac{1}{2} O_2 + 2H^+ + 2e^- \rightarrow H_2O$
PAFC	Hydrogen	No	Sulphur $CO < 1\%$ vol	$H_2 \rightarrow 2H^+ + 2e^-$ $\frac{1}{2} O_2 + 2H^+ + 2e^- \rightarrow H_2O$
MCFC	Hydrogen NG, Biogas	Yes	Sulphur	$H_2 + CO_3^{2-} \rightarrow H_2O + CO_2 + 2e^-$ $\frac{1}{2} O_2 + CO_2 + 2e^- \rightarrow CO_3^{2-}$
SOFC	Hydrogen NG, Biogas	Yes	Sulphur	$H_2 + O^{2-} \rightarrow H_2O + 2e^-$ $\frac{1}{2} O_2 + 2e^- \rightarrow O^{2-}$

classification and the main operating details of the Polymer Electrolyte Membrane Fuel Cell (PEMFC), the Phosphoric Acid Fuel Cell (PAFC), the Molten Carbonate Fuel Cell (MCFC) and the Solid Oxide Fuel Cell (SOFC) are illustrated.

The stationary FC sector represented more than 70% of global FC revenue in 2014, and it is expected to continue to lead the overall FC market in the coming years. This sector includes four different market segments (Roland Berger-FCH JU 2015, Garche and Jörissen 2015) as shown in Fig. 2 residential, commercial, industrial and power generation, and uninterruptible power supply (UPS).

The performance of a fuel cell is characterized by the polarization curve, which describes the relationship between the output cell voltage and the current density.

Figure 3 shows the polarization curves of fuel cells commercially available or in pre-commercial status, referring to the mentioned market segments. For PEMFC technology, low temperature (60–80°C) and high temperature (120–200°C) solutions have been considered.

Engen-2500 is a 2.5 kWe SOFC module manufactured by SOLIDpower S.p.A, fed by natural gas (courtesy of SOLIDpower), DFC300 is a 300 kWe MCFC base module manufactured by FuelCell Energy U.S. (Patel et al. 2009) fed by natural gas, Mark1030 is an 1 kWe water-cooled LT-PEMFC manufactured by Ballard Power System that can be fed by syngas with CO concentration of 1 ppm (Jannelli et al. 2013), PureCell400 is a 400 kWe PAFC base module produced by Doosan Fuel Cell America U.S. (Remick et al. 2009), fed by natural gas and Serenus 166/390 is an 1 kWe HT-PEMFC (High Temperature Pem Fuel Cell) manufactured by Serenergy (User Manual-Serenus 166/390) that is fed by a syngas with CO concentration of 1%.

The *residential segment* regards CHP units, sized between 0.5 kWe and 5 kWe, that use either PEMFC or SOFC technology.

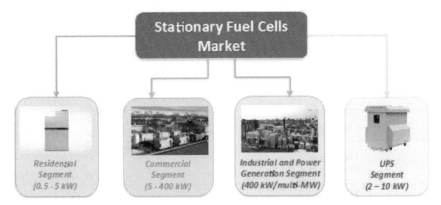

Fig. 2: Fuel cell systems market segments.

The residential segment comprises one-and two-family dwellings; here the fuel cell has the highest mass-market potential as standardized heating solutions typically target a large variety of buildings (i.e., both new buildings and the building stock). In the residential segment, the deployment of fuel cell micro-CHPs can realize substantial savings in primary energy, local emissions and energy costs. Stationary fuel cells primarily seek to replace existing gas heating solutions as integrated CHP applications, but could also operate as pure base-load micro power plants in addition to an existing heating solution like a gas condensing boiler (Roland Berger-FCH JU 2015).

The *commercial segment* regards CHP units sized between 5 kWe and 400 kWe using PEMFC, PAFC, SOFC or MCFC technologies that encompass both residential (i.e., apartment buildings) and non-residential buildings. In integrated fuel cell CHP applications, supplying heat is still the primary demand driver, but power-driven solutions have also a potential development. This segment shows high standardization potential especially for larger commercial buildings (Roland Berger-FCH JU 2015).

The *industrial and power generation segment* includes both large multi-MW$_e$ primary power units (< 10 MW$_e$), developed to replace the network grid where it is little or absent, and industrial facilities where there are heat and power demands (i.e., wastewater treatment facilities, chemical production facilities, breweries and data centres). Stationary fuel cells

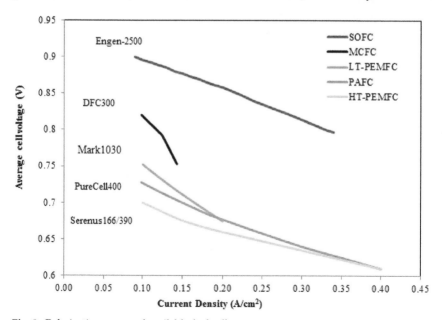

Fig. 3: Polarization curves of available fuel cells systems.

have to be designed to the specific needs of the utilities, but a standardized modular approach and the development of modular concepts can permit cost-efficient adaptation to specific energy requirements of different industries (Roland Berger-FCH JU 2015).

The *UPS segment* refers often to small power units (< 5 kW$_e$) that have to provide a guaranteed supply of power in the event of grid interruption. However, for data centers the UPS plants could be in multi-MW class.

UPS based-fuel cell development is led by two main technologies which are SOFC and PEMFC.

Finally, fuel cells can be arranged with other power engines in several configurations realizing integrated cycles or hybrid systems. These systems combine two or more power generating devices (Gas Turbine, Organic Rankine Cycle, Internal Combustion Engine) and make use of the synergism to generate maximum power and offer higher efficiencies.

2. Stationary Fuel Cells in Residential Segment Market

Residential CHP units produce heat and power mainly for single and multi-family residential dwellings. With respect to conventional CHP technologies (Internal Combustion Engine, Gas Turbine, Combined Cycle), fuel cells systems assure a significant reduction of CO_2 emissions and pollutant gasses. In particular, as reported in Dodds et al. (2015), some manufacturers declare 35–50% as CO_2 reduction (1.3–1.9 tCO_2/year in a four person household) considering small-size CHP systems (0.7–1 kWe), while higher percentages can be reached for larger systems (3 tCO_2/year for CFCL BlueGen device).

With respect to the pollutants, they can be only generated in the burner of the reforming system that is used when the fuel cell is fed with hydrocarbons (i.e., natural gas). In this case further fuel and the unreacted anode stream are combusted to supply thermal energy for the endothermic reactions of the reformer. Thus, the measured pollutants from fuel cells are around a tenth (NOx 1–4 g/MWh; CO 1–8 G/MWh, SO_2 0–2 g/MWh) of those from other gas-burning technologies like condensing boiler and CHP engines (Ellamla et al. 2015, Dodds et al. 2015). Two types of fuel cells are used for micro-CHP systems: PEMFC (80% used) and SOFC (20% used).

The market offers CHP systems in the power range 0.5–5 kW$_e$, having an electrical efficiency of 30–40% (PEMFC) and 40–60% (SOFC) and CHP efficiency > 85%.

PEMFCs are able to produce energy when needed, therefore, for example, they work during the day and are switched off at night. Differently, SOFC units, even if they have a higher maximum efficiency, need to operate continuously, because the start-up and shut-down times are too long and can reduce the life of the stack because of the thermal stress during the transient phases.

2.1 Polymer Electrolyte Membrane Fuel Cell Applications

PEMFCs are the most popular and versatile of all the types of fuel cells existing in the market. Due to their modularity, simple manufacturing, high power density, low operating temperature, and fast start-up and shutdown, residential CHP based on PEM fuel cells are widespread in the market and strong efforts are devoted to reduce their costs and to increase their life time.

PEMFC operates at low temperatures (65–85°C), uses a solid polymer membrane as electrolyte and platinum as electrodes catalyst. Due to the high cost of platinum and to its scarcity, a substitute catalyst is currently being researched by the main PEMFCs developers. An alternative is represented by graphene-based components which can operate either with hydrogen or with reformed fuel (Iwan et al. 2015, Graphene Science Handbook 2016).

PEMFCs function best with high purity hydrogen gas as fuel source. In this case the CHP system configuration is quite simple as shown in Fig. 4.

The anode off-gas contains water vapour mixed with hydrogen (the cell does not work in dead-end mode) according to the anode utilization

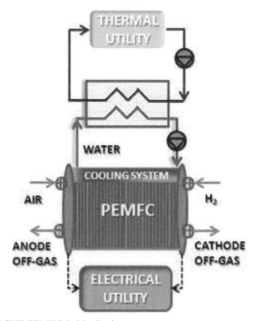

Fig. 4: Layout of CHP PEMFC fed by hydrogen.

factor, while the cathode off-gas contains oxygen, nitrogen and water vapour. The thermal energy for cogeneration purpose is recovered by the stack cooling system.

Because the pure hydrogen is unlikely to be the fuel source in the near future due to the lack of hydrogen infrastructure, a large use of PEMFC in CHP applications is possible by using syngas generated by reforming processes (Perna 2007a,b, Di Bona et al. 2011). The most economic source of hydrogen is the steam reforming of natural gas due to its widespread availability and low cost, but other fuels (LPG, kerosene, alcohols, etc.) and a variety of renewable sources such as landfill gas and sewer gas can also be used.

In this case the CHP system configuration is more complex because the PEMFC power unit is integrated with a reforming unit consisting of three components (Steam Reformer, Water Gas Shift Reactor, CO removal unit) as shown in Fig. 5. It can be noted that the conversion of natural gas in a hydrogen rich stream for PEMFC feeding is carried out in two steps, a high temperature endothermic step that takes place in the fuel reforming reactor (steam reforming reactor) in which the hydrocarbons are converted into a gaseous mixture of H_2, CO, CO_2, CH_4 and unreacted H_2O and a low temperature slightly exothermic step that occurs in the water gas shift reactor in which CO is reacted with H_2O towards H_2 and

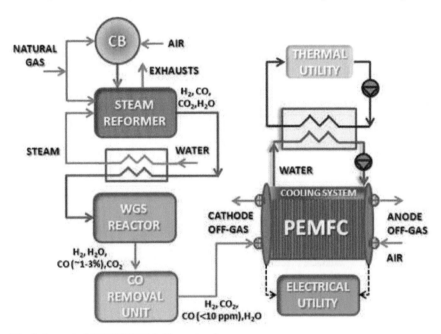

Fig. 5: Layout of CHP PEMFC fed by natural gas.

CO_2. Because the shift reaction is equilibrium-limited, CO conversion is not complete and an additional step of CO removal is necessary in order to reduce the CO concentration at the value (< 10 ppm) required by the low temperature PEM fuel cell. This can be achieved by using a preferential oxidation reactor where the syngas reacts with a controlled amount of air over an activated Ru catalyst (a novel Ru/Al_2O_3 catalyst) that allows to lower the CO content below 1 ppm (Di Bona et al. 2011).

The heat required for the steam reforming reaction is supplied by a catalytic burner (CB) fed with natural gas and the anodic exhausts from the PEMFC power unit. Finally, the syngas and the air are sent to the PEMFC where the electrochemical reactions occur. Water is used as stack cooling stream and the heat recovered is used for house heating purpose.

The regular operation of a PEMFC fed by syngas is assured if the CO concentration, that causes the poisoning of the platinum anode catalyst, is less than 10 ppm (Kim et al. 2001, Baschuk and Li 2001, Minutillo and Perna 2008). Therefore, considerable efforts in the development of PEMFC have been made to reduce the effect of CO.

An interesting approach to overcome the CO poisoning is the increasing of the fuel cell operating temperature at values higher than 100°C (High Temperature PEM fuel cell, HT-PEMFC), because, in this way, the CO adsorption onto the catalyst sites is greatly reduced. This approach implies that the conventional polymeric membrane has to be modified. Therefore, since many years, great attention has been paid to the research of other membrane materials such as PFSA membranes, sulfonated polyaromatic polymers and composite membranes (PEEK, PI, PSF and SPSF), polybenzimidazole-based (PBI) membrane, and aromatic polyether polymers or copolymers that bear pyridine units (Zhang et al. 2006, Oono et al. 2009, Geormezi et al. 2011).

Highest operating temperatures have an interesting advantage in developing CHP systems based on HT PEMFCs integrated with reforming systems because a minor complexity and a better compactness (Fig. 6) can be obtained thanks to some aspects:

- the CO removal unit is not required;
- reactants humidification systems are not required because the proton transport in the membrane occurs without water dragging.

From the operating point of view HT-PEMFCs require cell voltages of over 0.7 V to achieve system efficiencies higher than that of LT PEMFCs but their performances are currently still low; as a matter of fact, by analyzing the electric and cogeneration efficiencies of CHP systems based on LT-PEMFC and HT-PEMFC, it can be affirmed that 38% and 40%, 80% and 78% are attended, respectively (Jannelli et al. 2013).

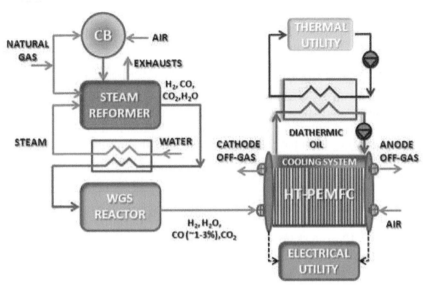

Fig. 6: Layout of CHP HT-PEMFC fed by natural gas.

2.2 Solid Oxide Fuel Cell Applications

CHP systems based on SOFC technology are very attractive, due to their higher efficiencies (electric and cogeneration efficiencies) and to the possibility of using the natural gas from the network as fuel, simplifying the integration into an existing heating system. Moreover, SOFC can also tolerate higher levels of sulphur than other types of fuel cells and this is another reason for which they are very interesting for the application in residential CHP systems.

The SOFC uses a solid, usually Y_2O_3-stabilized ZrO_2 (YSZ), as electrolyte but other electrolytes are under development. The operating temperatures are between 700°C and 1000°C. These high temperatures allow: (1) to operate without precious metals as catalysts; (2) to eventually avoid an external fuel reformer (internal reforming takes place); (3) to obtain high-quality waste heat; (4) to use different fuels.

As a consequence, SOFC-based CHP systems have a simpler plant configuration (a compact design), as shown in Fig. 7, so manufacturing costs could be reduced.

Other system configurations and operating conditions are analyzed in the scientific literature. In (Braun et al. 2006) the performance of an anode-supported SOFC in residential CHP system was evaluated. The authors estimated system performance by varying the feeding fuel (i.e., pure hydrogen or methane), the reforming configuration (i.e., internal or external reforming with or without anode off-gas recirculation), the

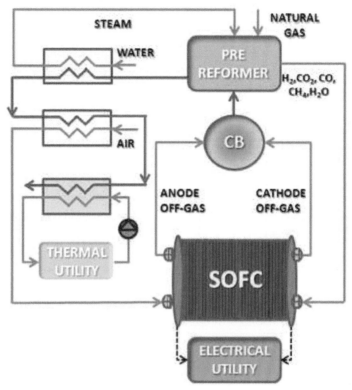

Fig. 7: Layout of CHP SOFC system.

oxidant processing with cathode recirculation, and the combination of recirculation and internal reforming. The results indicated that maximum efficiency was achieved when the cathode and anode gas recirculation was used along with the internal reforming of methane. The electric and cogeneration efficiencies were 40% HHV and 79%, respectively.

The above mentioned characteristics and performances have allowed to these systems entry in the residential CHP market and thus, to be installed in small residential houses in urban areas, apartment houses and condominiums.

The market of residential fuel cell CHP systems is large and there are numerous companies (about 20) actively involved. Some manufacturers and their products are illustrated in Table 2.

Figure 8 compares the performance of these systems (Ellamla et al. 2015, McPhail et al. 2017).

In this market domain, Japan is in a leading position. The Japanese Ene-Farm project that started in 2009 is the world's most successful program for CHPs development and installation. Major companies, like Panasonic,

Table 2: Micro-CHP systems manufacturers.

Manufacturer	Fuel Cell type	Electric power (W)	Thermal power (W)	Electric efficiency (%)	CHP efficiency (%)
Buderus (A)	SOFC	700	620	45	85
Solid Power (B)	SOFC	1500	610	60	85
Elcore (C)	HT-PEMFC	300	600	32	98
Hexis (D)	SOFC	1000	1800	35	95
Junkers (E)	SOFC	700	620	45	85
Solid Power (F)	SOFC	2500	2000	50	90
Vaillant (G)	SOFC	700	1300	33	93
SenerTec (H)	LT-PEMFC	700	950	37,6	89
Viessmann e Panansonic (I)	LT-PEMFC	750	1000	38,5	90
Toshiba (L)	LT-PEMFC	700	1000	38,5	91

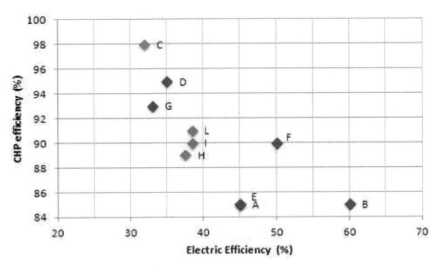

Fig. 8: Micro-CHP systems performance.

Toshiba and AISI-Seiki, successfully commercialized PEMFC and SOFC-based CHP systems and, with the help of government support, about 200,000 units (700–750 W each, CHP efficiency ≈ 95%) were installed by March 2017 (Ellamla et al. 2015, McPhail et al. 2017), as shown in Table 3.

Concerning the targets for residential applications, the Table 4 shows the state-of-the-art in 2012 and the main targets to be achieved for the

Table 3: ENE-FARM Project in Japan.

Year	2009	2010	2011	2012	2013	2014	2015	2016
N. systems	5,030	4,985	17,995	14,477	47,000	25,812	38,746	50,000
Cumulative	5,030	10,015	28,010	42,487	89,487	115,299	154,045	204,045
Financial support for the period 2009–2015: 964 M€								
Financial support in 2016: 88 M€								

Table 4: FCH JU Targets for residential application (FCH JU – Multi Annual Working Plan 2014).

Segment	Key Performance Indicator (KPI)	Unit	State of the Art (2012)	FCH_JU Target		
				2017	2020	2023
	CAPEX	€/kW	16,000	14,000	12,000	10,000
Residential – mCHP for single family homes and small buildings (0,5–5 kW)	Durability	Years of plant operation	10	12	13	14
	Availability	% of the plant	97	97	97	97
	Electrical efficiency	% LHV	30–60	33–60	35–60	35–60
	Thermal efficiency	% LHV	25–55	25–55	25–55	25–55
	LCOE	€ Ct/kWh	3*grid parity	2.5*grid parity	2*grid parity	< 2*grid parity
	Emissions	mg/kWh	NOx < 2 no SOx	NOx < 2 no SOx	NOx < 2 no SOx	NOx < 2 no SOx

generic stationary fuel cells applications across Europe (FCH JU–Multi Annual Working Plan 2014).

3. Stationary Fuel Cells in the Commercial Segment Market

Fuel cell systems intended for use in commercial applications are introduced by individual users (hotels, hospitals, restaurants, office buildings) for energy generation ranging from 5 kW$_e$ to 400 kW$_e$. The main technology has been MCFC in the size of 300 kW$_e$ (DFC 300, manufactured by Fuel Cell Energy), but currently this product is not on the market anymore, because the leading company decided to develop systems starting from 1.4 MW for industrial and power generation applications. There are also solutions with PAFC and SOFC which have some market share.

3.1 Molten Carbonate Fuel Cell Applications

Among the various fuel cell types, the MCFC has been mainly applied for stationary generation together with the production of highly valuable heat and it is thus suitable for many commercial applications as well as for distributed power supply.

MCFCs use a carbonate electrolyte, which is generally a mixture of lithium and potassium carbonates (salts). The electrolyte materials, which become molten at typical operating temperatures of 650°C, are usually supported and wicked to cover electrode surfaces through use of a ceramic electrolyte support mesh (often LiAlO2). Anode materials are typically Ni–Cr/Ni–Al alloys and cathode materials are composed of lithiated NiO.

The nickel anode is an excellent catalyst for the shift reaction which converts carbon species (ultimately carbon monoxide) and water into hydrogen that then releases the electrons generating the electric current. As a consequence, the MCFC can operate with both pure hydrogen as well as hydrocarbons, and water is formed at the anode side. As depicted in Fig. 1 and reported in Table 1, carbon dioxide is necessary as a closed-loop reagent: CO_2 is consumed at the cathode (together with oxygen) at the same rate at which it is released at the anode. This operation makes an interesting opportunity to use the MCFC as a CO_2 concentrator.

The MCFC's performance is influenced by many operating conditions depending on the application, such as load profile, availability, power-to-heat ratio required, fuel and oxidant chemical compositions. However, the MCFC's excellent capacity to work also at part load—with practically unchanging efficiency values—makes it a technology which comes out head and shoulders above conventional generators in the intermediate power range.

The MCFCs can be fed with any hydrogen-carbon mixture as fuel, therefore a variety of fuels such as natural gas, biogas, gasified biomass, syngas from coal or waste, but even liquid fuels such as ethanol can be adopted. The high operating temperature of the MCFC helps to process all these different fuels, but for safe and enduring operation of the MCFC a careful clean-up of the fuel is necessary beforehand (McPhail 2015).

The main producers of MCFC power plants are Fuel Cell Energy (FCE), United States, with its subsidiary FCES, Germany, and South Korean partner POSCO Energy.

In the commercial segment, the Fuel Cell Energy commercialized the DFC 300 (Direct Fuel Cell) power unit, a stack of 400 fuel cells, generating about 350 kWe of power at 480 V and 50–60 Hz. The net electrical efficiency of the DFC 300 is about 47% (based on LHV) when operating on natural gas.

Figure 9 shows the concept scheme of the DFC300 power plant that consists of the fuel cell stack and the equipment needed to provide the proper gas flows and power conversion.

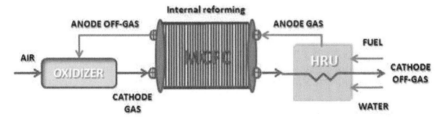

Fig. 9: Concept scheme of DFC300.

Fuel and water are heated to the required fuel cell temperature (about 650°C) in a heat recovery unit (HRU) and sent to the anode side of the fuel cell stack, where the fuel is converted to hydrogen by internal steam reforming. Anode off-gas is supplied to a catalytic reactor (oxidizer) where reacts with the incoming air to provide the cathode reactants (oxygen and carbon dioxide). Cathode exhaust exits the system at 340–400°C, so high quality waste heat for cogeneration and/or cooling purpose is available. However, this system, as mentioned before, is not commercialized by the company any more.

In Margalef and Samuelsen (2010), the authors studied the application of trigeneration power systems based on high-temperature fuel cells integrated with absorption chillers for the distributed generation (DG). As fuel cell power unit, the DFC300 was adopted and the 40RT Yazaki CH4040-KE was selected as the absorption chiller. This chiller is a dual exhaust and gas fired double-effect chiller which takes exhaust gases directly as the heat source and is supported by a high-temperature generator which burns natural gas in case of insufficient thermal energy. In addition, lithium bromide-water (LiBr–H_2O) solution is used as working fluid.

3.2 Phosphoric Acid Fuel Cell Applications

PAFCs are considered as the first generation of modern fuel cells, and are in a commercial stage for stationary electricity generation and hundreds of units have been installed over the world.

They use concentrated phosphoric acid as the electrolyte and the operating temperatures are between 150°C and 220°C. Like the PEMFC, they need to use platinum as a catalyst which makes them expensive.

The leading company producing PAFCs was for many years United Technologies Corp (UTC). In 2013, UTC was taken over by ClearEdge and subsequently by the Doosan Group in 2014. Doosan Corporation Fuel Cell has launched to the market (April 2014) two improved PAFCs (PureCell products) in two sizes: 5 kW and 400 kWe. The 400 kWe PureCell system has dimensions 8.74 m × 2.54 m × 3.02 m, with a total volume of 67.04 m³. The performance of this unit is summarized in Table 5 (Jason et al. 2012).

Another player that has emerged in developing PAFC power units is Fuji Electric. This company began developing phosphoric acid fuel cells in 1973, and has developed 50 kWe, 100 kWe and 500 kWe models for use in onsite applications, and in cooperation with gas companies and electric power companies, has field-tested more than 100 units. Fuji Electric developed a new phosphoric acid fuel cell known as the "FP-100i", and launched this model in 2009. Figure 10 illustrates the plant lay-out.

The main features of Fuji Electric's commercial-model 100 kWe phosphoric acid fuel cell are: (a) high electrical efficiency, from low output to high; (b) capable of utilizing multiple sources of fuel; (c) excellent environmental characteristics. This PAFC power system, fed by natural gas (22 m³/h), has an electrical efficiency of 42% (LHV) and the overall efficiencies (cogeneration efficiency) are 62% or 91% with a high temperature heat recovery of 50 kW (90°C) or medium temperature heat recovery of 123 kW (60°C), respectively. Moreover, FP-100i can also

Table 5: Performance of PureCell 400.

Characteristics	Units	Nominal data
Electric power	kW	400
Fuel consumption	Nm³/h	94.2
Electric efficiency	%	42
Cogeneration efficiency	%	90
High Grade Heat Output (120°C)	kW	188
Low Grade Heat Output (60°C)	kW	258

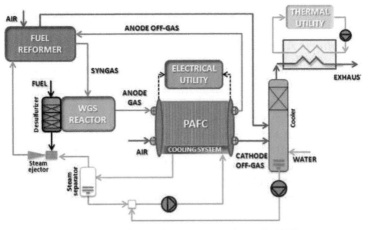

Fig. 10: Concept scheme of a PAFC cogeneration system (Kuroda 2011).

be fed by digester gas; in this case the fuel consumption is 44 m³/h, the medium temperature heat recovery is 130 kW (60°C), and the electric and cogeneration efficiencies are 40% and 90%, respectively (Kuroda 2011).

3.3 Solid Oxide Fuel Cell Applications

The first demonstration of SOFC technology at a commercially relevant scale was the Siemens-Westinghouse CHP100 cogeneration system. This system produces 109 kWe with 46% net electrical efficiency into the local grid and 64 kWt of hot water into the local district heating system (Siemens SOFCs 2010). It has a record of 99.1% availability with more than 36,900 hours (updated on June 2007) of operation.

The Siemens CHP100 (Cali' et al. 2005) power system mainly consists of the fuel supply system, the power module, the electrical control system, the thermal management system and the heat export system. Natural gas is supplied from high pressure pipeline at the site and is desulfurized in two pressure vessels. Humidification for natural gas reforming is obtained by re-circulating and mixing a portion of the spent fuel from stack with the feeding fuel. The control system includes a programmable logic controller (PLC), a power conditioning and a switching equipment, a DC power dissipater, an instrumentation interface, and an operator interface. The thermal management system, that mainly consists of two process air blowers, two regenerative heat exchangers, two electric air preheaters and a mass flow meter, allows to heat the air before entering in the anode side of the stack. Finally, the produced hot water for cogeneration purpose is generated in a further heat exchanger.

In Fontell et al. (2004) the authors studied a 250 kWe natural gas-fuelled SOFC plant. Figure 11 shows the proposed SOFC system.

A desulphurizer unit is included to treat natural gas before being fed to the pre-reformer; the reformed gas, before entering in the SOFC, is pre-heated by the anode off-gas that is then split into two parts; the first part is combusted with cathode off-gas in the catalytic burner (CB), whereas the second part is recirculated to the pre-reformer. The exhaust gas from the catalytic burner is applied to vaporizing the water stream and preheating the natural gas feed stream. The cogeneration system efficiency is 85%.

Bloom Energy commercializes the Energy Server power plants that are modules of different size. Table 6 reports the main performance of Energy Server 5.

Concerning the targets for commercial applications, Table 7 reports the state-of-the-art in 2012 and the main targets to be achieved for the generic stationary fuel cells applications across Europe (FCH JU–Multi Annual Working Plan 2014).

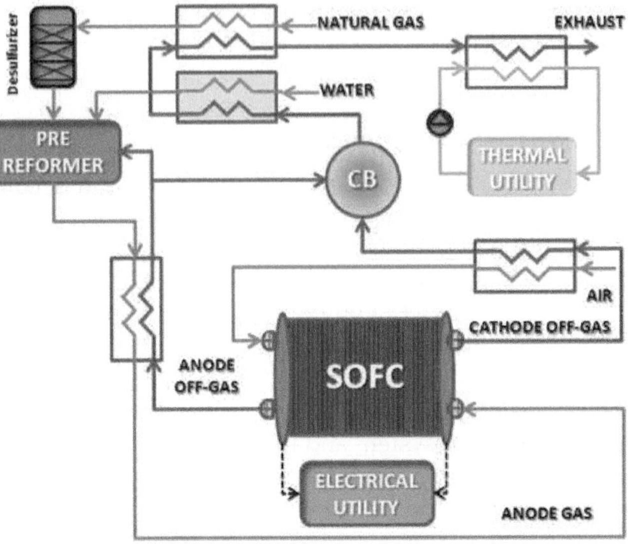

Fig. 11: Layout of CHP-SOFC system proposed by (Fontell et al. 2004).

Table 6: Performance of Energy Server 5.

Characteristics	Units	Nominal data
Electric power (net AC)	kW	250
Fuels		Natural gas (Directed Biogas)
Electric efficiency (LHV, net AC)	%	65 (53)
Heat rate (HHV)	Btu/kWh	5811 (7127)

4. Stationary Fuel Cells in the Industrial and Power Generation Segment Market

The main applications for the industrial use of FCs are prime power (3–60 MW), CHP, and tri-generation, mainly for new office builds, retail parks, hospitals, universities, or data centers. The fuel cells employed for these applications are the MCFCs, SOFCs and PAFCs, and the market is led by three companies—FuelCell Energy, manufacturer of SureSource products line, based on MCFC technology, Bloom Energy, manufacturer of Energy Server products line, based on SOFC technology, and Doosan Group, manufacturer of PureCell products line, based on PAFC technology.

4.1 Molten Carbonate Fuel Cell Applications

High-temperature MCFCs recently became the market leader for large stationary applications (Review FC industry 2015). MCFC power plants

Table 7: FCH JU Target (FCH JU–Multi Annual Working Plan 2014).

Segment	Key Performance Indicator (KPI)	Unit	State of the Art (2012)	FCH_JU Target		
				2017	2020	2023
	CAPEX	€/kW	6,000–10,000	5,000–8,500	4,500–7,500	3,500–6,500
Commercial– Mid-sized installations for commercial and larger buildings (5–400 kW)	Durability	Years of plant operation	2–20	6–20	8–20	8–20
	Availability	% of the plant	97	97	97	97
	Electrical efficiency	% LHV	40–45	41–50	42–55	42–55
	Thermal efficiency	% LHV	24–40	24–41	24–42	24–42
	LCOE	€ Ct/kWh	3*grid parity	2.5*grid parity	2*grid parity	2*grid parity
	Emissions	mg/kWh	NOx < 40	NOx < 40	NOx < 40	NOx < 40

are suited to food and drink processing applications and sewage gas which generate anaerobic digester gas. An additional benefit for the food and beverage industry is the CHP capabilities inherent to the stationary MCFC plants. Harvesting waste heat, steam can be produced for hot water and other heating needs, further increasing the efficiency of the power plant, up to double than that of grid-supplied power. Because most food and drink processing plants require 5 MW or less of power, fuel cell power plants can produce most, if not all, of the power requirements at these facilities. In places where digester gas production volume is variable, blending with natural gas can be carried out for reliable base-load power and heat (McPhail 2015).

MCFCs benefit from relatively low cost due to non-platinum catalysts, simpler ancillary systems, availability ratings of about 95% (offering base load power 24 hours a day), but suffer from low lifetime and power density (Staffell 2015).

Fuel Cell Energy is the main player in this market, delivering MCFC modules since 1969 when the company was called ERC. In 1999 it was renamed as Fuel Cell Energy Inc. and in 2007 this company started a cooperation with Posco Energy, a Korean company. The proposed power generation solutions, available in different sizes and configurations, achieve electric efficiencies ranging from 47% to 60% and cogeneration efficiencies up to 90%. The solutions delivered by Fuel Cell Energy are available in different sizes and configurations and follow the 2007

California Air Resources Board (CARB) standards. The available solutions are:

- SureSource 1500™ (1.4 MW) (https://www.fuelcellenergy.com/wp-content/uploads/2017/02/Product-Spec-SureSource-1500.pdf), ideal for on-site power generation for large installations requiring continuous power and value high-quality heat for facility heating and/or absorption chilling. The system is suitable for a wide range of applications, including wastewater treatment plants, manufacturing facilities, hospitals and universities;

- SureSource 3000™ (2.8 MW) (https://www.fuelcellenergy.com/wp-content/uploads/2017/02/Product-Spec-SureSource-3000.pdf), comprised of two 1.4 megawatt (MW) modules, the SureSource 3000 generates 2.8 MW of ultra-clean power. Similar to the SureSource 1500, the system is well-suited for on-site applications and caters to customers with an even greater power load requirement. Ideal applications include large universities, manufacturing facilities, wastewater treatment plants, or for multi-plant fuel cell parks to support the electric grid;

- SureSource 4000™ (3.7 MW) (https://www.fuelcellenergy.com/wp-content/uploads/2017/02/Product-Spec-SureSource-4000.pdf), the largest in the SureSource power plant fleet and generates 3.7 megawatts (MW) of ultra-clean power with an industry-leading electrical efficiency of approximately 60%. This enhanced-efficiency fuel cell system is designed for utilities, large industrial users, data centers, and other customers focused on clean and affordable power driven by the benefits and economics of high system electrical efficiency.

Figure 12 shows the SureSource 3000™ plant configuration that supplies 2.8 MW of electric power with 47% electrical efficiency.

The SureSource 3000 power plant is comprised of two 1.4 MW fuel cell modules that operate in parallel. Each module comprises four fuel cell stacks (DFC300). The fuel cell module is combined with the electrical and mechanical balance of plant (BoP) to complete the power plant. The mechanical BoP processes the incoming fuel for the fuel cells (i.e., clean natural gas or renewable biogas), whereas the electrical BoP converts the direct current (DC) generated by the fuel cells into alternating current (AC) for use by the customer.

The SureSource™ power plants are scalable, so that multiple plants can be combined to create multi-megawatt fuel cell parks. In Fig. 13 the plant configuration of 3.7 MW is depicted. With respect to the 2.8 MW

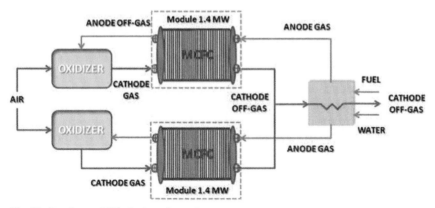

Fig. 12: SureSource3000 plant configuration.

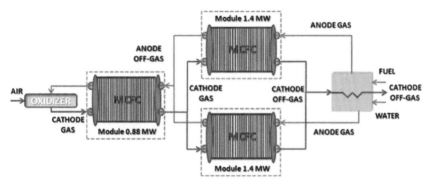

Fig. 13: SureSource4000 plant configuration.

solution, a third module of 0.88 MW, fed by the anode off-gasses from the other two modules, is added. In this configuration the electrical efficiency achieves 60% (LHV of natural gas).

A number of multi-megawatt fuel cell parks are operating around the world including a 14.9 MW fuel cell park in Connecticut (USA), multiple fuel cell parks ranging between 10–20 MW in South Korea, and the world's largest fuel cell park, a 59 MW facility composed of 21 SureSource 3000 power plants in Hwasung City, South Korea (POSCO Energy in partnership with FCE).

The manufacturing cost of a MCFC power plant is approximately $2,500–$3,000/kWe, but the goal is to reach $1,500/kWe by increasing the module lifetime (> 5 years) and by lowering the cost of fuel processing (Garche and Jörissen 2015).

4.2 *Solid Oxide Fuel Cell Applications*

Large SOFC systems are realized only by Bloom Energy, after Siemens-Westinghouse stopped their activities at the end of the 2000s due to high costs (> $17,000/kW) and limited lifetime.

Their products based Energy Server modules (ES-5000, ES-5400, and ES-5700 systems generates 100 kWe, 105 kWe, and 210 kWe, respectively) are combined to realize large power installation. Specific system costs are between $7,000–$8,000/kW (@ 100 kWe). In 2013, a 6 MW (30 Bloom 200 kWe systems) CHP plant was opened at e-Bay's data center in Utah (U.S.) (Garche and Jörissen 2015).

4.3 *Phosphoric Acid Fuel Cell Applications*

Large PAFC power systems are realized by Doosan Fuel Cell America with PureCell products. This company working with Samsung C&T Corp. and Korea Hydro and Nuclear Power, announced a deal to supply 70 of its PureCell 400 PAFC power plants for a 30.8 MW facility for a new residential complex in Busan. They have previously delivered more than 200 units of the 200 kWe ONSI PAFC.

Concerning the targets for industrial applications, Table 8 shows the state-of-the-art in 2012 and the main targets to be achieved for the generic stationary fuel cells applications across Europe (FCH JU–Multi Annual Working Plan 2014).

Finally, in the large FC power plants segment, Ballard commercializes ClearGen® fuel cell system based on the PEMFC technology. This system is a complete fuel cell power solution, designed to generate clean energy from hydrogen. The system is a CHP unit and is based on modular PowerBanks that are combined to produce multiple megawatts of zero-emission electricity.

5. Fuel Cells for UPS Application

An uninterruptible power supply (UPS) can be full time (Primary Power) or standby (Backup Power). In the first case the system runs to ensure that the supply of electrical power to the device it is protecting is constant and uninterrupted; otherwise, a standby UPS system only works when the primary source fails. Thus, the main difference is the fuel tank capacity (back-up power systems for emergency use need lower fuel capacity).

Fuel Cells offer several advantages with respect to conventional batteries and diesel generators that, at present, are used in the UPS market sector. These include:

Table 8: FCH JU Target (FCH JU–Multi Annual Working Plan 2014).

Segment	Key Performance Indicator (KPI)	Unit	State of the Art (2012)	FCH_JU Target		
				2017	2020	2023
	CAPEX	€/kW	3,000–4,000	3,000–3,500	2,000–3,000	1,500–2,500
Industrial–Large scale installations for industrial use larger buildings (0.3–10 MW)	Durability	Years of plant operation	20	21	22	22
	Availability	% of the plant	98	98	98	98
	Electrical efficiency	% LHV	45	45	45	45
	Thermal efficiency	% LHV	20	20	22	22
	LCOE	€ Ct/kWh	1.8*grid parity	1.3*grid parity	grid parity	<grid parity
	Emissions	mg/kWh	NOx < 5, SOx < 0.05, CO_2< 5k	NOx < 5, SOx < 0.05 CO_2 < 5k	NOx < 5 SOx < 0.05 CO_2 < 5k	NOx < 5 SOx < 0.05 CO_2 < 5k

- High efficiencies and low emissions
- Fast dynamic response
- Lower maintenance costs and longer maintenance-free lifetime
- High power density
- No limitations in start and stop cycles
- Scalable systems according to the specific applications

Moreover, by using an external hydrogen supply, the operation is continuous, limited only by the amount of fuel storage. A typical UPS configuration consists of: a fuel cell power module, a control system (it has to handle the battery management and to control the stack), an energy storage system (the ESS is generally a battery that allows both to overcome the transients, such as instantaneous power fluctuations, dynamics of fuel cell auxiliaries, and to satisfy overload conditions), a battery charger (it is used to prevent the battery discharging eventually due to a long stop), a DC/AC converter (it converts the power from the fuel cell output voltage to 230 VAC at the frequency of 50 Hz). Figure 14 shows the electric wiring flow diagram of a UPS system.

Typical applications for UPS-FC based systems include telecommunications, data and video services, financial institutions, medical centres, hospitals, government buildings, utilities and public safety networks. The on-site fuel capacity of some of these applications must be large enough to cover the required autonomous operation time that, sometimes, means several days of continuous operation. Most UPS applications require a dedicated fuel supply or storage unit, but some FC systems also incorporate a fuel reformer so that the unit can operate not only with hydrogen, but also with other conventional fuels. Hydrogen-UPS solutions are primarily used for backup power generation, while reformate-fuelled are preferred for off-grid power generation or in regions with frequent power outages.

Ballard Power Systems offers UPS power solutions, ElectraGen™ fuel cell systems, based on different fuel solutions: (a) ElectraGenTM-H2 (2 kWe output power, 40% efficiency) that uses direct hydrogen; (b) ElectraGen™-ME (5 kWe output power, 31% efficiency) that is fed by blended methanol and water (62% methanol and 38% de-ionized water).

The ElectraGen™-ME fuel cell systems are designed for reliability, long autonomy and minimal maintenance and are, at least, 20% more efficient than conventional diesel generators.

Ballard has shipped more than 3,000 ElectraGenTM fuel cell systems to customers, equivalent to 12.6 MW of power (http://ballard.com).

Several commercially FC UPS systems, in the range 2 to 10 kWe, are now available, above all for communications and telecom applications.

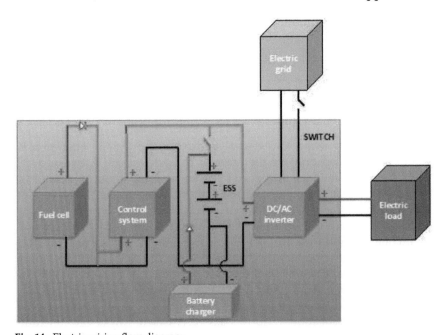

Fig. 14: Electric wiring flow diagram.

Commercially available PEMFC systems include (Garche and Jörissen 2015): Axane (0.5–10 kWe), Power Cell (3 kWe–100 kWe units), Electro Power Systems (1.5–10 kWe), Heliocentris (1.2–20 kWe), Horizon (0.1–25 kWe), Hydrogenics (2–200 kWe), and ReliOn-Plug Power (0.2–17.5 kWe). Systems generating H_2 on site from renewable sources and electrolyzers are also under development, with examples including ElectroSelf TF (1.5–12 kWe) and the MF-UEH Series (1–3 kWe).

6. Fuel Cells in Hybrid Power Plant Solutions

SOFCs and MCFCs (High Temperature Fuel Cells, HTFCs) can be easily combined with bottoming thermodynamic cycle (e.g., Rankine cycle, Brayton cycle) (Song et al. 2006, Zabihian and Fung 2009, Zhe et al. 2010, McPhail et al. 2011, Minutillo and Perna 2013) thanks to their high temperature exhaust gases. In this case, the overall electric efficiency of the hybrid cycle can theoretically be higher than 70% (Buonomano et al. 2015).

In general, hybridization of fuel cells with other power systems (i.e., Gas Turbine, GT) leads to a plurality of hybrid power plants (Zabihian and Fung 2009, Zhe et al. 2010, McPhail et al. 2011, Desideri and Barelli 2015, Barelli et al. 2017). HTFC/GT power plants can be theoretically fed by a variety of fuels other than natural gas. In particular, very attractive is the possibility to use HTFC/GT systems fed by gasified biomass in order to include the utilization of a renewable energy source.

SOFC is the most attractive fuel cell technology for possible system hybridization, since its working conditions, high temperature and pressurized or atmospheric configurations, allow the integration with steam cycles and/or Brayton cycles achieving very high efficiency. According to the SOFC operating pressure, the GT cycle can be externally or internally fired. In Figs. 15 and 16 a hybrid SOFC-GT externally fired (atmospheric SOFC) and a hybrid SOFC-GT internally fired (pressurized SOFC) configurations are depicted, respectively.

In these layouts, a pre-reformer is used to enhance the natural gas conversion to hydrogen and thus to improve the system efficiency because, when no external reformer is employed, the inlet sections of the SOFC typically just convert methane into hydrogen, whereas the electrochemical rate of reaction is very scarce, due to the low availability of hydrogen (Buonomano et al. 2015).

Since the SOFC operating pressure impacts on the system efficiency and reliability, techno-economic considerations drive the choice of the hybrid plant configuration. The majority of SOFC/GT power plants are based on the pressurized configuration. This layout allows achieving higher conversion efficiencies and lower capital costs, but the system

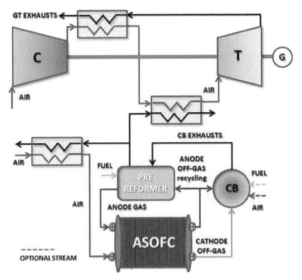

Fig. 15: SOFC-GT atmospheric hybrid plant configuration.

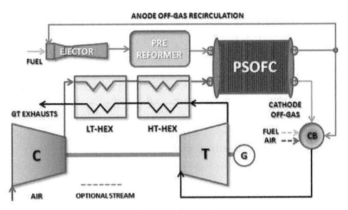

Fig. 16: SOFC-GT pressurized hybrid plant configuration.

management is more complex and restricted. The test trials showed that the direct coupled (pressurized) SOFC/GT is very complex (Gengo et al. 2008).

On the contrary, the operation of SOFC under atmospheric pressure should be preferred (Zhang et al. 2011) when easy management and operation are the design specifications. In this case, the integration between the SOFC topping section and the GT bottoming section is made by heat exchangers, so that the operation of the SOFC is completely independent from the GT power plant. This permits the management of both the FC and the GT, in the whole operating range, with high efficiencies (in part-load conditions the efficiencies of turbomachineries dramatically decrease).

However, the atmospheric configuration is characterized by lower electrical efficiencies than the pressurized one that is also expected to be cheaper since no expensive heat exchanger is needed.

Hybrid SOFC systems have been under development for some time in Japan, USA and UK. The world's first SOFC/GT hybrid system was delivered to the National Fuel Cell Research Center in 2000 in California, for operation and testing, in cooperation with Siemens Westinghouse Power Corporation. The hybrid system includes a pressurized Siemens-Westinghouse SOFC module (SPGI 120 modified for pressurized operation at 3 atm) integrated with a micro-GT/generator supplied by Ingersoll-Rand Energy. The system provided a total output of 220 kWe, with an output of about 180 kWe from the SOFC and about 40 kWe from the micro-GT generator. This pilot plant has already demonstrated the high efficiency feature of the hybrid systems, with a conversion efficiency of approximately 53%—that is very high for this size class (Brouwer 2006).

LG Fuel Cell Systems, with operations in Ohio USA, Derby UK, and Korea, ran large scale pressurized SOFC/micro-GT unit trials at the 200 kWe scale. In 2016 the company announced to move to a fully integrated 250 kWe system test, and then towards commercial deployment.

Mitsubishi Hitachi Power Systems has delivered a 250 kWe unit to Kyushu University, complete with an integrated micro-GT supplied by Toyota. It is expecting to install hundreds of megawatts of power generation at efficiencies higher than 70% (Fuel Cell Industry Review 2015). The unit is operating in a pre-commercial verification programme at the Next Generation Fuel Cell Centre at Kyushu University.

In 2014 General Electric announced a return to the fuel cell market with a 50 kWe hybrid SOFC system, coupling with a reciprocating gas engine, thanks to a novel SOFC manufacturing technique, which will dramatically reduce cost.

With respect to the SOFC/GT hybrid systems, the technology based on MCFCs was considered less competitive and, even if it was widely studied (Lunghi et al. 2003, Chen et al. 2006, Huang et al. 2015), it is still at the conceptual level. However, thanks to the MCFC technology development and to its potential for effective and systematic CO_2 capture with very high efficiency, MCFC/GT systems are currently considered more attractive and a viable option, even if, currently, only Fuel Cell Energy produces these systems (Wee 2011).

Similarly to the SOFC/GT systems, there are two configurations for the combination of a MCFC and a GT according with the operating pressure (Liu and Weng 2010). Figure 17 shows the configuration of the atmospheric MCFC/GT system (Wee 2011). The system includes an atmospheric

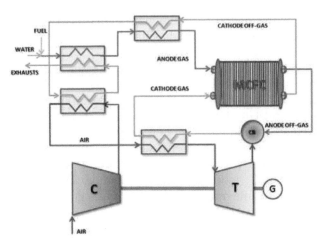

Fig. 17: MCFC-GT atmospheric hybrid plant configuration.

MCFC module with internal reforming, a GT, a catalytic burner and four heat exchangers. Due to the internal reforming configuration of the MCFC stack, water and fuel are mixed and pre-heated before entering in the fuel cell. The thermal energy required for heating the air entering the turbine is supplied by the cathode off-gas and the exhausts of the catalytic burner. These are generated by the combustion of the anode off-gas and the expanded air leaving the turbine (hot air). In this configuration the GT operating pressure can be chosen independently from the cell operating pressure, so that it can work very efficiently over a wide range (3–15) of pressure ratios (Wee 2011).

The pressurized configuration is illustrated in Fig. 18. The system includes a pressurized MCFC module, a GT, a catalytic burner, a fuel compressor, a steam reforming and three heat exchangers.

It is possible to note that the oxidant (air) is delivered to the MCFC as part of the cathode gas at high pressure according to the GT pressure ratio, whereas the outgoing high pressure exhausts from the catalytic burner drives the turbine. The cathode off-gas is divided in two streams that are sent to the GT combustor and to the catalytic burner. The cathode off-gas recirculation to the cathode inlet can also be realized (dashed stream in Fig. 18).

The higher operating pressure guarantees higher cell performance. Some basic studies have concluded that, by assuming equivalent design parameters, a pressurized system may have higher system efficiency over an atmospheric pressure system (Liu and Weng 2010).

Finally, a hybrid MCFC-GT prototype (fed by a digester gas), with a nominal power of 300 kWe (250 kWe MCFC) and an electric efficiency of 52%, was presented at the Aichi World Exposition, held in 2005 in Japan.

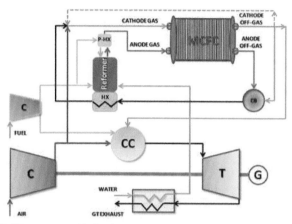

Fig. 18: MCFC-GT pressurized hybrid plant configuration.

The MCFC was manufactured by Ishikawajima-Harima Heavy Industries (currently they have stopped their activity on MCFC) whereas the GT was a modified version of a turbine commercialized by Toyota Turbine and Systems for cogeneration systems (Azegami 2003).

7. Expanding Market for Stationary Fuel Cell Applications

Recently, great attention has been paid to new market segments for stationary fuel cell applications.

The first application concerns "poly-generation" with MCFC technology. In this field, FuelCell Energy has developed a plant for the tri-generation of hydrogen, electrical power and heat, based on MCFCs fed by renewable biogas, generated from wastewater treatment facilities, as fuel source (see Fig. 19). This plant has received a contingent certification from the Low Carbon Fuel Standard Agency (administrated by the California Air Resources Board) attesting the production of a transportation fuel (renewable hydrogen) by a carbon-neutral and non-polluting process. Fuel Cell Energy's hydrogen-co-production system, utilizing a DFC3000 plant, generates approximately 1200 kg/day of hydrogen, which can service approximately 300 cars/day or 50 buses/day. The hydrogen production results in a modest reduction of electrical output (2 MW vs. 2.8) in the tri-generation configuration compared to the power/heat-only configuration (McPhail 2015).

Another new market segment for stationary fuel cells application is the energy storage of renewable energy sources such as wind and solar that are significantly increased in the last decade, and a further growing is expected in the next years. Their integration in the power grid is problematic and presents a great challenge in energy generation and load

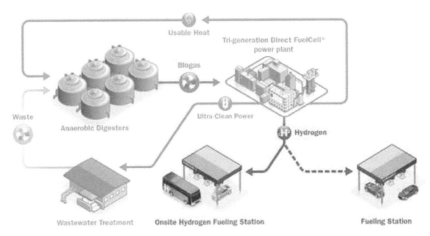

Fig. 19: Renewable hydrogen from tri-generation fuel cells included under California Low Carbon Fuel Standard.

balance maintenance to ensure power network stability and reliability. Amongst all the possible solutions, Electrical Energy Storage (EES) has been recognized as one of the most promising approaches. Recent studies demonstrate that an EES system based on fuel cell technology offers a potentially high efficiency, low cost, and scalable distributed energy resource. Thus, a new market segment can be found in this application field for stationary fuel cells. In particular, PEMFC and SOFC, operating either as a fuel cell (generating electric power) to electrochemically oxidize fuel species (i.e., hydrogen or syngas) or as an electrolysis cell (consuming electric power) to electrochemically reduce reactant species (i.e., water or water and carbon dioxide) are under R&D as EES systems. If steam is the only reactant, the process is typically referred to as electrolysis, while co-electrolysis refers to simultaneous reduction of H_2O and CO_2 to produce syngas. A ReSOC energy storage system is well suited for energy management applications because it can operate over a wide range of energy-to-power ratios by sizing the energy and power ratings independently, and is expected to have high energy storage efficiency and energy capacity suitable for storage duration in the order of hours to days.

Looking at the waste management market segment, another application concerns Microbial Fuel Cell (MFC) technology, which is considered a tool for energy recovery and organic load removal of a variety of biodegradable substrates as fuel, such as wastewater and organic fraction of municipal solid waste. The conventional MFC consisted of anode and cathode compartments. MFCs could be utilized as power generators in small devices such as biosensors. MFCs represent a valid alternative to achieve small-scale, distributed and efficient conversion of biodegradable substrate into electricity. Their main features are the direct electrical power production, the conversion of the chemical energy contained in any form of biomass, low environmental impact, low operating temperatures and simple architectures.

MFCs performances are heavily affected by operating parameters such as reactor configurations and scales, electrode materials, electrode surface areas and the nature of electron donors, if present, by the types of microbe in the anodic chamber. Several data about specific parameters are actually available in lab-scale experiment and a few studies about the dynamics of the wide range of biotic and abiotic parameters affecting MFCs power production in waste treatment have been carried out. Thus, the main issues actually limiting the performances of scaled MFCs seem to be the interactions among the above-cited factors (Nastro et al. 2015).

Finally, the Direct Carbon Fuel Cells (DCFC), that are still under research and development, represent an emerging technology for stationary FC application. These FCs use solid carbon as fuel and convert the chemical energy of the solid carbon directly into electricity through its direct electrochemical oxidation, with a theoretical efficiency of around 100%.

The combination of these two factors leads to a DCFC electric efficiency close to 80% (approximately twice the efficiency of current generation coal fired power plants), thus involving a 50% reduction in greenhouse gas emissions. The amount of CO_2 for storage/sequestration is also halved. Moreover, the exit gas is an almost pure CO_2 stream, requiring little or no gas separation before compression for sequestration. Therefore, the energy and cost penalties to capture the CO_2 will also be significantly less than for other technologies. Furthermore, a variety of abundant fuels such as coal, coke, tar, biomass and organic waste can be used. Despite these advantages, the technology is at an early stage of development requiring solutions to many complex challenges related to materials degradation, fuel delivery, reaction kinetics, stack fabrication and system design, before it can be considered for commercialization (Giddey 2012).

8. Conclusion

The decarbonization of our society and the realization of a clean energy system represent the challenging goals for the next few decades (2050); it means to produce energy not only without CO_2 emissions but also without any pollutants. Concisely speaking, we have to build zero emissions energy systems.

There is no doubt, fuel cells systems will play an important role in this contest, they have all potentialities to replace incumbent technologies, i.e., all energy systems based on combustion engines, in all sectors from the portable generation, to micro generation, to power generation. It is expected that in the near future power plants, even in co-generation mode, in the range of ten up to hundreds MW will realize where the energy is needed, i.e., at the customer cite producing electricity and heat to be supplied to district heating system.

Almost all the FC systems mentioned in this chapter have reached a very high technology readiness level. They are ready for the market but there is still a problem: the "market is not ready". There is the need for a clear policy that moves citizens towards the decentralized production, towards the co-generative systems with very high electrical and thermal efficiency with the aim to reduce emissions. If countries start to define clear policies to promote a clean and decarbonised society, fuel cell systems have the potential to be the most adopted technologies. This will be the only way to start market penetration. Moreover mass production is the only way to overcome the main hurdle that today prevents the FC entry into the market which is the cost per kWe.

It will be sufficient to move from an annual production of a few thousand systems per year to a mass production to lower the capex costs and to make FC systems reach the grid parity that indicates economic sustainability.

A good example is given by Japan where, with a funding of above 1 billion of € for the period from 2009 to 2016, almost 200.000 CHP systems were put in operation at customer cities resulting in a cost reduction of more than 60% within the first 6 years. A further reduction of 20–30%, achievable mainly thanks to wider applications and sales, is needed in order to enter the "free market".

References

Aliofkhazraei, M., N. Ali, W.I. Milne, C.S. Ozkan, S. Mitura and J.L. Gervasoni. 2016. Graphene Science Handbook, applications and Industrialization. Taylor & Francis Group.

Azegami, O. 2003. MCFC/MGT hybrid generation system. Technical Journal/R&D, Review of Toyata CRDL 41(1): 36.

Barelli, L., G. Bidini and A. Ottaviano. 2017. Integration of SOFC/GT hybrid systems in micro-grids. Energy 118: 716–728.

Baschuk, J.J. and X. Li. 2001. Carbon monoxide poisoning of proton exchange membrane fuel cells. Int. J. Energ. Res. 25: 695–713 (behaviour modelling).

Braun, J., S.A. Klein and D.T. Reindl. 2006. Evaluation of system configurations for solid oxide fuel cell-based micro-combined heat and power generators in residential applications. Journal of Power Sources 158: 1290–1305.

Brouwer, J. 2006. Hybrid gas turbine fuel cell systems, Chapter 4. In: Richard A. Dennis (ed.). The Gas Turbine Handbook. U.S. Department of Energy, DOE/NETL-2006/1230, Morgantown, West Virginia.

Buonomano, A. and F. Calise, M. Dentice d'Accadia, A. Palombo and M. Vicidomini. 2015. Hybrid solid oxide fuel cells–gas turbine systems for combined heat and power: A review. App. Energ. 156: 32–85.

Cali', M., E. Fontana, V. Giaretto, G. Orsello and M. Santarelli. 2005. The EOS Project: a SOFC Pilot Plant in Italy Safety Aspects, HySafe—International Conference on Hydrogen Safety, 8–9 September 2005, Pisa, Italy.

Chen, Q., Y. Weng, X. Zhu and S. Weng. 2006. Design and Partial Load Performance of a Hybrid System Based on a Molten Carbonate Fuel Cell and a Gas Turbine, UEL CELLS 06(6): 460–465.

Desideri, U. and L. Barelli. 2015. Solid Oxide Fuel Cell (SOFC). Handbook of Clean Energy Systems 1–14.

Di Bona, D., E. Jannelli, M. Minutillo and A. Perna. 2011. Investigations on the behaviour of 2 kW natural gas fuel processor. Int. J. Hydrogen Energy 36(13): 7763–7770.

Dodds, P.E., I. Staffell, A.D. Hawkes, F. Li, P. Grunewald, W. McDowall and P. Ekins. 2015. Hydrogen and fuel cell technologies for heating: A review. Int. J. Hydrogen Energy 40: 2065–83.

Ellamla, H.R., I. Staffell, P. Bujlo, B.G. Pollet and S. Pasupathi. 2015. Current status of fuel cell based combined heat and power systems for residential sector. Journal of Power Sources 293: 312–328.

ElMubarak, E.S. and A.M. Ali. 2016. Distributed generation: Definitions, benefits, technologies & challenges. International Journal of Science and Research 5(7): 1941–1948.

Fontell, E., T. Kivisaari, N. Christiansen, J.B. Hansen and J. Palsson. 2004. Conceptual study of a 250 kW planar SOFC system for CHP application. Journal of Power Sources 131: 49–56.

Fuel Cell and Hydrogen Joint Undertaking, FCH JU. Multi-Annual Work Plan 2014–2020. 2014; Fuel Cell Industry Review 2015. http://www.fuelcellindustryreview.com.

Garche, J. and L. Jörissen. 2015. Applications of Fuel Cell Technology: Status and Perspectives The Electrochemical Society Interface 2015: 39–43.

Gengo, T., Y. Kobayashi, Y. Ando, N. Hisatome, T. Kabata and K. Kosaka. 2008. Development of 200 kW class SOFC combined cycle system and future view. Technical Review. Mitsubishi Heavy Industries, Ltd. 2008: 45.

Geormezi, M., C.L. Chochos, N. Gourdoupi, S.G. Neophytides and J.K. Kallitsis. 2011. High performance polymer electrolytes based on main and side chain pyridine aromatic polyethers for high and medium temperature proton exchange membrane fuel cells. J. Power Sources 196: 9382–90R.

Giddey, S., S.P.S. Badwal, A. Kulkarni and C. Munnings. 2012. A comprehensive review of direct carbon fuel cell technology. Progress in Energy and Combustion Science 38(3): 360–399.

Hidayatullah, N.A., B. Stojcevski and A. Kalam. 2001. Analysis of distributed generation systems, smart grid technologies and future motivators influencing change in the electricity sector. Smart Grid and Renewable Energy 2: 216–229.

Huang, H., J. Li, Z. He, T. Zeng, N. Kobayashi and M. Kubota. 2015. Performance analysis of a MCFC/MGT hybrid power system Bi-fueled by city gas and biogas. Energies 8: 5661–5677.

Iwan, A., M. Malinowski and G. Pasciak. 2015. Polymer fuel cell components modified by graphene: Electrodes, electrolytes and bipolar plates. Renewable and Sustainable Energy Reviews 49: 954–967.

Jannelli, E., M. Minutillo and A. Perna. 2013. Analyzing microcogeneration systems based on LT-PEMFC and HT-PEMFC by energy balances. App. Energy 108: 82–91.

Kim, J.-D., Y.-I. Park, K. Kobayashi and M. Nagai. 2001. Effect of CO gas and anode-metal loading on H_2 oxidation in proton exchange membrane fuel cell. J. Power Sources 103: 127–33.

Kuroda, K. 2011. Present and Future of PAFC at Fuji Electric, 4th IPHE Workshop—Stationary Fuel Cells, 1st March 2011, Tokyo, Japan.

Liu, A. and Y. Weng. 2010. Performance analysis of a pressurized molten carbonate fuel cell/ micro-gas turbine hybrid system. Journal of Power Sources 195: 204–213.

Lunghi, P., R. Bove and U. Desideri. 2003. Analysis and optimization of hybrid MCFC gas turbines plants. Journal of Power Sources 118: 108–117.

Margalef, P. and S. Scott. 2010. Integration of a molten carbonate fuel cell with a direct exhaust absorption chiller. Journal of Power Sources 195: 5674–5685.

McPhail, S.J., A. Aarva, H. Devianto, R. Bove and A. Moreno. 2011. SOFC and MCFC: commonalities and opportunities for integrated research. Int. J. Hydrogen Energy 36: 10337–45.

McPhail, S.J., J. Kiviaho and B. Conti. 2017. The Yellow Pages of SOFC Technology, International Status of SOFC deployment. Implementing Agreement Advanced Fuel Cells Annex 32 – SOFC, http://www.ieafuelcell.com/documents/The_yellow_pages_of_SOFC_technology%202017.pdf (accessed 28/06/17).

McPhail, S.J., L. Leto, M. Della Pietra, V. Cigolotti and A. Moreno. International Status of Molten Carbonate Fuel Cells Technology 2015, IEA Advanced Fuel Cells Implementing Agreement, Annex 23-MCFC, 2015, http://www.enea.it/it/pubblicazioni/pdf-dossier/2015_MCFCinternationalstatus.pdf (accessed 28/06/2017).

Minutillo, M. and A. Perna. 2014. Renewable energy storage system via coal hydrogasification with co-production of electricity and synthetic natural gas. Int. J. Hydrogen Energy 39(11): 5793–5803.

Minutillo, M. and A. Perna. 2008. Behaviour modelling of a PEMFC operating on diluted hydrogen feed. International Journal of Energy Research 32: 1297–1308, DOI: 10.1002/er.1424.

Minutillo, M., A. Perna and E. Jannelli. 2014. SOFC and MCFC system level modelling for hybrid plant performance prediction. Int. J. Hydrogen Energy 39(36): 21688–21699.

Momoh, J.A. 2014. Centralized and Distributed Generated Power Systems—A Comparison Approach; Prepared for the Project. The Future Grid to Enable Sustainable Energy Systems. Funded by the U.S. Department of Energy.

Nastro, R.A., G. Falcucci, M. Minutillo and E. Jannelli. 2017. Microbial fuel cells in solid waste valorization: Trends and applications. Springer Book, Modelling Trends in Solid and Hazardous Waste Management pp. 159–171.

O'Hayre, R.P., S.–W. Cha, W.G. Colella and F.B. Prinz. 2009. Fuel Cell Fundamentals, John Wiley & Sons Inc., Hoboken, New Jersey.

Oono, Y., A. Sounai and M. Hori. 2009. Influence of the phosphoric acid-doping level in a polybenzimidazole membrane on the cell performance of high-temperature proton exchange membrane fuel cells. J. Power Sources 189: 943–9.

Owens, B. and J. Mcguinness. 2015. GE-Fuel Cells: The power of tomorrow, General Electric Company.

Patel, P., A. Lipp, F. Jahnke, E. Heydorn and F. Holcomb. 2009. Co-production of renewable hydrogen and electricity. ECS Transaction 17(1): 569–580.

Perna, A. 2007a. Hydrogen from ethanol: theoretical optimization of a PEMFC system integrated with a steam reforming processor. Int. J. Hydrogen Energy 32(12): 1811–1819.

Perna, A. 2007b. Theoretical analysis on the autothermal reforming process of ethanol as fuel for a PEMFC system. Journal of Fuel Cell Science and Technology.

Remick, R.J., D. Wheeler and P. Singh. 2009. MCFC and PAFC R&D Workshop, U.S. Department of Energy Summary Report.

Roland Berger Strategy Consultants. 2015. Advancing Europe's energy systems: Stationary fuel cells in distributed generation. A study for the Fuel Cells and Hydrogen Joint Undertaking.

Shaffer, B., B. Tarroja and S. Samuelsen. 2015. Dispatch of fuel cells as transmission integrated grid energy resources to support renewables and reduce emissions. Applied Energy 148: 178–186.

Song, T.W., J.L. Sohn, T.S. Kim and S.T. Ro. 2006. Performance characteristics of a MW-class SOFC/GT hybrid system based on a commercially available gas turbine. J. Power Sources 158: 361–7.

Staffell, I. 2015. Zero carbon infinite COP heat from fuel cell CHP. App. Ener. 147: 373–385.

User Manual Serenus 166/390 Air C v2.5, v1.10410.

Wee, J.H. 2011. Molten carbonate fuel cell and gas turbine hybrid systems as distributed energy resources. Applied Energy 88: 4252–4263.

Wiser, J.R., P.E. James, W. Schettler, P.E. John and L. Willis. 2012. Evaluation of Combined Heat and Power Technologies for Wastewater Facilities, September 25, 2012, Columbus Water Works, Columbus, Georgia.

Zabihian, F. and A. Fung. 2009. A review on modeling of hybrid solid oxide fuel cell systems. Int. J. Eng. 3: 85–119.

Zhang, J., Z. Xie, J. Zhang, Y. Tang, C. Song, T. Navessin, Z. Shi, D. Song, H. Wang, D.P. Wilkinson, Z.-S. Liu and S. Holdcroft. 2006. High temperature PEM fuel cells. J. Power Sources 160: 872–91.

Zhang, X., S. Su, J. Chen, Y. Zhao and N. Brandon. 2011. A new analytical approach to evaluate and optimize the performance of an irreversible solid oxide fuel cell-gas turbine hybrid system. Int. J. Hydrogen Energy 36: 15304–12.

Zhe, Y., L. Qizhao and B. Zhu. 2010. Thermodynamic analysis of ITSOFC hybrid system for polygenerations. Int. J. Hydrogen Energy 35: 2824–8.

Chapter 10

Biopower Technologies

D. Chiaramonti,* M. Prussi and A.M. Rizzo

1. Introduction to Bioenergy

Energy generation from biomass has certainly been one of the main routes-to-energy for mankind: human beings have been using biological feedstock for heating, lighting, mechanical energy since ever, including— more recently—power and heat cogeneration and transport fuels.

A large variety of feedstocks can be collected and converted into energy and other bioproducts, such as lignocellulosic biomasses, sugar and starch crops, oil crops, organic residues, aquatic biomasses.

Depending on the scale of the application, as well as on the end use (e.g., decentralized energy generation in developing or industrialized countries, power generation in industrial sites, and other applications) the technical characteristics of the bioenergy system will differ, as the skills needed to operate the plant will change. Key element of biomass sourcing is the environmentally sound production and supply of the bio-feedstock, a necessary pre-condition for the sustainability of the overall chain, together with its use in modern clean technologies.

This chapter will deal with power generation from biomass pathways, therefore we will not address other uses, as conversion into conventional or advanced biofuels for transport, or heat generation, despite these

RE-CORD and CREAR, Department of Industrial Enineering, University of Florence, Viale Morgagni 40, 50134 Florence (ITALY).
Emails: matteo.prussi@unifi.it; andreamaria.rizzo@unifi.it
* Corresponding author: david.chiaramonti@unifi.it

applications also representing major destinations of the resource. We will focus on the process and technological options to generate electricity (and heat), either through thermochemical or biochemical pathways.

Basic biomass can be converted into power through various routes. Focusing on the main and most used options, we could probably define the following list:

- Direct combustion of biomass or biomass-derived products (torrefied biomass, charcoal, hydrochar, produced through torrefaction, pyrolysis or hydrothermal carbonisation) and generation of power and heat in steam or ORC cycles. This route includes co-combustion with fossil fuels, e.g., coal, oil and natural gas.

- Thermochemical conversion of lignocellulosic biomass into a liquid (pyrolysis or hydrothermal liquefaction oil) or gaseous (gasification) intermediate fuel, and power generation in reciprocating engines and gas turbines.

- Biological conversion of biodegradable organic feedstock into biogas, rich in biomethane, and use of the gas for power generation in engines and turbines. This route also includes possible separation of natural gas from the biogas stream (biomethane).

- Extraction, cleaning and utilization of pure vegetable oils in engines and turbines for power and heat generation.

In this chapter, we will not focus on hydrothermal liquefaction and anaerobic digestion to biomethane, as these two routes—being also at a very different Technology Readiness Levels—find their main scope in the production of a transport fuel, and not for power generation. Nevertheless, these pathways also generate side streams that can be directly (in case of combustible gaseous streams) or indirectly (in case, for instance, of CO_2 converted into CH_4 through thermochemical methanation) used again for power and heat generation.

As regards the thermochemical conversion processes, the main biomass characteristics that play a key role in the combustion, pyrolysis and gasification process are:

- Carbon, Hydrogen, Nitrogen, Sulfur, Oxygen and Ash content;
- Ash composition, and ash sintering/melting characteristics;
- Moisture content, physical shape/dimension of biomass and its size statistical distribution (e.g., content of fines).

As regards the Anaerobic Digestion process, the content of biodegradable material (i.e., Volatile Matter) is among the major parameters to be examined, as discussed later in the chapter.

2. Pretreatment of Lignocellulosic Feedstock for Energy Conversion

Solid biomass is normally subject to different kinds of pretreatment before being fed to plants. It ranges from chipping at collection, moisture reduction, grinding and compactation, to more complex processes as torrefaction.

2.1 Chipping, Drying and Compacting

Reduction of biomass trees/logs into small chips is normally the first pretreatment step done on the feedstock, often at the site itself of biomass collection.

Chipping reduces wood into small fragment (chips) of a length ranging between 2 and 5 cm, a width of max 2 cm and a thickness of a few millimetres. The main benefits achieved through chipping are the following: increasing surface-to-volume ratio (which accelerate the chemical reactions), fluidizing the fuel flow, increasing the amount of biomass recovered and made available, and compacting very bulky feedstock. Drawbacks of chipping includes some risks of fermentation (due to microorganisms) and not uniform chips' size. In fact, the microbial activity on fresh chips generates a certain amount of dry matter losses, which in a temperate climate often vary between 2 to 4% per month. Moreover, storage of chips in large amount requires appropriate measures to prevent possible safety and health risks.

The operating principle of wood chippers may differ, from disc-wheel cutters to drum chippers, auger chippers, and other solutions, as well as powered by mechanical power or electric power. In fact, they can be built as mobile unit (working on site) or stationary plant. The first one can be even tractor-mounted ones, while stationary ones show higher throughput capacities and higher efficiencies, and therefore lower production costs. However, stationary chippers are recommended only in case the transport distance of wood is very short, otherwise on-site chipping should be preferred. Chip quality is essentially defined by fibre content, particle size distribution and moisture content. Particle size distribution depends on several factors, mainly: the raw material, the chipping machine used, the conditions of the chipper knives, and the eventual use of refining devices in the chipper. Finally, knife wear and tear has a major impact on chip quality as well as on chipper productivity.

Moisture is normally reduced to a level compatible with the downstream technologies: thus, biomass can be fed at very different level of water content based on the type of thermochemical route adopted. For instance, typical biomass combustion plants normally receive biomass in the range of 20–40% w/w (on a wet basis), most frequently 25–35% w/w, but some combustion technologies allow being fed with feedstock at up to 55–60% w/w moisture. Dried biomass in boilers contributes to reduce boiler size, as well as increase combustion temperature and reduce CO emissions. On the opposite, as for any pretreatment method, drying will increase biomass cost, while high temperatures of combustion will require careful control to avoid ash-fusion related problems in the combustion chamber and NOx emissions.

On the contrary, moisture conditions for biomass at inlet are normally more stringent in the case of gasification systems (excluding the updraft configuration) and even more for fast pyrolysis plants. Various methods can be used for moisture reduction: the selection of the appropriate solution depends on the type of feedstock, the dimension of the installation (typically, biomass to be processed per hour), the system configuration, etc. Dewatering removes the water form biomass as a liquid, through open air storage or mechanical dewatering. Drying, which removes water as vapour, can be done through open air solar drying (including seasoning), batch or kiln drying (circulating hot gases by natural or forced draft), continuous drying by stack gases, and steam dryers.

On dried biomass, compacting is also sometimes carried out. Most frequent compacting approach distinguishes between briquetting and pelletizing. In the case of briquettes, small biomass cylinders of various diameters are generated through screw or piston press and extruders, where finely grinded biomass is forced to pass, while in the case of pellet biomass the process is conceptually similar, with fine biomass forced to go through a die by press rolls. Biomass moisture at inlet is normally in the range of 8–10% w/w. The main benefits from biomass compacting comprise increased bulk density and volumetric energy content, standardization of feedstock (allowing optimized automatic feeding), lower moisture and thus easier ignition and higher combustion temperature, and others: the withdraw is the higher cost of the feedstock at plant inlet.

2.2 Torrefaction

Torrefaction (Bergman et al. 2005a,b) is a mild thermochemical process which aims to make biomass more easily grindable, hydrophobic, and increase its energy density (in particular if compacted through pelletisation).

The process is carried out in absence of Oxygen (anaerobic conditions), and consists of heating the lignocellulosic feedstock in a typical range of 200–300°C. It can be actually seen as an extremely low pyrolysis process, in which the main product is solid (i.e., the torrefied material) and some gas/vapour phase. The typical mass balance gives a 70% yield, while the energy balance (referred to the solid product) offer an interesting 90% yield, thus achieving an average 0.9/0,7 = 1.3 energy densification effect (before compating).

The gas phase is composed by a small amount (order of 3–4%) of permanent gases (as CO and CO_2), and a large part of condensable components, dominated by water (approx. 90% of the condensable part) and other chemicals (such as acetic acid, furfural, methanol, formic acid). Moisture content of torrefied material at outlet is typically in the range of 1.5–3%, sometimes up to 4–7%.

During the heating phase, lignin is subject to a glass transition and softening, while hemicellulose is the first polymer to decompose-recondensate and in small part devolatize and carbonize, followed by cellulose at higher temperatures (up to 300°C, when conventionally slow pyrolysis is assumed to begin).

As said, the main effect of torrefaction is making the material more easily grindable, and resistant to biological decomposition (hydrophobic material), in addition to the energy densification effect. For this reason, the process of torrefaction has also been used by the timber industry to improve the characteristics of construction wood (Finnish Thermo Wood Association 2003) and have a more durable material in exterior use. The process can be carried out in steam atmosphere.

The following SEM figure show the changes occurring in the biomass structure after torrefaction. The fibrous and tenacious lignocellulosic biomass becomes friable and less fibrous, more homogeneous, with a lower contaminant level and more easily storable.

 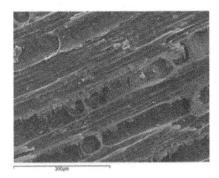

Fig. 1: SEM images of raw (left) and torrefied (right) lignocellulosic biomass.

3. Thermochemical and Biochemical Conversion of Biomass for Biopower

3.1 Pyrolysis

3.1.1 Generalities

Pyrolysis is the process of chemically decomposing organic materials by heating in the absence of oxygen; the process is carried out at near atmospheric pressure and in the temperature range between 300 and 600°C. Products of pyrolysis are a solid, also called char, a mixture of condensable vapours, and some permanent gases (mainly CO_2, CO, CH_4). Pyrolysis process is significantly flexible, as products distribution, their relative abundance and properties can be influenced (steered) by properly tuning the process parameters or conditioning the feedstock, according to the desired final goal. A resume of products spectrum, in terms or relative abundance of liquid, solid and gas according to process condition is reported in Table 1.

When fast and intermediate pyrolysis is applied, the rapid condensation of pyrolysis vapours and aerosols originates a dark brown mobile liquid, the bio-oil, that has a heating value about half that of conventional fuel oil; synonyms for bio-oil include pyrolysis oils, pyrolysis liquids, bio-crude oil (BCO), wood liquids, wood oil, liquid smoke, wood distillates, pyroligneous acid, and liquid wood (Mohan et al. 2006). The interest in pyrolysis for the production of liquid fuels resides in the possibility to convert a solid biomass into a liquid intermediate with higher energy density, improving transportation cost, and simplifying logistics and further processing, either upgrading or use.

Table 1: Yield of products (%w/w) according to biomass pyrolysis mode (Bridgwater 2012).

Mode	Conditions	Liquid %w/w	Solid %w/w	Gas %w/w
Fast	≈ 500°C, short hot vapours residence time ≈ 1 s	75	12	13
intermediate	≈ 500°C, hot vapours residence time 10–30 s	50[a]	25	25
slow (*carbonization*)	≈ 400°C, long vapours residence (hours up to days)	30	35	35
torrefaction	≈ 290°C, solids residence time ≈ 10–60 min	0[b]	80	20

[a]: in 2 phases. [b]: unless condensed, then up to 5%.

3.1.2 Exploiting the Potential of Pyrolysis Oil for Biopower

Electricity generation from pyrolysis oil is an attractive opportunity, because of the higher market value of electricity compared to process heat, the possibility to benefit from feed-in tariff for renewable energy sources, and the ease of distribution. The challenges associated with the use of bio-oil in energy conversion plants have been extensively reviewed by Shaddix and Hardesty (1999), Czernik and Bridgwater (2004), Chiaramonti et al. (2007) and Venderbosch and van Helden (2001). In the following sections, a brief introduction on the historical development of the use of bio-oil in turbines and internal combustion engines will be given.

3.1.3 Bio-oil in Gas Turbines

A limited number of independent research works have been carried out in the past on bio-oil combustion in gas turbines. The first reported successful demonstration of bio-oil combustion in a gas turbine combustor test rig was the work of Kasper et al. (1983) in the '80s at Teledyne (USA). Authors fed bio-oil from forest and agricultural residues to the annular combustor of a J69-T-29 turbine. The measured combustion efficiency in the rig using pyrolysis oil as fuel was 95%, emissions of CO were higher compared to fossil fuel, but CH and NO_x were within the limits observed for petroleum fuels. Also a slag build-up in the exhaust section resulting from ash in bio-oil was identified as a potential problem.

In the '90s Ardy et al. (1995) at ENEL CRT (Italy) reported to have tested pure fast pyrolysis bio-oil produced by ENSYN and bio-oil/ethanol mixtures in a pressurized gas turbine combustor test rig, rated at 40 kWe. Authors found that bio-oil atomization and combustion quality were extremely sensible to the crude bio-oil feeding temperature; exhaust temperature had a strong influence on the combustion efficiency; CO concentration was relatively low only for high exhaust temperatures. Several problems arose in the fuel pumping section, especially in the throttling section of the valve. The combustor performance was influenced by gas residence time in the combustor, when air/fuel ratios and average exhaust outlet temperatures remained unchanged. Smoke (*Bacharach index*, *nda*) and CO were relatively low only for low loads, when increasing the air flow rate, smoke and CO rate increased by 1 order of magnitude. At the time, bio-oil stability was poor, owing to noticeable differences in test results over time under the same conditions.

In the '90s, Orenda Aerospace Corporation (Canada) and Zorya-Mashproekt (Ukraine) started a still ongoing R&D project on a 2.5 MWe industrial gas turbine, that was modified in its hot section and combustion system for alternative fuels applications and fed with fast pyrolysis bio-oils from ENSYN and Dynamotive (Andrews et al. 1997a,b).

Lupandin et al. (2005) reported that emissions data generated during bio-oil operation showed that NO_x, CO and SO_x emissions were below the Ontario Emissions limits, and compared to measurements with #2 diesel fuel, NO_x and SO_x concentrations were lower, while CO concentration was higher when feeding both bio-oils.

Strenziok et al. (2001) at University of Rostock (Germany) modified a T216 gas turbine, rated at 75 kWe. The combustion chamber of the turbine was adapted and equipped with a dual fuel system that included an ignition nozzle for diesel fuel and a main nozzle for bio-oil. The engine operated in a dual fuel mode at 73% of the full power that would be generated in a standard fuel mode, with about 40% of total power produced from bio-oil and 60% from diesel. Compared to the operation on diesel fuel, CO and HC emissions were significantly higher and NOx less for dual fuel operation. The use of bio-oil in the turbine resulted in deposits in the combustion chamber and on the blades.

At the University of Florence, tests with pyrolysis oil and blends of ethanol/pyrolysis oil were initiated in 2010; the combustion chamber of a Garrett GTP 30-67 gas turbine was modified to accommodate the requirements of bio-oil combustion and a new fuel line was set-up. Preliminary results gave encouraging feedback, but to date a sustained run on pure pyrolysis oil has not been achieved yet. Details of the combustion chamber development and testing activities can be found in references (Cappelletti et al. 2013, Chiaramonti et al. 2013).

A recent development in the field was the work of Beran and Axelsson (2014) at OPRA Turbine (The Netherlands) in the framework of a R&D project in partnership with Biomass Technology Group (BTG, The Netherland) and University of Twente; Authors developed and tested a tubular combustor for low-calorific fuels to equip their OPRA OP16 radial gas turbine rated at 1.9 MW. The experiments have been performed in an atmospheric combustion test rig. Authors reported to have burned pure pyrolysis oil in the load range between 70% and 100% with a combustion efficiency exceeding 99% and without creation of sediments on the combustor inner wall. They found that NOx emissions were similar for pyrolysis oil and diesel, whereas the CO emissions were twice as high for pyrolysis oil. They found that an air blast nozzle was found to be more suitable than a pressure nozzle, due to its better performance over a wider operating range and its higher resistance to erosion and abrasion. Moreover, they found that the maximum allowed droplet size of the pyrolysis oil spray had to be 50–70% of the droplet size for diesel fuel.

3.1.4 Bio-oil in Internal Combustion Engines

The possibility to use bio-oil in internal combustion engines was investigated worldwide by several research institutes (VTT, MIT, Aston

University, University of Rostock and University of Florence, CNR-IM), sometimes in partnership with private companies (Wartsila Diesel, Carterpillar, Ormrod Diesels) (Van de Beld et al. 2013, Bertoli et al. 2000, Chiaramonti et al. 2003a, Shihadeh 1998, Jay et al. 1995); however, internal combustion engines present a larger number of technical barriers to overcome for the use of bio-oil, compared to gas turbines, mainly due to (1) limited time available for the combustion, which might be insufficient to accommodate the peculiar behaviour of bio-oil during combustion, and (2) material of construction of "wetted" parts, which are generally not tolerant toward harsh fuels.

R&D focused on using FPBO in the engine, rather than CHP: several works considered the addition of a pilot injector to the diesel engine, fed with conventional diesel, to keep the combustion stable, and or run with only few piston fed with FPBO. Others investigated the use of diesel/FPBO emulsions in engines (David Chiaramonti et al. 2003b). Recently, BTG extensively tested a micro-scale (10 kWe) one-cylinder diesel engine, with very promising results (Van de Beld et al. 2013).

Calabria et al. (2007), who performed several experiments in a single droplet combustion chamber, have addressed the fundamentals of bio-oil combustion recently. They reported that after an initial heat up phase, the bio-oil droplet starts swelling, volatiles are released and start burning, but the combustion time is dictated by the heterogeneous combustion of cenospheres, i.e., the solid carbonaceous particle formed by liquid phase pyrolysis during the last stages of droplet combustion.

According to Venderbosch and Prins (ibid.), the lack of available amounts of bio-oil was a key factor at that time for testing and development activities; however, in several tests, severe wear and erosion/corrosion phenomena were observed in the injector needles, due to fuel acidity and particulates; high viscosity and loss of stability with increasing temperature were major problems, and a carbon deposit build-up was reported in the combustion chamber and exhaust valve.

Chiaramonti et al. (2007) clearly pointed out that injector and fuel pump material as well as emissions are critical factors, and that development of effective and reliable pumping and injection systems and good combustion to avoid deposits on the hot parts (cylinder, piston, injector) are the main R&D needs.

Few researchers have also investigated the use of FPBO/alcohols blend, usually with the addition of Cetane improvers to compensate for the exceedingly low Cetane number of FPBO. Kim and Lee (2015) investigated the combustion and emission characteristics of FPBO/n-Butanol/polyethylene glycol 400/2-ethylhexyl nitrate blends up to

30/50/15/5% w/w, % in a single cylinder, diesel cycle, direct injection test engine with increased compression ratio. At higher FPBO content, they observed a decrease in nitrogen oxides and particulate matter at increasing engine load compared to diesel, while carbon monoxide increased. Authors attributed these effects to (1) for PM: micro-explosions that promotes mixing of air and fuel molecules and formation of local lean air-fuel mixture that enhances the oxidation of particulate matter, (2) for NOx: presence of water and longer fuel injection, and (3) for CO: the high viscosity of the fuels, abundant carbon-oxygen bonds, and low self-ignitability.

3.1.5 Analysis of Different GT System Configurations versus CHP Plant Size

Large-scale industrial gas turbine (including the above mentioned Orenda/Magellan and Opra) are non-regenerated (or recuperated): thus, the electrical efficiency, depending on the type of GT, normally stays in the range of 20–25% for the larger ones studied for FPBO feeding. Their thermodynamic cycle, Joule cycle is named "simple cycle". These systems are designed to run at full scale or very near to design conditions, as efficiency at part load rapidly decreases. At this scale regeneration is not possible (regeneration, or recuperation, consists in heating the compressed air available at the compressor outlet with the hot flue gases at the exhaust of the GT). So, electrical efficiency is not very high and a lot of high temperature heat is available. Large scale industrial GTs, in order to increase electrical (only) efficiency, are used as "top-cycles" in combined cycles, where the hot exhaust gases from the upper GT cycle are directed to a so called HRSG (Heat Recovery Steam Generator), that provide steam which is fed to the steam-based "bottom cycle", generating additional power and increasing the overall efficiency. In large thermal power stations this CC cycles achieve 55% and higher electrical efficiencies.

However, while micro GT and large scale GT in power stations are optimized in this way, i.e., either through regeneration (micro/small GTs, such as Capstone or Turbec, for instance) or adding a bottom steam cycle to develop a GT-CC (large plants), at the size of interest for biomass-based CHP it is not possible to adopt this solution at reasonable costs, since small scale steam cycles using steam turbines/engines are not very efficient nor low cost. Sometimes ORC cycles are used at this scale, since they are more practical than steam-based systems (no need for continuous 24/7 presence of operator qualified for running steam cycle is needed), but costs still remains in the high end. Summarizing, where regeneration is not possible and developing a CC (Combined Cycle) is not yet economically convenient (thus GTs operated in Simple Cycle), the GT CHP system must run at design load/speed (so to ensure optimum efficiency), and thus continuous consumption of large amount of high temperature heat must be found at end users' sites.

Smaller-medium scale GTs can be regenerated instead, but their design electrical efficiency is only around 30–33%. Moreover, most of these GT manufacturers normally provide containerized packages of multiple systems, all regenerated, connected in Master/Slave mode, providing multiple power in a single system, controlled by a master control regulating all GTs or single GTs and keeping the others at design rate, depending on the local conditions. So, very efficient, clean and compact systems suitable for continuous operation can be designed at almost at any size in the range of MWs scale.

3.1.6 Key Differences Between GT and ICE with Respect to FPBO Use

In the following Table 2 key differences between GT and ICE are highlighted in reference to the FPBO use. Specifically the comparison is made with regards to fuel injection, combustion, efficiency and typical size.

Table 2: GT and ICE comparison with respect to FPBO use.

Parameter	Gas turbine	Reciprocating engines
Fuel injection	Either with pressure swirl or air-assisted nozzles; does not require exceedingly high pressure level (<< 50 bar). Air-assisted atomizers have shown better performance than pressure swirl with FPBO. In principle no fuel recirculation is needed	Fuel atomization in diesel cycle engine leverages very high pressure differences between fuel pump/mains[1] and cylinder. In case of fuel recirculation, loop temperature have to be kept below critical temperature (< 80°C to avoid repolymerisation)
Combustion	Steady combustion process. Combustor length/volume can be adjusted (at least in principle) to accommodate FPBO requirements	Cyclic combustion. Unsteady process to be completed in very short time (few ms)
Plant efficiency	<< than diesel engines for simple cycle systems < for regenerated cycle systems	Largely higher than GT in simple cycle
Typical Size (single unit)	Microturbines (regenerated cycle): up to 4–500 kWe Industrial (simple cycle): above 0.5 MW	From several MW to tens of MW and groups of these in utility mode

NOTES
[1]: depending on the specific arrangement: *injector-pump* or *common rail*

3.2 Hydrothermal Carbonization

Fuel conditioning to attain a certain moisture content is an essential step in solid combustion process, and the original moisture content is of paramount importance when dealing with fuels with a relatively low specific energy density, e.g., biomass. In wood-fired boilers, for example, it has been found that the combustion process cannot be maintained if the wood moisture content exceeds 60% w/w on a wet basis (Van Loo and Koppejan 2008). The problem is therefore exacerbated when residual organic feedstock with high moisture content are being considered, as in the case of the organic fraction of municipal solid waste. High-moisture fuels can be beneficially processed in a hydrothermal environment to produce a solid carbonaceous material with a higher calorific value, lower volatiles and higher fixed carbon, which can be co-fired in power plants.

In hydrothermal carbonization (HTC) organic materials are processed in a water-rich environment in order to produce a solid carbonaceous material also known as hydrochar, that exhibits a carbon content similar to lignite in a yield range between 35 and 60% w/w (Kruse et al. 2013). In hydrothermal conversion, water has multiple roles: medium of heat transfer, solvent, reactant and product. The solid product can be exploited as a renewable fuel or to produce valuable carbon-based materials with specific properties and applications such as carbon fixation, water purification, fuel-cell catalyst, energy storage, CO_2 sequestration (Hu et al. 2010). Side-products of the conversion are a Carbon-laden water, retaining a significant amount of Carbon in the form of dissolved organics (sugars and derivatives, organic acids, furanoid and phenolic compounds), and an off-gas, mainly composed of CO_2, some CH_4 and CO, traces of H_2 and C_nH_m (Funke and Ziegler 2010).

Among the three hydrothermal processes, hydrothermal carbonization is the technology where the less severe conditions are required: temperature generally ranges from 160°C to 250°C and often pressure is autogenous, i.e., generated by the process itself (batch process). The main characteristic of HTC is the long biomass residence time at reaction conditions, which is needed to form the solid carbonaceous product. In fact, in order to attain good char yields, the slurry has to be processed from few hours to several days (Pavlovič et al. 2013).

As a process, key general features of hydrothermal carbonization are reported below, while relevant range of process parameters can be found in Table 3:

- Avoids the drying step
- Can process high-ash feedstock
- Carried out at moderate pressure and temperature

Table 3: Process parameters of hydrothermal carbonization.

Parameter	Range (typical)	Unit	Relevance[1]
Pressure	10–40	bar	++
Temperature	160–250	C	+++
Moisture content	Only technical limits		+
Residence time	0.5–12	h	++
Note. [1]: (+) low, (++) medium, (+++) high.			

- Hydrophobicity of the solid product
- Increased energy density of the solid product compared to the feed

HTC entails a complex set of reactions, that have been summarized by Funke and Ziegler (Funke and Ziegler 2010):

- Hydrolysis
- Dehydration
- Aromatization
- Decarboxylation
- Polymerization

From a process perspective, it is difficult to determine the exact influence of process parameters on the distribution and quality of the conversion products, as the dependence on the feed composition is very large. Recirculation of Carbon-laden water was shown to have beneficial effect on the process Carbon and energy yield, as recirculated reactive substances polymerized and formed additional solid substance and *dewaterability* of HTC solid product was enhanced (Stemann et al. 2013). High biomass to water ratio will result in better polymerization and higher overall solid product yields (Stemann et al. 2013), but also in less energy consumption, i.e., sensible heat and losses. As process severity increases, solid yields will decrease; however, H/C and O/C ratios will also decrease, resulting in greater energy densification and higher heating values (Sevilla and Fuertes 2009). Also the amount of water being produced during HTC is influenced by the process temperature, and an increase of water production was noticed at increased process temperature (Hoekman et al. 2011).

3.2.1 Applications

Nowadays the production of solid carbonaceous fuel for co-firing with coal is the main industrial application of hydrothermal carbonization, though activated carbon and biochar manufacturing are being evaluated in pilot trials. A brief review of companies operating pilot or demo HTC installations is presented below.

Ingelia is a Spanish company located in Valencia tha is operating an industrial demo HTC plant since 2010. Their modular design leverages a continuous process operating at about 20 bar and 200C, with a residence time of approximately 8 h. The standard module offered by the company is capable of converting 1 ton/h of wet feedstock (55% w/w moisture) in about 350 kg of hydrochar at 7% w/w moisture, 80 kg of P- and N-rich fertilizer at 50% w/w moisture and other by-products (water, CO_2). The solid products obtained from the Ingelia process are lignite-like products with a LCV (low calorific value) between 21 and 23 MJ/kg.

AVA-CO2 is a Swiss company founded in 2009, which is a pioneer of the hydrothermal carbonization technology. In 2009 they installed the HTC pilot plant K3-335 in collaboration with the Karlsruhe Institute of Technology (KIT). In 2010 AVA-CO2 and KIT put into operation the first demonstrative HTC plant in Karlsruhe (Germany), HTC-0, consisting of parallel batch reactors. In 2012 AVA-CO2 commissioned the HTC-1, the world first commercial-scale hydrothermal carbonization plant, set-up in Relzow (Germany), with a production capacity of 8200 tons/year (dry matter) of "AVA cleancoal", corresponding to about 10800 dry ton/year of biomass. AVA cleancoal has a 65% carbon content, a HHV between 25 and 30 MJ/kg and a very small particle size (99% < 300 μm). Carbon efficiency and net energy efficiency of the process have been reported to be higher than 90% and higher than 70% respectively.

TerraNova Energy is a German company which developed the TerraNova Ultra process: a continuous HTC for sewage sludge treatment, offering a feasible and efficient alternative to expensive disposal. In 2010 a pilot plant was implemented in the Kaiserslautern (Germany) wastewater central treatment plant. This facility has been active until the end of 2012. In 2014 two more plants have been set-up and started operation in Düsseldorf, Germany and in Maribor, Slovenia. In 2016 the first large scale plant (14000 tons of sewage sludge) has been inaugurated in Jining, China; an expansion to the capacity of 40000 tons has been planned. TerraNova process can be integrated with an anaerobic sludge digestion plant, fulfilling the HTC heat demand and recycling the excess water from the HTC to the digester. This process-coupling would allow to increase the biogas yield by about 10%.

Artec Biotechnology is a German company located in Bad Königshofen and founded in 2009. The first continuous pilot plant they developed is the MOLE I (tubular reactor, 180 l, 9 m length, 150 mm internal diameter), which has accomplished more than 3000 hours of operation. The next project was the design and construction of a downsized version of the MOLE I (20 l, 3 m length and 120 mm internal diameter). The biggest-scale plant developed by the company is the Artcoal 3000 k, a 3 m³ reactor with 250 mm internal diameter and 40 m long, installed in the waste management facility of Halle, Germany. The plant is able to process 2500 ton/year of input material and produce about 1000 ton/year of hydrochar.

Suncoal Industries is a German Company founded in 2007, located at Ludwigsfelde, that patented in 2007 the CarbonREN process. In 2008 the first continuous HTC pilot plant was built and operated in Koenigs Wusterhausen. In 2011 the company submitted the permit application for the first industrial plant with an annual capacity of about 60 kton of biomass for a production of approximately 20 kton of hydrochar. In the CarbonREN process a wide range of biomass feedstocks can be converted, such as straw, wood chips, rice husks, waste from palm oil and fruit juice mills, bagasse and so on, with a moisture content from 20 to 75%. Valmet presently commercializes the CarbonREN HTC process developed by SunCoal Industries for organic sludge treatment and technical carbons production.

Shinko Tecnos Co. is a Japanese company founded in 1997, which designed and developed a HTC plant together with the Tokyo Institute of Technology. Since 2007 it has built several plants in Asia (Japan, China, Indonesia, Thailand, Sri Lanka) for fertilizer or solid fuel production, processing several kind of feedstock: sewage sludge, hospital waste, MSW, palm empty fruit bunch.

3.3 Gasification

3.3.1 Introduction

Among the various thermochemical conversion processes, gasification aims to maximize the production of a combustible gas stream. Gasification occurs at high temperatures and in presence of an oxidising agent. The oxidant used depends to the scale and type of the process: it can be pure oxygen, while in small scale decentralized applications it is more often air; steam gasification has also been developed and is implemented at commercial scale.

The gasification process has been studied for centuries; the first reported application of a gasification process, in the modern concept, can be attributed to the inventor Robert Garner in 1788. In 1792 the first syngas applications for public lighting were introduced; the feedstock was coal. In 1823 several British cities were illuminated by coal derived syngas.

The following development was the connection of gasification reactors to energy generation systems. The first commercial reactors appeared in France in 1840, but only in 1878 there was a real successful demonstration of syngas use in an internal combustion engine.

Fig. 2: A vehicle fuelled by producer gas (1940).

At the beginning of the 20th century this technology had a rapid diffusion, especially in Germany, that developed an innovative system to use coal and char as fuels. During the Second World War gasifiers were largely used: for instance, in Sweden in 1939 more than 250,000 gasification-based vehicles were registered, of which 90% were using producer gas, and 20,000 tractors were powered with the technology (Knapp and Goss 1981).

With the end of the Second World War, and with the emergence of a wide availability of liquid fossil fuels, the interest for this technology rapidly faded. The energy crisis of 1970 gave instead a new opportunity to wood-based energy systems and already in the 1980s some tens of gasifier producers were back on the market. Today small and medium scale gasifiers are of great interest for decentralized renewable energy generation from lignocellulosic biomass, while larger projects are more focused on the advanced biofuel production.

3.3.2 The Thermochemical Process

The reactions taking place in the gasification zone are both heterogeneous (solid-gas) as well as homogenous phases (gas-gas). Steam is generated by the water fed with biomass to the high-temperature gasification reactor: steam reacts with biomass carbon, to produce Carbon monoxide and Hydrogen:

$$H_2O + C \rightarrow CO + H_2 \qquad +118.5 \text{ kJ/mol}$$

A second fundamental C-reducing step is the Boudouard reaction:

$$C + CO_2 \rightarrow 2CO \qquad +159.9 \text{ kJ/mol}$$

Among all the other reactions occurring in the reactor, the *water gas shift* and the methanation reactions represent two key steps:

$$H_2 + CO_2 \rightarrow CO + H_2O \quad +40.9 \text{ kJ/mol}$$
$$H_2 + C \rightarrow CH_4 \qquad -87.5 \text{ kJ/mol}$$

The majority of the reactions that take place in the gasification zone are endothermic. In order to provide the energy needed and thus maintain the desired temperature, part of the biomass carbon must be used in combustion and partial oxidation reactions:

$$C + O_2 \rightarrow CO_2 \qquad -393.8 \text{ kJ/mol}$$
$$C + \frac{1}{2}O_2 \rightarrow CO \qquad -123.1 \text{ kJ/mol}$$

The presence of an oxidizing agent in the reaction zone determines also the oxidation of some of the gasification products:

$$CO + \frac{1}{2}O_2 \rightarrow CO_2 \qquad -283.9 \text{ kJ/mol}$$

$$H_2 + \frac{1}{2}O_2 \rightarrow H_2O \qquad -285.9 \text{ kJ/mol}$$

Despite that these reactions produce the required heat for support of the process, they have the drawback to reduce the calorific value of the producer gas.

The final composition of the gas mixture (*producer gas*) is determined by the quality of the reactions that take place, that are influenced by the actual reaction temperature and by other parameters, among which is decisive the available time to complete the reactions.

The partial completion of the reactions determines some negative particularities of the system, such as:

- a different gas composition from the one expected;
- a lower calorific value;
- a larger production of tars;
- a lower carbon conversion efficiency.

Tars are one of the major pollutant for producer gas, that need to be removed (cleaning) or converted in to gas (tar cracking) to make the produced gas usable in conversion systems.

3.3.3 Temperature Influence

The reaction temperature has a direct effect on the chemical equilibrium of the reactions, and thus on the final gas composition. The control of the reactor temperature is a key parameter for the quality of the process, especially for the gasification zone: low concentrations of H_2 and CO and high tars concentration can be expected for temperatures below 850–900°C.

Increasing the reaction temperature modify the gas composition, with a shift toward higher H_2 and CO concentration and a lower tars content. There are however some elements that limit the reachable temperatures in the gasification zone:

- At temperatures typically higher than 900–1000°C, and depending on the feedstock, ashes could start melting and sintering
- More expensive materials must be adopted to build the reactor
- The overall energy balance of the gasification system is affected, as well as heat losses

3.3.4 Impact of the Gasification Agent

The energy needed for the gasification process is often supplied through partial combustion of a small amount of biomass derived products (as pyrogas, char and biomass). The use of different oxidizing agents affect the quality of the producer gas is shown in Table 4.

The use of oxygen allows high temperatures in the reactor, absence of N_2 (diluting the gas heating value), and a higher production of Hydrogen and Carbon monoxide, with a resulting gas heating value ranging from 10–15 MJ/Nm³. The use of oxygen in gasification plants, however, has safety implications that have limited its use in small scale units.

The use of steam allows obtaining a gas with a calorific value of 12–20 MJ/Nm³ but, for oxygen, the plant configuration becomes more complex, which fact favors its applications to medium-large scale installations.

The most used gasification agent in small scale plants is air; however, as said, a significant reduction of the gas calorific value (down to 3–6 MJ/Nm³) due to the presence of nitrogen is determined.

Table 4: Syngas composition as function of the oxidant used for the process (composition %vol) (Couto et al. 2013, Rapagna et al. 2000).

Oxidant	H_2	CO	CO_2	CH_4	N_2	LHV (MJ/Nm³)
Air	9–10	12–15	14–17	2–4	56–59	3–6
Oxygen	30–34	30–37	25–29	4–6	-	10–15
Steam/CO_2	24–50	30–45	10–19	5–12	-	12–20

3.3.5 Pressure Effect

Most of the biomass gasifiers are operated at ambient pressure. Increasing the pressure leads to improving the quality of the gas, and thus its further use in the downstream energy conversion systems. However, the complexity of realizing airtight biomass supply systems limits the use of pressurized reactors.

3.3.6 Influence of the Equivalent Ratio

Aiming to achieve chemical balance in the reaction zone, the gas composition mostly depends on the oxygen introduced into the reactor. The main parameter that is used to tune the reactions in the gasifier is thus the equivalence ratio, defined as the ratio between the amounts of Oxygen introduced in the reactor to the Oxygen required for complete combustion of all the biomass carbon (stoichiometric Oxygen). An overall equivalence ratio higher than one is adopted for biomass furnaces and combustors, even if the equivalence ration in primary combustion zone is normally below 1; in gasifiers, this parameter is instead always below 1. In particular, the energy density of the gas is maximized at around 0.25. Below 0.25, part of the carbon remains unconverted in the ashes.

3.3.7 Superficial Velocity

A typical assumption in several gasification models is that chemical equilibrium is reached in the gasification zone. In an actual plant, this condition is however strongly dependent on superficial gas velocity. This parameter represents the velocity of the gas flow in the reaction zone. It is calculated as the ratio between the gas flow rate and the section of the reactor.

At nominal conditions and Imbert-type reactors, the superficial gas velocity stays between 0.8–2.5 m/s. Upon variation of the load required to the reactor this value is changed, and the consequent unbalanced conditions may also create important modification in gas composition. These effects limit the operability of gasification systems at partial loads.

3.3.8 Small Scale Gasification

There are many possible ways to classify gasification reactors. Although there are technological differences depending on the plant scale (which will be presented below), the definition of "scale" itself is often questionable, as it is connected to different regulatory requirements or even economic support levels. These classes were thus not related to technical issues, but have defined the distinction between "small and large scale plant".

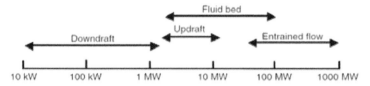

Fig. 3: Electrical power range of applicability for biomass gasifier types.

From a purely technological point of view, a relevant distinction can be carried out between small and large scale on the basis of the type of bed reactor (Fig. 3), distinguishing between:

- **fixed bed reactors:** typically in a range of 100–1000 kW$_{el}$
- **fluidized bed reactors:** suitable for plants capacity higher than 1 MW$_{el}$.

In the fixed-bed reactors the biomass is substantially stationary while the gasifying agent passes through it.

In fluidized bed reactors, biomass is suspended by a gas flow, strongly mixed with an inert material; there is thus an intimate mixing and a greater homogeneity in temperature distribution. The increased complexity in plant management and in the biomass preparation make this solution more suitable for larger scale plants.

Today, small-scale gasifiers are of considerable interest since they allow the production of electricity from solid biomass at scales otherwise normally not covered, in an economically sustainable way, from other technologies.

The use of biomass in small-scale gasifiers require the compliance of certain minimum quality specifications, dictated by the process itself; of fundamental importance are biomass humidity and size uniformity.

In general gasification, compared to biomass combustion, have significant limitations with respect to biomass versatility and partials load.

3.3.8.1 Fixed bed reactors

Fixed-bed reactors can be divided into three classes, depending on the relative motion between biomass and gas: *Updraft*, *Downdraft* and *Crossdraft*.

In countercurrent fixed bed reactor (*Updraft*), biomass and gas move in opposite direction, this allows a preheat of the fuel. In this type of reactor, while the char produced by pyrolysis continues the descent towards the reduction zone, part of the produced tar vapors come in contact with the solid particles while a large amount of them is dragged out from the reactor with the syngas. Because of this high tar content, the producer gas has a high calorific value, but must be largely cleaned before entering a conversion system. The major advantages of this type of reactors are the constructional simplicity, the tolerance to high moisture content (up to 50%) and the ability to accept finely grinded biomass. The main drawback lies instead in the high presence of tars in the gas. This problem is limited if gas is directly used for combustion in external combustion systems.

In co-current fixed bed reactors (*downdraft*) (Fig. 4) biomass is loaded from the top and both char and producer gas exit the reactor from the bottom. The producer gas, rich in tars passes through the char bed, allowing tars reduction and thus a better gas quality (in average of about 500 mg tars/Nm^3 but even lower than 100 mg tars/Nm^3 if optimized). Biomass bed is sustained by a shrinkage (throat) of the reactor (*Imbert cone*).

In these type of reactors, the combustion zone is placed above that of reduction; this layout allows that the drying zone and the pyrolysis are mainly heated by radiation, and that a good part of the pyrolysis vapors is burned or cracked in the high-temperature combustion area. The path of the gas allows to achieve a high conversion rate and thus lower unconverted compounds, from which it follows a small presence of residual tar in the producer gas. The difficulty in obtaining steady gas quality depends on ability to ensure biomass homogeneity in the reactor and to control superficial gas velocity in a way to ensure that all the produced gases have a sufficient residence time to complete reactions. Despite the low tars concentration, the gas produced by *downdraft* gasifier still requires a cleaning section before being used in a reciprocating engine or a gas turbine.

The main drawbacks of the *downdraft* reactors are:

- the gas passes through the char bed, which positively impact on tars reduction but, at the same time, generates the need of separators to clean the gas by the transported solids;

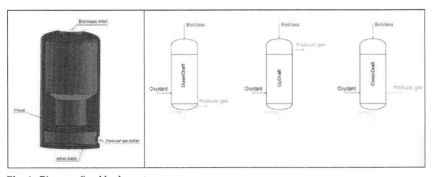

Fig. 4: Biomass fixed bed reactors.

- the biomass needs to be pre-treated, as the moisture content at inlet should typically fall in the range of 10–15%, and the average size must be between 4 and 10 cm in order to maintain sufficiently porous the throat of the reactor;
- the relatively high temperature of the gas leaving the reactor reduce the overall energy conversion efficiency, as it must be cooled before entering in the engine.

Another kind of fixed bed reactor is the cross-flow (*crossdraft*) that, even though it has a similar structure of the *updraft*, it has an oxidant injection in the cross section. This configuration allows reaching very high temperatures in the core zone of the reaction, resulting in low concentration of tars in the outlet gas.

The main advantage of this solution is the theoretically smaller gas cleaning section even for small scale applications. The quality of the biomass feedstock is fundamental for *crossdraft* reactors too.

The *Open-Core* are also fixed-bed gasification systems, specifically developed to allow the use of fine particle size fuels, characterized by a low density, unsuitable for the other fixed bed reactors.

To achieve this goal, the open-core does not present the throat that supports the solid bed since, given the characteristics of the fuel, it would obstruct the flow downwards and/or create problems like bridging or channeling of the biomass in the reactor. This technology, addressing fuels with high ash content, such as rice husk, can also be equipped with rotating grates for continuous fuel mix and ashes removal.

The following table compares the main and typical characteristics of the described technologies.

Table 5: Characteristics of fixed bed reactors.

Characteristics	Downdraft	Updraft	Open-Core
Allowed Moisture content (%wb)	12 (max 25)	40 (max 60)	7–15 (max 15)
Ashes (%db)	0.5 (max 6)	1.4 (max 25)	1–2 (max 20)
Biomass size (mm)	20–100	5–100	1–5
Outlet gas temperature (°C)	700	200–400	250–500
Tars (g/Nm^3)	0.015–0.500	30–150	2–10
Hot gas efficiency at full load (%)	85–90	90–95	70–80
Cold gas efficiency at full load (%)	65–75	40–60	35–50
Syngas LHV (kJ/Nm^3)	4.5–5.0	5.0–6.0	5.5–6.0

3.3.8.2 Fluidized bed reactors

Contrariwise to fixed bed reactors, in fluidized beds the solid biomass particulate is suspended by a gaseous fluid. Here the fuel is mixed together with an inert bed material, acting like a fluid.

The fluid behavior is obtained by forcing a gas through the bed; examples of typical fluidization media are air, steam or steam/oxygen mixtures. Even if the fluidization is a function of several parameters, such as particles size and shape, material density, bed geometries, etc., the velocity of the fluid flows has a dramatic impact on the behavior of the fluidization process (Gómez-Barea et al. 2013).

Depending on the velocity of the fluidization medium, the bed reactors can be divided into bubbling fluidized beds (BFB) and circulating fluidized beds (CFB).

Bubbling beds are operated at relatively low gas velocities (below 1 m s^{-1}), while in circulating fluidized beds higher gas velocities are expected (3–10 m s^{-1}). In circulating beds, solid particles are dragged upwards by the gas flow and they must be recovered to be reintroduced in the reactor (Siedlecki et al. 2011). A cyclone is usually used to separate and then recycle the biomass particles.

To be successfully used in fluidized bed gasification, the biomass needs to be pre-treated, as it needs to be grinded to obtain a fine and uniform fuel for the reactor. A delicate issue in fluidized bed technology is again that great attention must be given to fuels with high content of ash, and alkali metals (Marrero et al. 2004). Such elements can sinter/melt at process temperature, causing sticking of the particles, eventually leading to the formation of bigger agglomerates, affecting the hydrodynamics of the reactor and ultimately leading to bed de-fluidization.

Together with biomass an inert material is used: silica sand is the most used bed material. The inert material has the aim to enhance the heat

Fig. 5: Principle of fluid bed gasification.

exchange, thus allow operating approximately at isothermal conditions. Maximum operating temperatures typically lie between 800 and 900°C, practically limited by the melting point of the bed material. In these conditions, and due to the short gas residence times, gasification reactions cannot reach chemical equilibrium if a catalyst is not used (Basu et al. 2006). The main goal of catalysts in fluidized bed regards tars decomposition. Metal oxides from natural rock minerals can be successfully used in the fluidized bed, and they can be even more advantageous than the use of commercial Ni-based catalysts (Constantinou et al. 2009). Dolomites $(CaMg(CO_3)_2)$, calcites $(CaCO_3)$, magnesites $(MgCO_3)$ and olivines $((Mg,Fe)_2SiO_4)$ can also be potentially used as bed materials because their catalytic properties.

When properly operated, tar content in fluidized bed is lower than fixed bed (such as downdraft gasifier) of about one order of magnitude (Siedlecki et al. 2011).

Another major characteristic of fluidized bed gasifiers with respect to fixed beds is that due to their geometry and excellent mixing properties, fluidized beds are well suited for scaling up. On the other hand, the plant complexity is considerable compared to fixed bed.

3.3.8.3 Energy conversion via producer gas: reciprocating engines

The most diffused solution for generating heat and power from producer gas is represented by reciprocating engines. These are usually combined with downdraft gasifiers, thanks to the gas quality achievable with this reactor system. Diesel engines or gas (i.e., spark ignition) engines can be used in small-scale applications: sometimes diesel engines are equipped with pilot injection to facilitate ignition (typically using fossil diesel or vegetable oil/biodiesel). For markets with stringent emission regulations, engines are typically of very high technical standard, ensuring high efficiency, while where emission standards are less strict, lower technology levels have been adopted in the past to reduce capital costs.

Feeding a gas engine requires a clean dust-free gas, with tars dew point below the engine inlet temperature (or even lower in case of a turbo-charged engine) in order to prevent tar condensation and fouling. The

Table 6: Typical gas limits for feeding producer gas to gas engines (source: Le Coq et al. 2012).

Component	Unit	Acceptable
Tar content	mg/Nm³	< 50
Dust content	mg/Nm³	< 50
Particle size	µm	< 10

removal of tars, ashes and corrosive gaseous compounds is needed before the engine, to achieve the gas quality required by the manufacturers.

A possible strategy to limit potential damages to engine is to work with a gas feeding temperature above 80°C; this temperature, higher than light tar dew point, prevents the condensation of water and any remaining harmful tar components. This can help to mitigate the need for more intensive gas treatment, but it also reduces the engine performances due to the low gas density. A turbocharged engine can also be considered, to increase the gas pressure and therefore the gas energy density, compensating the effect of the high temperature gas feeding.

If optimized for low LHV gas feeding, the electrical efficiency of reciprocating engines coupled with gasifiers can reach 35–45%.

3.3.8.4 Energy conversion via producer gas: gas turbine

Micro gas turbine (MGT) technology has rapidly penetrated the market in recent years, thanks to its high performances in terms of efficiency and the low maintenance requirements. Today commercial MGTs cover a typical power range of 30–400 kWel.

One of the most interesting potential benefits of coupling MGT with downdraft gasifier is represented by the possibility of using part of the high temperature exhausts to heat the gasifier, so as to recover part of the heat to supply endothermic reactions.

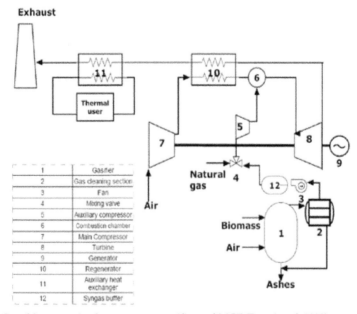

1	Gasifier
2	Gas cleaning section
3	Fan
4	Mixing valve
5	Auxiliary compressor
6	Combustion chamber
7	Main Compressor
8	Turbine
9	Generator
10	Regenerator
11	Auxiliary heat exchanger
12	Syngas buffer

Fig. 6: Possible connection layout among gasifier and MGT (Prussi et al. 2008).

The main challenge for the gasification unit is to deliver clean gas to the combustion chamber of the gas turbine, the main contaminants being: tars, particulate, alkalis and heavy metal content. As regards tars, they tend to condensate on cold surfaces, causing problems in piping or clogging in valves and filters.

There are several estimations regarding the maximum allowable tar level for a gas turbine, and manufacturers do not indicate tar limits in their specifications; in case of an optimized downdraft gasifier, a mean value of $0.015-0.5\,g/Nm^3$ can be expected. As regards particulates, this contaminant must be strictly controlled since it can rapidly cause significant erosion in nozzles and blades (Consonni et al. 1996). Alkalis, such as Na and K (contained in the biomass feedstock at different amounts), cause corrosion of turbine blades and for this reason the maximum allowable content is rather low: for a 30 kWel commercial machine the limit is set at 0.5 ppm when operated by natural gas (CAPSTONE 2005).

Knoef et al. (2005) suggest that turbines are less sensitive to tars compared to reciprocating engines, since the feeding gas temperature can be high and tars maintained in vapors form. This can be easily implemented only for pressurized gasifiers (Kitzler et al. 2011), while in case of atmospheric gasifiers (almost all small scale commercial installations today on the market), the energy cost of hot gas compression is generally unacceptable.

Despite the needs of managing tars and contaminants, given the low LHV of the gas, a larger flow is fed to the gas turbine combustion chamber, increasing the compression ratio, with risk of surge. In order to limit this risk, a de-rating strategy can be adopted. De-rating consists of imposing a lower gas turbine burning temperature (TIT) and thus, lowering the compressor pressure ratio, but this also reduces machine efficiency; co-combustion of low LHV fuels and natural gas can be successfully adopted to limit this problem.

3.4 Combustion of Lignocellulosic Biomass

3.4.1 Basic of Biomass Combustion

The combustion of lignocellulosic biomass is always composed of the following process: drying, pyrolysis and gasification, char oxidation and gas phase oxidation (Van Loo et al. 2002, Nussbaumer 2002, 2003). In fact, under typical conditions, almost 80% of biomass is subject to devolatilization and burn as a gas phase, while the remaining char is oxidized through heterogeneous reaction.

Drying already starts at T being slightly lower than 100°C: the evaporation of water reduces the temperature in the combustion chamber. Thus, the furnace and plant design will have to meet the maximum moisture content of biomass at inlet. Devolatilization (pyrolysis) follows the drying phase, in a temperature range between 200°C and 400°C, with limited devolatilization still present until approximately 500°C. The devolatilization rate of hemicellulose, cellulose and lignin (the main constituent of lignocellulosic biomass, together with ashes and extractives) differ, with hemicellulose being the first to depolymerize into the gas phase, followed by cellulose at higher temperature, and finally by lignin. This behavior is well shown by the Thermo Gravimetric Analysis (TGA). Combustion is then completed by gas phase oxidation of gas/vapour from devolatilization, and sold phase char oxidation.

The main factors affecting efficiency and emissions from biomass combustion in furnaces are—among others—the heat transfer mechanism, the furnace design (excess air ratio, air staging and distribution, materials/insulation, air preheating, fuel load, draught), the combustion temperature, and the fuel characteristics (elemental composition CHNSO, ash and moisture content).

The fuel composition determines its calorific value and—with its physical shape and bulk density—the energy density (energy per unit volume), is a very important element for plant design and specifically for the design of the feeding section. Biomass can be fed in a variety of forms, from chips to bales (square or rounded), to briquettes and pellets.

Ashes also play a key role in biomass combustion, being responsible for slagging, agglomeration, fouling, corrosion and erosion phenomena.

3.4.2 Industrial-scale Biomass Combustion

Biomass combustion is typically carried out in fixed bed, fluidized bed or pulverized reactors. In the first case, biomass is placed on a grate (with or without movement between the bed embers and the grate depending on the grate type), while in fluidized systems (either bubbling or circulating) well grinded biomass is inserted in a bed of hot inert material, normally sand, which is suspended by an air preheated flow. Cyclones then separate the inert, which is recirculated in the combustion chamber.

Biomass combustion can be performed in co-current, counter current or cross-current systems, depending on the flow of hot gases/flame versus the fuel flow. The co-current solution is better suited for dry fuels, while the counter current works well with wet materials thanks to the increased heat exchange by convection (in addition to radiation).

Air is normally fed in at least two or more different positions in the furnace: first, in the grate area where devolatilization of biomass and char oxidation occur, called primary zone, and then in the secondary combustion zone where gas phase oxidation is completed.

In order to improve the combustion of biomass and control the emissions, excess air (lambda value) is normally limited to substoichiometric values on the grate (primary zone), while the overall amount of air introduced in the furnace is above 1, usually in the range of 1.5–3. Exhaust gas recirculation is also widely adopted (together with air preheating), in order to control the temperature in the primary zone as well as the oxygen content. Movement of the grate and the bed of ember further accelerates the biomass devolatilization phase.

A non-exhaustive list of industrial biomass combustion systems based on fixed bed includes: travelling grates, inclined moving grates, horizontally moving grates, vibrating grates, cigar burners, underfeed rotating grates, and underfeed stokers. Fluidised beds however, differentiate, as said, between a bubbling and circulating fluid bed.

3.4.3 Power Generation from Biomass Combustion

The generation of electricity and heat from biomass combustion does not differ from external combustion systems based on fossil fuels. Thus, differently from gasification and pyrolysis, where a combustible liquid or gaseous fuel is produced, the typical system to generate power from solid biomass is still represented by the well-known steam cycle.

Other external combustion systems, especially at small and micro scale at various levels of maturity, have been tested, such as Stirling engines, externally fired gas turbines and in particular Organic Rankine Cycles (ORC). Among these systems, ORC systems achieved a full commercial status in combination with biomass fuels, and were widely adopted in Europe and abroad.

3.5 Anaerobic Digestion

3.5.1 Introduction

Biogas technologies aim at converting organic feedstock in a gas stream suitable for direct energy conversion, as well as for upgrading it to biomethane. Biogas is produced under anaerobic conditions: it is a mixture of gases, with a variable composition (Table 7) depending on the feedstock employed.

Biogas can be obtained from a wide variety of organic materials. Some possible supply chains and related economic sectors are as follows:

- Agricultural sector: digestion of manure, agricultural residues, and energy crops;
- Waste water treatment plants: conversion of the organic load of the sludge and, sometimes, the influent, with relevant energy recovery in the process;
- Industrial sector: digestion of waste streams, such as those from food industries and food/feed processing;

Table 7: Average biogas composition.

Components	Agricultural wastes and energy crops	Waste of agrifood industry	Waste water treatment
CH_4 %vol	60–75	68	60–75
CO_2 %vol	33–19	26	33–19
N_2 %vol	1–0	-	1–0
O_2 %vol	< 0,5	-	< 0,5
H_2O %vol	6 (@ 40°C)	6 (@ 40°C)	6 (@ 40°C)
Total %vol	100		100
H_2 %vol	< 1	< 1	< 1
H_2S mg/m$_3$	3000–10000	400	1000–4000
NH_3 mg/m$_3$	< 1	< 1	< 1

- Municipality level: digestion of urban and food wastes, as well as other types of biowastes;
- Landfills: recovery of landfill gas for energy production.

Europe has a relevant number of plants and technology providers in the field of biogas/anaerobic digestions; the number of installed biogas plant has grown rapidly in the last decade. According to available data, Europe has 17.300 biogas plants, 367 of them are biomethane plants (European Biogas 2015). Germany is leading the scene of biomethane with

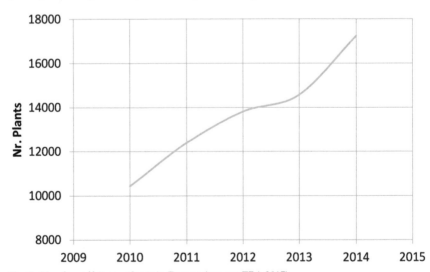

Fig. 7: Number of biogas plants in Europe (source: EBA 2015).

almost 200 plants and a production potential of about 120.000 Nm^3/h (FNR 2016).

According to recent data from CIP (Consorzio Italiano Biogas), Italy has built more than 1700 biogas plant built and more than 1300 MWe ws installed in 2016, which is equivalent to approximately 3 billion Nm^3 of biomethane equivalent utilized each year.

Biogas can be used on-site, in combined heat and power (CHP), units or upgraded to biomethane. Biomethane can be used as vehicle fuel or injected in to the grid (Dzene et al. 2013). Usually biogas plants are located far from houses or other heat demands, thus the majority of the installations target only electricity production. When heat is dissipated, the average net efficiency of a biogas plant is around 30–35%. If the biogas is transported via natural gas pipeline and used in efficient CHPs or modern domestic boiler, more than 90% of the primary energy of the biomass can be recovered (Ryckebosch et al. 2011).

As said, biomethane can be obtained from biogas upgrading, which mainly consist in methane separation from the wet biogas stream. Other forms of upgrading are also possible, as thermochemical methanation or *in situ* biometanation. The main product of this processes is methane that can be directly used in existing infrastructures without any modification. There is also a stream of concentrated CO_2 from the process, that can be diverted to C sequestration or other chemical or agro-industrial processes.

3.5.2 Biogas Production Process

In the anaerobic digestion process, organic materials are decomposed by a population of anaerobic microorganisms. Three main steps compose the global process that convert wet biomass in biogas (De Baere et al. 2000):

- hydrolysis and acidogenesis,
- acetogenesis;
- methanogenesis.

In the hydrolysis and acidogenesis step, the complex organic molecules of the biomass are broken down into smaller molecules (sugars, amino-acids and fatty acids) by bacteria. These molecules are then further decomposed during the acetogenesis phase, to obtain mainly acetic acid. The intermediate products of the acetogenesis step are then are converted in methane, carbon dioxide and water by methanogenic organisms.

The process works at different temperatures and thus the steps can be carried out in different reactors. The majority of the plants operate in a single stage, at an average temperature that can be in the range of 30–45°C (mesophilic) or 45–60°C (thermophilic). The time required to convert the biomass varies from few weeks up to three months.

3.5.3 Feedstock

Any biodegradable non-woody plant or animal material is potentially suitable as feedstocks for anaerobic digestion. Lignocellulosic biomasses cannot be completely converted as micro-organisms cannot break down lignin, consequently feedstock such as paper, straw, etc., need strong pretreatments or will slow the digester work.

To be economically viable, biogas plants require cheap feedstocks with high fermentable carbohydrate content, such as corn or other energy crops.

Feedstock can be typically distinguished on the base of:

• dry matter content;
• storability;
• pre-treatment required;
• time required to be digested;
• volatile solids (VS).

For instance, livestock manures are widely used in biogas plants, even if these are low-energy feedstocks. Manure is used since it has a generally neutral pH and shows a high pH-buffering capacity, provide nutrients and is easily pumpable. Moreover manure naturally contains a mix of microorganisms able to perform anaerobic degradation.

Table 8: Feedstock biogas potential.

Feedstock	Biogas Yield (m³/t)
Cattleslurry	15–25 (10% DM)
Pigslurry	15–25 (8% DM)
Poultry	30–100 (20% DM)
Whole wheatcrop	185 (33% DM)
Crude glycerine	580–1000 (80% DM)
Wheatgrain	610 (85% DM)
Miscanthus	179–218
Potatoes	276–400
Maizesilage	200–220 (33% DM)
Maizegrain	560 (80% DM)
Sorghum	295–372
Grass	298–467
Triticale	337–555
Straw	242–324

Securing a reliable feedstock supply is fundamental to profitably run a biogas plant: for example, blending of energy crops with livestock manure is a common practice to maximize biogas production, with a positive economical balance.

The rate of biogas production is influenced by the feedstock total solid content (TS). The total solid load affects the pH, and the performances of the microorganisms in the AD process (Khanal 2008).

3.5.4 The Biogas Plant

A biogas plant is very a flexible plant, able to process a wide range of feedstock. Biogas plants are made of several sections:

1. storage and feedstock leading;
2. feedstock pretreatment;
3. reactors;
4. separation of digestate;
5. biogas cleaning;
6. biogas use (CHP) or upgrading (biomethane).

The definition of a properly sized storage is a fundamental element in plant design, as this part guarantee the operability of the unit. A complex storage allows several substrates mixing, in order to realize the proper "diet" for the reactors. Moreover, storage can be also used for a basic pretreatment of the biomass, as it allows for the production of silages (e.g., corn silage).

Fig. 8: Biogas plant components.

Once the proper diet has been prepared, the loading section feeds the digester. This section is able to mix various feedstocks in the proper amount. Water is usually added here, in order to obtain a concentration of substrate suitable for the Anaerobic Digestion (AD) process.

Pretreatments can be necessary on some kinds of substrates: this is particularly true for lignocellulosic ones. The aim of the pretreatment step is to open the structure of the material, in order to increase the surface/volume ratio, allow easier penetration by microorganisms, and improve the conversion in biogas.

The reactor typically has a large volume (usually of several thousand of m³), where mixing is guaranteed by propellers or recirculating pumps; the operational temperature is maintained by a heat exchanger placed around the reactor, eventually heated by the water warm from the CHP. In the majority of installed plants, reactions can occur in a single reactor (single stage plant), but two stage plants can also be realized (separating the acetogenesis and acidogenesis phases from the methanogenic one). By increasing the number of stages theoretically the performances but also the management complexity is increased. AD reactors usually operate at the following conditions:

Table 9: Reactor operational conditions.

Temperature	Mesophilic	35–45°C
	Thermophilic	45–60°C
pH	Hydrolysis/acidogenesis	4.5–7.0
	Acetogenesis/methanogenesis	6.5–8.0
Average hydraulic retention time	Single stage	20–90 days
	Multistage	37–200 days
Power demand		7–8% of the total electrical production
Heat demand		20–30% of the total electrical production

Raw biogas is collected from the upper space of the reactor and needs to be cleaned before being used, in order to remove contaminants such as halogenates, hydrocarbons, Sulphur compounds, ammonia and dust particles. Raw biogas is mostly cleaned using chilled water scrubber; this system allows to obtain a dry clean gas, suitable for further utilization.

Biogas can be fed to CHPs: however, due to the usual locations where AD plants are installed (i.e., agricultural areas), the heat from the engine can only partially be used to lack of demand for heat, and therefore it is in part dissipated.

The product outflowing the AD digester is the digestate that is typically separated into a solid and a liquid fraction. The solid fraction is directly distributed on the soil, in order to recover the N and P. The liquid fraction is also rich in nutrients, but in several countries its use is highly regulated because of the potential leakage of nitrogen compound in the deep water layer.

3.5.5 Biogas Upgrading

Biogas can be upgraded to separate biomethane from carbon dioxide and water vapour. Once the biomethane is in turn upgraded to meet natural gas specifications, it can be injected into the natural gas grid. This technological pathway offers considerable opportunities in terms of increasing the renewable energy share within the natural gas-fired systems (Kovacs 2013): biomethane can be considered as one of the most viable renewable substitutes for natural gas (Adelt et al. 2011). Moreover, biomethane, either separated in a gaseous or liquefied form, can serve a variety of uses: heat generation, power and combined heat and power, and biofuel for transport in Natural Gas Vehicles.

The upgrading process can be performed by several chemical-physical techniques: water scrubbing, organic solvents and membrane separation (see Table 10). An interesting physical technique for upgrading biogas is the Pressure Swing Adsorption (PSA). PSA is performed by several steps of gas pressurization, which facilitates the CO_2 adsorption on molecular sieves.

The methane content of the gas can reach to 95%–98% (IEA 2007).

In order to be suitable for grid injection or vehicle utilization, the final quality of the biomethane has to meet the technical specifications at Country level. Since biomethane is a commodity which can be physically exchanged by countries, there is the issue in adapting these standards to a common EU level. EU specifications, for instance, have not yet been set, nonetheless the single Countries have already initiated biomethane production and distribution. The Italian standard, for example, has been recently issued and set the limits for the major biomethane parameter (see Table 11).

4. Vegetable Oil for Power and CHP

4.1 Vegetable Oil Production Chain and Market

In the last decades, EU consumption of biofuels (mainly biodiesel produced from vegetable oils) has grown rapidly, especially thanks to the targets set by the Renewable Energy Directive of the European Union (EU-RED,

Table 10: Technologies for biomethane separation and upgrade.

		Pressure swing adsorption	Water Scurbbing	Physical Absorption	Chemical Absorption	Membrane	Cryogenic process
Process heat temperature	°C	-	-	55–80	110–160	-	-
Process Pressure	Bar	4–7	5–10	4–7	0.1–4	5–10	-
Power Consumption	kWh/Nm³	0.20–0.25	0.20–0.30	0.23–0.33	0.06–0.15	0.18–0.25	0.18–0.33
Heat consumption	kWh/Nm³	0	0	0.3	0.5–0.8	0	0
Methane loss	%	1–5	0.5–2	1–4	0.1	2–8	-
Demand for Chemical	-	no	no	yes	yes	no	no
Water Demand	-	no	yes	no	yes	no	no

Table 11: Italian specifications for biomethane use and distribution (UNI-TS 11537).

Parameter	Symbol	Unit	Limits
Higher heating value	HHV	MJ/Sm^3	$34.95 \leq$ HHV ≤ 45.28
Wobbe index	WI	MJ/Sm^3	$47.31 \leq$ WI ≤ 52.33
Relative density	ρ	-	$0.5548 \leq \rho \leq 0.8$
Hydrocarbon condensation point			$\leq 0°C$ @ 100 kPa $\leq$ P $\leq$ 7000 kPa r.
Oxygen content	O_2	%mol	≤ 0.6
Carbon dioxide content	CO_2	%mol	≤ 3
Hydrogen sulfide content	H_2S	mg/ Sm^3	≤ 6.6
Sulphur content by mercaptans	-	mg/ Sm^3	≤ 15.5
Total Sulphur content		mg/ Sm^3	≤ 150

Panichelli et al. 2015, ICCT 2013). In particular, the world consumption of vegetable oil is still increasing (Issariyakul et al. 2014).

On a large scale, such as for biodiesel or hydroprocessed lipids, the use of palm oil represented the largest part of the world demand of vegetable oils (Mosarof 2015). However, the use of lipid-based feedstock as bioliquid for power and CHP generation is largely diffused, also in the context of small scale decentralised production and use of locally-grown lipidic feedstock. Short bioenergy chains, with high-added value for the farmers, has shown positive socio-economic impacts (Chiaramonti and Prussi 2009), even if the use of land for energy crop has been subject of a strong debate and confrontation in Europe.

4.2 Vegetable Oil Characteristic and Production

Industrial oil extraction is normally based on the use of solvents; this technique allows for extracting practically all the oil contained in the seeds. The co-product of industrial extraction is a flour that is widely used as animal feed. The vegetable oil obtained from solvent extraction needs to be treated before being used as a fuel.

In small scale decentralized plants, the most common technology to recover oil from seeds is based on cold mechanical pressing (i.e., screw-press). Oil seeds need to be pre-conditioned before extraction: moisture content needs to range in 8–9%. If the vegetable oil is extracted by mechanical pressing, screw velocity results to be a fundamental parameter affecting the final vegetable oil quality: the higher speed and thus the temperatures, the higher the oil yields, but also the higher the contamination. In fact,

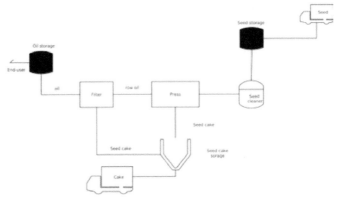

Fig. 9: The vegetable oil decentralised chain.

operating at high extraction speed lead to a significant contaminants presence (e.g., phosphorus content, solid contents, etc.) (Sidibé et al. 2010). Temperature should be limited to 75°C. This temperature level allows to obtain an oil suitable for use in engines and systems converted for pure vegetable oil fueling.

Compared to diesel fuel, straight vegetable oil shows different chemical-physical characteristics. The direct use of vegetable oil as liquid biofuel is possible in adapted reciprocating engines (Bohl et al. 2014, Corsini et al. 2015, Pasyniuk et al. 2013) or microturbines.

For the direct use as fuel, vegetable oil is not subjected to a specific standard. There are several standards issued by engine manufacturer and a German norm, the DIN 51623 (DIN 2015), which sets specifications, providing reference values and testing methods. Straight (or pure) vegetable oil is a generic term used to refer to a wide class of vegetable oils: actually, as regards their use as liquid fuel, the characteristics of each vegetable oil can largely differ. Cleaning strategies and technology adaptations are thus required to feed engines with pure vegetable oil.

4.3 Conventional Conversion Technologies

As said, conventional energy generation technologies—such as reciprocating engines—must be modified in order to accept pure vegetable oil, since they have to guarantee a high reliability, low consumption, and limited emissions over a long-term operation. The first element to be considered in using pure vegetable oil instead of Diesel fuel is the viscosity. The viscosity of vegetable oil at ambient temperature is one order of magnitude higher than Diesel (Vegetable Oil: 40 mm²/s a 40°C; Diesel: 3–5). Furthermore, the heating value of vegetable oil is lower, even though its higher density limits the difference with Diesel oil on a volume base (Vegetable Oil: 36.8 MJ/l; Diesel fuel: 35.8 MJ/l).

Pre-heating the oil contributes to reduce the viscosity, and consequently limits the negative effects on engine performance, especially as regards atomization and vaporization: this is a measure normally taken when engines are adapted for energy production. Nevertheless, heating the fuel only is not sufficient to ensure low emissions and component life in the long-term; the fueling system and the operating parameters (injection pressures, timing, etc.) must be appropriately modified.

4.4 Innovative Conversion Technologies

Power systems are generally based on reciprocating engines, a fully-mature technology that today offers high reliability and high performances. However, it is characterized by rather high maintenance costs and considerable high emissions (in terms of NO_x and CO). Compared to small scale diesel engines, micro gas turbines are able to reduce both emissions and maintenance costs, while they still offer excellent performances, and a larger share of heat available. Pure vegetable oils can be used to fuel modified MGT, without significantly affecting the engine performances and emission (Prussi et al. 2012).

5. Conclusions and Future Perspectives

Biomass has been able to satisfy energy needs of human beings since ancient times. Today, new and modern technologies and processes can exploit biomass products in a very efficient and clean way.

Solid biomasses, as lignocellulosic feedstocks, or lipid-based one (oil seeds) can be converted into power and heat in a very effective way, often also at a rather small or even micro scale, depending on the technology and the local context.

In fact, bioenergy systems have to be designed on the base of the specific conditions existing on-site, such as the type of biomass locally available, the skills of operators, the supply chain, etc.

Environmentally sound techniques for biomass collection and provision at the plant gate must be implemented, to ensure the renewable character of the biomass feed.

Under these circumstances, the sustainable exploitation of biomass resources through decentralized schemes involving local communities in a circular economy approach bring socio-economic development to rural communities.

On the technological side, a variety of conversion systems can be adopted, from diesel engines to gas turbines and steam cycles. Downscaling of these solutions is possible, the economic sustainability depending on the market of competing fossil fuels and therefore on the presence of specific incentives for renewable energy production. Future developments are expected to drive the whole bioenergy sector towards further lowering of emissions, improvement of efficiency, easier operability and compactness of the plants.

References

Adelt, M., D. Wolf and A. Vogel. 2011. LCA of biomethane. Journal of Natural Gas Science and Engineering 3(5): 646–650.

Andrews, R.G., S. Zukowski and P.C. Patnaik. 1997a. Feasibility of firing an industrial gas turbine using a bio-mass derived fuel. pp. 495–506. In: Bridgwater, A.V. and D.G.B. Boocock (eds.). Developments in Thermochemical Biomass Conversion SE-39. Dordrecht: Springer Netherlands. doi:10.1007/978-94-009-1559-6.

Andrews, R.G., D. Fuleki, S. Zukowski and P.C. Patnaik. 1997b. Results of industrial gas turbine tests using a biomass-derived fuel. pp. 425–435. In: Overend, R.P. and E. Chornet (eds.). Making a Business from Biomass in Energy, Environment, Chemicals, Fibers, and Materials. New York: Elsevier Science Inc.

Ardy, P.L., P. Barbucci, G. Benelli, C. Rossi and G. Zanforlin. 1995. Development of gas turbine combustor fed with bio-fuel oil. In Second Biomass Conference of the Americas: Energy, Environment, Agriculture, and Industry, 429–38. Portland, OR (USA): NREL.

Basu, P. 2006. Combustion and Gasification in Fluidized Beds; CRC Press: Boca Raton, FL, USA.

Beld, B. Van de, E. Holle and J. Florijn. 2013. The Use of Pyrolysis Oil and Pyrolysis Oil Derived Fuels in Diesel Engines for CHP Applications. Applied Energy 102 (February). Elsevier Ltd: 190–97. doi:10.1016/j.apenergy.2012.05.047.

Beran, M. and L.-U. Axelsson. 2014. Development and experimental investigation of a tubular combustor for pyrolysis oil burning. Journal of Engineering for Gas Turbines and Power 137(3): 31508. doi:10.1115/1.4028450.

Bergman, P.C.A., A.R. Boersma, R.W.R. Zwart and J.H.A. Kiel. 2005a. Torrefaction for Biomass co-firing in existing Coal-fired Power Stations. ECN-C-05-013, July 2005(a). http://www.ecn.nl/biomass.

Bergman, P.C.A., A.R. Boersma, J.H.A. Kiel, M.J. Prins, K.J. Ptasinski and F.J.J.G. Janssen. 2005b. Torrefaction for Entrained-flow Gasification of Biomass. ECN-C-05-067. http://www.ecn.nl/biomass.

Bertoli, C., J.D. Alessio, N. Del Giacomo, M. Lazzaro, P. Massoli, V. Moccia and C.N.R. Istituto Motori. 2000. Running Light-Duty DI Diesel Engines with Wood Pyrolysis Oil. Combustion, no. 724.

Bohl, T., G. Tian, W. Zeng, X. He and A. Roskilly. 2014. Optical investigation on diesel engine fuelled by vegetable oils. Energy Procedia 61: 670–674 [online] http://linkinghub. elsevier.com/retrieve/pii/S1876610214027696 (accessed June 30, 2015).

Bridgwater, A.V. 2012. Review of fast pyrolysis of biomass and product upgrading. Biomass and Bioenergy 38 (March): 68–94. doi:10.1016/j.biombioe.2011.01.048.

Calabria, R., F. Chiariello and P. Massoli. 2007. Combustion fundamentals of pyrolysis oil based fuels. Experimental Thermal and Fluid Science 31(5): 413–20. doi:10.1016/j. expthermflusci.2006.04.010.

Cappelletti, A., A.M. Rizzo, D. Chiaramonti and F. Martelli. 2013. CFD redesign of micro gas turbine combustor for bio-fuels fueling. In: XXI International Symposium on Air Breathing Engines (ISABE), 1199–1206. Busan, Korea. doi:10.13140/2.1.4096.1601.

CAPSTONE Turbine Corp. 2005. Capstone micro turbine fuel requirements. Technical Reference.

Chiaramonti, D. and M. Prussi. 2009. Pure vegetable oil for energy and transport. Int. J. Oil, Gas Coal Technol. 2(2): 186–198.

Chiaramonti, D., A.M. Rizzo, A. Spadi and M. Prussi. 2013. Testing of Pyrolysis Oil, Vegetable Oil and Biodiesel in a Modified Micro Gas Turbine: Preliminary Results. Oral Presentation 2BO10.5, 21st European Biomass Conference, Copenhgen.

Chiaramonti, D., A. Oasmaa and Y. Solantausta. 2007. Power generation using fast pyrolysis liquids from biomass. Renewable and Sustainable Energy Reviews 11(6): 1056–86. doi:10.1016/j.rser.2005.07.008.

Chiaramonti, D., M. Bonini, E. Fratini, G. Tondi, K. Gartner, A.V. Bridgwater, H.P. Grimm, I. Soldaini, A. Webster and P. Baglioni. 2003a. Development of emulsions from biomass pyrolysis liquid and diesel and their use in engines—Part 1: emulsion production. Biomass and Bioenergy 25(1): 85–99. doi:10.1016/S0961-9534(02)00183-6.

Chiaramonti, D., M. Bonini, E. Fratini, G. Tondi, K. Gartner, A.V. Bridgwater, H.P. Grimm, I. Soldaini, A. Webster and P. Baglioni. 2003b. Development of emulsions from biomass pyrolysis liquid and diesel and their use in engines—Part 2: tests in diesel engines. Biomass and Bioenergy 25(1): 101–11. doi:10.1016/S0961-9534(02)00184-8.

Consonni, S. and E.D. Larson. 1996. Biomass-gasifier/aeroderivative gas turbine combined cycles. Part A—technologies and performance modeling. Journal of Engineering for Gas Turbines and Power.

Constantinou, D.A., J.L.G. Fierro and A.M. Efstathiou. 2009. The Phenol steam reforming reaction toward H_2 production on natural calcite. Appl. Catal. B.

Corsini, A., A. Marchegiani, F. Rispoli, F. Sciulli and P. Venturini. 2015. Vegetable oils as fuels in diesel engine. Engine performance and emissions. Energy Procedia 81: 942–949.

Couto, N., A. Rouboa, V. Silva, E. Monteiro and K. Bouziane. 2013. Influence of the biomass gasification processes on the final composition of syngas. Energy Procedia 36: 596–606.

Czernik, S. and A.V. Bridgwater. 2004. Overview of applications of biomass fast pyrolysis oil. Energy & Fuels 18(2): 590–98. doi:10.1021/ef034067u.

De Bere, L. 2000. Anaerobic digestion of solid waste: state-of-the-art. Water Science and Technology 41(3): 283–290.

DIN 51623: Fuels for Vegetable Oil-Compatible Combustion Engines. Fuel from Rapeseed Oil-Requirements and Test Methods, Berlin, February 2015, S. 1–25.

Dzene, I. and L. Slotiņa. 2013. Efficient heat use from biogas CHP plants. Case studies from biogas plants in Latvia. Environmental and Climate Technologies 1: 45–48.

Finnish Thermo Wood Association. 2003. Thermowood Handbook. April 2003.

FNR. according dena. 2016. www.fnr.de.

Funke, A. and F. Ziegler. 2010. Hydrothermal carbonization of biomass: a summary and discussion of chemical mechanisms for process engineering. Biofuels, Bioproducts and Biorefining 4(2): 160–77. doi:10.1002/bbb.198.

Gómez-Barea, A., B. Leckner, A.V. Perales S. Nilsson and D.F. Cano. 2013. Improving the performance of fluidized bed biomass/waste gasifiers for distributed electricity: a new three-stage gasification system. Applied Thermal Engineering 50(2): 1453–1462.

Hoekman, S.K., A. Broch and C. Robbins. 2011. Hydrothermal carbonization (HTC) of lignocellulosic biomass. Energy & Fuels 25(4): 1802–10. doi:10.1021/ef101745n.

Hu, B., K. Wang, L. Wu, S.-H. Yu, M. Antonietti and M.-M. Titirici. 2010. Engineering carbon Materials from the hydrothermal carbonization process of biomass. Advanced Materials (Deerfield Beach, Fla.) 22(7): 813–28. doi:10.1002/adma.200902812.

ICCT. 2013. Vegetable Oil Markets and the EU Biofuel Mandate, Washington, USA [online] http://www.theicct.org.

IEA. Task 37 Project biogasmax. Biogas as vehicle fuel—market expansion to 2020 air quality report on technological applicability.

Issariyakul, T. and A.K. Dalai. 2014. Biodiesel from vegetable oils. Renew. Sustain. Energy Rev. 31: 446–471 [online] ttp://dx.doi.org/10.1016/j.rser.2013.11.001.

Jay, D.C.O., A. Rantanen, K. Sipilä and N.O. Nylund. 1995. Wood pyrolysis oil for diesel engines. In: Proceedings of the ASME Fall Technical Conference. New York: American Society for Mechanical Engineers (ASME).

Kasper, J.M., G.B. Jasas and R.L. Trauth. 1983. Use of pyrolysis-derived fuel in a gas turbine engine. *In*: Volume 3: Coal, Biomass and Alternative Fuels; Combustion and Fuels; Oil and Gas Applications; Cycle Innovations, V003T06A016. ASME. doi:10.1115/83-GT-96.

Kaupp, A. and J.R. Goss. 1981. State-of-the-art report for small scale (to 50 kw) gas producer-engine systems (No. PB-85-102002/XAD). California Univ., Davis (USA). Dept. of Agricultural Engineering.

Khanal, S.K. 2008. Bioenergy generation from residues of biofuel industries. Anaerobic biotechnology for bioenergy production: Principles and Applications 161–188.

Kim, T.Y. and S.H. Lee. 2015. Combustion and emission characteristics of wood pyrolysis oil-butanol blended fuels in a DI diesel engine. International Journal of Automotive Technology 16(6): 903–12. doi:10.1007/s12239-015-0092-4.

Kitzler, H., C. Pfeifer and H. Hofbauer. 2011. Pressurized gasification of woody biomass—variation of parameter. Fuel Processing Technology 92(5): 908–914.

Knoef, H.A.M. 2005. Biomass Gasification Handbook. Handbook, published by BTG group pp. 189–191.

Knoef, H.A.M. 2000. A review of fixed bed gasification systems for biomass. School of Energy Studies for Agriculture, India.

Kovacs, A. 2013. Proposal for a European biomethane roadmap. European Biogas Association, Brussels, Belgium.

Kruse, A., A. Funke and M.M. Titirici. 2013. Hydrothermal conversion of biomass to fuels and energetic materials. Current Opinion in Chemical Biology 17(3). Elsevier Ltd: 515–21. doi:10.1016/j.cbpa.2013.05.004.

Le Coq, L. and A. Duga. 2012. Syngas treatment unit for small scale gasification-application to IC engine gas quality requirement. Journal of Applied Fluid Mechanics 5(1): 95–103.

Loo, S.V. and J. Koppejan. 2008. Biomass fuel properties and basic principles of biomass combustion. *In*: The Handbook of Biomass Combustion and Co-Firing. EARTHSCAN.

Lupandin, V., R. Thamburaj and A. Nikolayev. 2005. Test results of the GT2500 Gas turbine engine running on alternative fuels: bio oil, ethanol, bio diesel and heavy oil. In Proc. of the ASME TURBO EXPO 05, 2005: 421–26. ASME. doi:10.1115/GT2005-68488.

Marrero, B.P., W.R. McAuley, J.S. Sutterlin, S. Morris and E. Manahan. 2004. Fate of heavy metals and radioactive metals in gasification of sewage sludge. Elsevier, Waste Management 24: 193–198.

Mohan, D., C.U. Pittman and P.H. Steele. 2006. Pyrolysis of wood/biomass for bio-oil: a critical review. Energy & Fuels. American Chemical Society 20(3): 848–89. doi:10.1021/ef0502397.

Mosarof, M.H., M.A. Kalam, H.H. Masjuki, A.M. Ashraful, M.M. Rashed, H.K. Imdadul and I.M. Monirul. 2015. Implementation of palm biodiesel based on economic aspects, performance, emission, and wear characteristics. Energy Convers. Manag. 105: 617–629 [online] http://dx.doi.org/10.1016/j.enconman.2015.08.020.

Nussbaumer, T. 2003. Combustion and co-combustion of biomass. Energy & Fuels 17: 1510–1521.

Nussbaumer, T. 2002. Combustion and co-combustion of biomass. Proceedings of the 12th European Conference and Technology Exhibition on Biomass for Energy, Industry and Climate Protection, 17–21 June 2002, Amsterdam (NL).

Panichelli, L. and E. Gnansounou. 2015. Impact of agricultural-based biofuel production on greenhouse gas emissions from land-use change: key modelling choices. Renew. Sustain. Energy Rev. 42: 344–360 [online] http://dx.doi.org/10.1016/j.rser.2014.10.026.

Pasyniuk, P., W. Golimowski, L. Sciences and L. Sciences. 2013. The study of rapeseed oil production technology of reduced macronutrients content as an engine fuel. J. Res. Appl. Agric. Eng. 58(1): 143–146.

Pavlovič, I., Ž. Knez and M. Škerget. 2013. Hydrothermal reactions of agricultural and food processing wastes in sub- and supercritical water: a review of fundamentals, mechanisms, and state of research. Journal of Agricultural and Food Chemistry 61(34): 8003–25. doi:10.1021/jf401008a.

Prussi, M., D. Chiaramonti, G. Riccio, F. Martelli and L. Pari. 2012. Straight vegetable oil use in micro-gas turbines: system adaptation and testing. Appl. Energy 89(1): 287–295 [online] http://linkinghub.elsevier.com/retrieve/pii/S0306261911004764 (accessed 27 April 2015).

Prussi, M., G. Riccio, D. Chiaramonti and F. Martelli. 2008. Evaluation of a micro gas turbine fed by blends of biomass producer gas and natural gas. In ASME Turbo Expo 2008: Power for Land, Sea, and Air. January (pp. 595–604). American Society of Mechanical Engineers.

Rapagna, S., N. Jand, A. Kiennemann and P.U. Foscolo. 2000. Steam-gasification of biomass in a fluidised-bed of olivine particles. Biomass and Bioenergy 19(3): 187–197.

Ryckebosch, E., M. Drouillon and H. Vervaeren. 2011. Techniques for transformation of biogas to biomethane. Biomass and Bioenergy 35(5): 1633–1645.

Sevilla, M. and A.B. Fuertes. 2009. The production of carbon materials by hydrothermal carbonization of cellulose. Carbon 47(9): 2281–89. doi:10.1016/j.carbon.2009.04.026.

Shaddix, C.R. and D.R. Hardesty. 1999. Combustion Properties of Biomass Flash Pyrolysis Oils. Sandia Report SAND99–8238.

Shihadeh, A.L. 1998. Rural Electrification from Local Resources: Biomass Pyrolysis Oil Combustion in a Direct Injection Diesel Engine. Massachusetts Institute of Technology. http://hdl.handle.net/1721.1/43601.

Sidibé, S.S., J. Blin, G. Vaitilingom and Y. Azoumah. 2010. Use of crude filtered vegetable oil as a fuel in diesel engines state of the art: literature review. Renew. Sustain. Energy Rev. 14(9): 2748–2759 [online] http://linkinghub.elsevier.com/retrieve/pii/S1364032110001656 (accessed 8 April 2015).

Siedlecki, M., W. De Jong and A.H. Verkooijen. 2011. Fluidized bed gasification as a mature and reliable technology for the production of bio-syngas and applied in the production of liquid transportation fuels—a review. Energies 4(3): 389–434.

Stemann, J., A. Putschew and F. Ziegler. 2013. Hydrothermal carbonization: process water characterization and effects of water recirculation. Bioresource Technology 143. Elsevier Ltd: 139–46. doi:10.1016/j.biortech.2013.05.098.

Strenziok, R., U. Hansen and H. Kunster. 2001. Progress in Thermochemical Biomass Conversion. Edited by A.V. Bridgwater. Progress in Thermochemical Biomass Conversion. Oxford, UK: Blackwell Science Ltd. doi:10.1002/9780470694954.

Van Loo, S. and J. Koppejan (eds.). Handbook of Biomass Combustion and Co-Firing, Twente University Press, Enschede 2002, ISBN 9036517737.

Venderbosch, R.H. and M. van Helden. 2001. Diesel engines on bio-oil: A Review.

Venderbosch, R.H. and W. Prins. 2010. Fast pyrolysis technology development. Biofuels, Bioproducts and Biorefining 4(2): 178–208, Copyright © 2010 Society of Chemical Industry and John Wiley & Sons, Ltd.

Web sites

http://european-biogas.eu/2015/12/16/biogasreport2015.

CHAPTER 11

Energy Plantations Value-Added Options

Gheorghe Lazaroiu,[1,*] *Lucian Mihaescu,*[2]
Gabriel Negreanu[2] *and Ionel Pisa*[2]

1. Introduction

In Europe, there are several financial support actions for the use of renewable fuels, such as oilseed crops for biodiesel in Germany and France, successive plantings of young forests with low growth periods for the production of heat and electricity in the United Kingdom, or energy willow plantations for the production of heat and electricity in Sweden. Denmark, Finland, Sweden and England are the only countries in the European Union where an increase in the use of biomass for the production of electricity is noticed.

According to the data presented in Romania's Energy Strategy (RES) 2016–2030 (Romania Ministry of Energy 2016) concerning the management of renewable resources, it resulted that:

- Romania fulfilled its European commitment for 2020 which implies increase of RES's share up to 24% of the gross final energy consumption, achieving the level of 26.3% for this indicator in 2015, at a not at all negligible cost for the end user (Commission of the European Communities 2006, 1997, European Renewable Energy Council).

[1] Department of Energy Production and Use, University Politehnica of Bucharest, Splaiul Independentei 313, 060042, Bucharest, Romania.
[2] Department of Thermotechnics, Engines, Thermal and Frigorific Equipments, University Politehnica of Bucharest, Splaiul Independentei 313, 060042, Bucharest, Romania.
* Corresponding author: glazaroiu@yahoo.com

- Three factors will determine the share of RES in Romania in 2030: (1) the capital cost for the funding of RES, (2) evolution of the consumption of biomass for heating and (3) reaching the biofuel target in 2020 in transportation and development of electro-mobility and electric heating (Stern 2007, Directive 2001/77/EC).
- Between 2017 and 2030, life quality for residents from rural areas and the improvement of the forest fund management will be among the national priorities.
- For 2030, Romania aims to equitably contribute in reaching the common European targets meaning increasing the share of RES in the gross final energy consumption up to 27%, increasing energy efficiency by 27% and reducing the emissions of greenhouse gases by 40% comparison to 1990.

In order to produce thermal energy with the help of boilers, in Romania, besides woody waste, the use of cereal straw and other agricultural waste, such as: corn cobs (cores), corn husks and stalks, soybean stalks, pea stalks, sunflower stalks, grapevine chords, etc., shall also be pursued.

Out of a total area of 23.8 million ha, Romania's agricultural area is almost 14.9 million hectares (approximately 62% from total). The European Union average is 41%. Arable land accounts for approximately 62%, permanent crops 3% and permanent pastures 33% of the agricultural area. In addition, 28% of Romania's area is wooded.

Romania's wooded surface is about 6.6 million ha, accounting about 27% of the country's area (Romania occupies the 13th rank in Europe by forest volume and the 26th rank for its territory share). In the EU, the forest represents about 180 million hectares, covering over 42% of the area (since 1990, the area has increased by 0.4% through afforestation actions) (Forestry resources in EU 2014).

According to data compiled by the ICAS (Institute of Forestry Research and Arrangements) the current annual potential for harvesting woody biomass in Romania is of 18.5 mill. m^3 and it is estimated to reach about 20 mill. m^3 in subsequent years.

In Romania, the renewable energy is currently produced in the field of hydro power, solar, wind, and biomass. The share of renewable energy in recent years has ranged between 25–45% of the gross domestic electricity consumption considering the amount of water available annually and the wind potential. The share of solar energy and biomass is still low, but is growing (Nistorescu et al. 2011).

Solid biomass is represented in particular by wood waste, agricultural waste, energy willow, household waste and some energy crops.

The energy potential of biomass is about 7,594 thousand tep/year, divided into the following fuel categories: residues from forestry work and firewood (1,175 thousand tep), wood waste (487 thousand tep), agricultural wastes resulted from grains (4,799 thousand tep), biogas (588 thousand tep) (Leca 2011).

Other woody wastes such as: tree bark, branches, sawdust and other processed wood wastes that are used with the same purpose.

Another definite source of renewable fuel are corn kernels due to their high heating value and bulk density. However, the use of this fuel determines the appearance of an ethical problem (Leca and Muşatescu 2010).

Currently, in Romania, vegetable oils (canola, sunflower, soybean, corn) are conducive to be used as liquid fuel. Newer researches aim to extend the use of these fuels (Mihaescu et al. 2012a).

Research on co-combustion of corn kernels and coal has been conducted. For a total area of approximately 2,665,000 thousand hectares, where straw cereal crops are planted (wheat, rye, barley, two-row barley, oats), an average of 1.5–2.0 tons of straw/ha, or about 4,200,000 tons of straw are obtained and this represent a great energy potential. Of this amount, a ratio of 25–34% of the collected and stored straw is used intrinsically in agriculture (feed, bedding, etc.) (Romanian Statistical Yearbook 2007).

Growers of biomass have the following benefits (Mihaescu et al. 2012a, Ciocea 2010):

- Expenses reduction pertaining the warehouses space due to its multiple use (agricultural and energy);
- Environmental pollution reduction and thereby, avoiding the sanitation costs of the adjacent area;
- Gaining income through the sale of biomass as eco fuel.

Other woody wastes such as: tree bark, branches, sawdust and other wood processing scraps are used for the same purpose.

On a surface of approximately 3,000,000 hectares where corn is cultivated, the production of stalks is 5.8 t/ha, or approximately 18,000,000 tons. For energy purposes, we can add 0.7 t/ha or 2,130,000 tons, production of corn cobs. In the future, a part of the production of corn kernels intended for energy production may be considered as well (Mihaescu et al. 2013a, Mihaescu et al. 2012b).

On the surface of about 870,000 ha where sunflower is cultivated, the resulted production of stalks is about 1.8 t/ha, which represents about 1,600,000 tons. Also, remnants of grapevine cords on a surface of 250,000 ha generate about 220,000 energy fuel tons/year, to which can be added the pruning from 260,000 orchards hectares, which generate about 240,000 energy fuel tons/year.

The arguments that favors the use of biomass energy are related to both environmental and a socio-economic contribution by increasing employment and stabilizing the labor force in some agricultural areas.

An important feature is the need to harness the biomass energy in the vicinity of its places of origin, imposed by the need to control the level of heat and electricity costs by reducing as much as possible the transport, distribution and storage costs.

The distribution of the agricultural biomass production as an energy resource per regions (Romanian Statistical Yearbook 2007), exhibits the highest values for:

- the North-East development region, 14.63%;
- the South-East development region, 23.71%;
- the South development region, 19.38%;
- the South-West development region, 10.85%;
- the West development region, 14.03%;

Obviously, in Romania, the purpose of biomass development is to utilize energy in order to obtain thermal and electrical energy. In the following, it is studied, which are the most used technologies for the recovery of biomass energy.

2. Biomass Energy Utilization Technologies

The main types of combustion plants for biomass are:

- Fixed bed combustion plants:
    - with fixed grate
    - with mobile-rolling grate
        - with pushback
        - with sideways push
    - with bottom feeding for pellet-type fuel
- Fluidized bed combustion plants
    - stationary
    - recirculating
- Air current (or pulverized state) combustion plants

Grate stokers, regardless of their construction and actuation, have a power of less than 50 thermal MW and can be used for burning various biomass sizes (but not less than 1 mm) with relatively high moisture and ash content.

Wood-based or coal-based biomass blends can also be burned, but not thereof blends with straw due to a significant difference between their burning characteristics, ash melting temperature and its decisive adherence.

Burning in stages or gasifying represents an indirect combustion technology, the predominant one is the secondary reaction of carbon dioxide reduction on an embers layer, the final combustion is exhibiting the removal of ash deposits.

Except in some cases where direct combustion is advantageous, as a rule, the gross biomass requires transformation into solid, liquid or gaseous fuels that may be used more easily and more efficiently for energy production. This conversion is accomplished by mechanical, thermal and biological processes. Mechanical processes do not represent a transformation of the nature of the fuel, but of its state. The main mechanical processes for the transformation of renewable solid fuels include:

- waste sorting and compacting;
- chopping of straw and cobs;
- processing chopped residues pressing in bales, briquettes and pellets;
- pressing of oilseeds.

Depending on the ratio of the amount of oxygen that enters into the reaction and the stoichiometric one, called the equivalent ratio, the characteristics of the gas produced can be determined. For a ratio of less than 0.1, the process is called pyrolysis and only a small fraction of the chemical energy of the biomass is found in the produced gas, the rest is found in the residual carbon and bio-oil. If the ratio is between 0.2 and 0.4, the gasification process is necessary.

The gas produced by gasification is called generator gas and has a low heating value (contains CO, CO_2, H_2, CH_4, and small quantities of hydrocarbons, water and nitrogen (if air is used as the oxidizing agent)). Various pollutants such as small particles of ash, coke, tar and oils also remain.

Solid biomass gasification with air leads to a gas with low heating value (4,000–7,000 kJ/m³N), which can be used effectively in combustion for energy purposes.

Gasification with oxygen produces a gas with higher heating value, which can be used in chemical synthesis. Generator gas can also result through pyrolytic or steam gasification.

Synthesis gas conversion efficiency in bio-methanol is about 85%, and the efficiency of the whole process (starting with biomass conversion) is 40–45%.

Taking into account the characteristics and production of biomass in Romania correlated with the energy recovery technologies, we focused our research efforts on two directions:

- the use of agricultural biomass in the form of cereal straw;
- the use of rapid growth biomass in sorghum and energy yeast.

The results of the researches from the two areas are presented in subchapters 3 and 4.

3. Achievements in the Field of Energy Recovery of Cereal Straw

Straw have a high heating value (14,000–15,000 kJ/kg), similar to wood, low humidity ($W^i_t = 11$–14.5%), and an ash quantity somewhat higher than wood ($A^i = 4.5$–6.5%).

As ash contains a relatively high proportion of phosphorus and potassium, it follows that in the end this may represent a light fertilizer.

The main problem that prevented an explosion in the straw energy use is the rough and adhering character of ash for the combustion area, which leads to definite technological disadvantages. This requires frequent and difficult cleanup interventions on the stoker, as well as the convective part.

In general, two combustion systems have been developed:

- burning straw bales;
- burning straw as briquettes or hash.

Burning straw bales is a discontinuous feed combustion technology, with special issues regarding the management of energy conversion process of the working fluid. The technology was developed in Western Europe and has not been adopted in our country, which has opted for continuous feed variants. Straw briquetting represents the most advanced technology developed in Romania.

To eliminate issues pertaining to the hardness of ash in the stoker, plants that maintain the temperature at the end of the stoker below the critical temperature of 850°C (in which case the ash remains friable and easily removed mechanically) have been developed in Romania.

In the period of 2010–2012, the company E. MORĂRIT HUȘI has built 80 and 100 kW boilers with power cooled stokers, rectangular water chambers located at the boiler's terminal area has been carried out, which represent an extension of the water volume framing the convective system. The fuel consists of straw briquettes of relatively large size (Mihaescu et al. 2012b, Mihaescu et al. 2013a, Mihaescu et al. 2013b, Mihaescu et al. 2012c).

Laboratory tests revealed the following average energy characteristics for straw from the Moldova North region (Mihaescu et al. 2012b):

- Reported elemental analysis:

  $C^i = 41{,}4\%$, $H^i = 6{,}3\%$, $N^i = 0{,}61\%$, $O^i = 34{,}1\%$, $W^i_t = 11{,}5\%$, $A^i = 6{,}09\%$
- Low heating value $Q^i_i = 16500$ kJ/kg
- Ash composition: $SiO_2 = 55{-}60\%$, $Al_2O_3 = 4{-}5{,}5\%$, $Fe_2O_3 = 0{,}9{-}1{,}3\%$, $MgO = 2{,}6{-}3\%$, $CaO = 3{,}1{-}3{,}9\%$, $Zn = 12{-}16$ ppm, $Pb = 9{,}1{-}10{,}2$ ppm.

  Briquettes were obtained by cold pressing.

The tested briquettes obtained at E. MORĂRIT Huşi, have the following dimensions (Mihaescu et al. 2013b):

- length –130 mm
- diameter –85 mm
- volume –0.03 m³

From the point of view of the pressing process, as well as those relating to combustion, briquettes with a central hole are recommended.

At the periphery, the briquettes were made with round or polygonal surfaces. The weight of wheat straw briquettes was 0.55–0.63 kg, their density being very high, up to 930–1000 kg/m³.

It was found that the dense biomass, between 255–250°C, leads to a greater density than that at lower temperatures.

Collecting the straw from the field leads to the formation of bales of varying sizes. In order to carry out straw briquetting, further shredding of these bales is necessary, up to a size of 40–60 mm. Figure 1 shows a mobile chopping plant with a rate of 1.2–1.4 t/h with a power of 18 kW, used at the briquetting station from which the was collected.

The briquette dimensions, have diameters ranging from 60 to 120 mm and a height of 60–270 mm (Mihaescu et al. 2012d).

The briquetting technology is shown in Fig. 2, the plant has a production of 50–60 kg/h for each line.

The briquettes produced in the piston pressure (with the pressure range of 60–200 bars) have a density of 900–1000 kg/m³. Briquette size is 80–100/80–160 mm (length/diameter) (Toader 2013).

In order to achieve a good compaction of the straw hash, each line is equipped with an electric heating system with a maximum power of 5 kW.

Figure 2 shows the fitting of the electric heater at the beginning of the briquette discharge zone. The optimum heating temperature has been determined and has the value of 250°C (Toader 2014).

The energy consumption for the operation of the piston was 2 kWh/line. For a production of 60 kg, the specific energy consumption was 0.033 kWh/kg (Toader 2015).

Figure 3 shows a sketch of the 80–100 kW boiler, highlighting the stepped grate stoker, with gradual size water chambers located at the final combustion area, surrounding the convective system consisting of horizontal pipes (Barta et al. 2011).

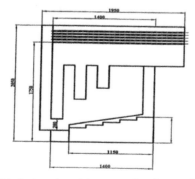

Fig. 1: Mobile chopping system for straw bales.

Fig. 2: Briquetting station heating plant.

Fig. 3: 100 kW boiler, with the location of the water chambers at the end of the stoker.

In 2013, based on the experience gained, the range of boilers with flame tube stokers with outputs of up to 450 kW was launched. The boilers were built by the E. Morărit Company from Huşi.

Figure 4 shows an overview of the 150 kW boiler plant with a flame tube, the positioning of the convective heat exchanger is above the cylindrical mantle (Toader 2015).

The fuel is consisted out of straw briquettes that have identical sizes to those from previously manufactured boilers. The boiler has fully automatic operation, in terms of fuel feeding and ash disposal, with continuous adjustment of the thermal load, with air excess, in order to achieve the best possible economic operation.

The boiler was a superior achievement to those with water chambers. The combustion in the flame tube is controlled through a pusher system attached to the grate. The grate type is flat stage, the first stage is fixed, the second is mobile with a forward push and the third is also fixed. Fixed stages are cooled with water from the boiler mantle, in order to increase the cooling in the maximum temperature zone. The actuation of the mobile stage is done with a grate-track rack system. The fuel pusher device is operated by the second gate-track rack, independently of the grate operation and it intends to advance the fuel bed concomitant with the burning speed in order to achieve full combustion at the end of the flame tube. Figure 5 shows the combustion bed at the end of the flame tube.

The vertical convective heat exchanger is made out of smoke stacks disposed in a three-way flue gas system, the vertical construction is allowing the recovery of ash deposits. A metal coil is fitted inside the smoke stacks with the purpose to increase the heat exchange by inducing a swirling motion in the flue gas flow. The same coil, by alternative axial displacement or by rotation removes the ash deposits.

Figure 6 shows the location of the coil in the convective system pipes.

The boiler has in nominal regime an efficiency of 91%. This efficiency is achieved for an excess of air of 1.6 and a stack temperature of 125°C. For an effective combustion and a control of polluting emission, the boiler has two air circuits:

- primary air circuit (blown under the grate bars);
- secondary air circuit (blown above the grate bars).

The secondary air circuit is designed to complete the combustion and maintain the emission of NO_x and CO below the allowable limits.

For a 11% oxygen content in the flue gas, pollutant emissions were:

➢ CO = 1300–2500 mg/m^3_N
➢ NO_x = 28–125 mg/m^3_N
➢ powders below 30 mg/m^3_N

Fig. 4: Overview of the 150 kW flame tube boiler.

Fig. 5: Aspect of burning in the flame tube.

Based on data obtained from the constructed plants, thermal power expansion in a 450 kW boiler was obtained. The Boiler is equipped with a combustion plant for a wide range of renewable fuels.

Figure 7 shows an overview of the boiler (Mihaescu et al. 2011).

The considered fuel is straw briquettes from Moldova area, with the following characteristics:

Elementary analysis: $C^i = 42.1\%$, $H^i = 6.6\%$, $N^i = 0.6\%$, $O^i = 33.4\%$, $W^i_t = 11.9\%$, $A^i = 5.4\%$.

Low heating value: 16.980 kJ/kg.

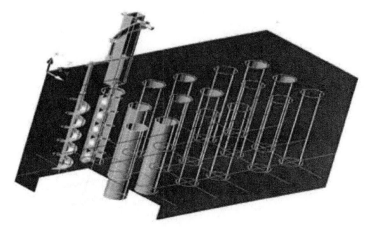

Fig. 6: Ash cleaner.

Fig. 7: Overview of the 450 kW boiler body.

Average composition of ash: $SiO_2 = 56.9\%$, $Al_2O_3 = 5.4\%$, $MgO = 3.1\%$, $CaO = 4.1\%$.

The boiler can also use (Negreanu et al. 2014):

- wood waste with a maximum humidity of 20%;
- sunflower stalk hash;
- corn stalk hash;
- briquettes made from mixtures of agricultural waste hash.

The grate is mounted inside the flame tube with an active volume of 0.5 m³, which includes the following elements mounted successively:

- a 270 mm fixed width bar, water-cooled;
- a row of mobile cast iron bars with a length of 270 mm and width of 70 mm;
- a 56 mm fixed width bar, water-cooled;
- a row of mobile cast iron bars;
- a 56 mm fixed width bar, water-cooled;
- a row of mobile cast iron bars.

Fixed and mobile bars have the dimensions shown above. Fixed bars have lateral clearances for primary air admission. Figure 8 shows the location of the grate in the flame tube.

Mobile bars are operated with a grate-track rack system. Another grate-track rack actuates the fuel pusher that is located after the grate as shown in Fig. 8.

Based on data obtained from the operating parameters, the pusher's stroke is adjusted to control the combustion at the end of the flame tube.

The convective heat exchanger is rectangular, with three paths for the combustion gases and is located above the boiler mantle, as shown in Fig. 9. The figure also highlights the actuating system which eliminates and cleans the ash deposit.

For thermal powers greater than 1,000 kW, suspended biomass burning is a future solution.

The analytical study of the dynamics of the oxygen consumption for the briquette bed recommends the use of a maximum flow rate of 2 m/s for the air that is traversing the fuel bed. Thus, for an air flow rate of 2 m/s, it is recommended that the height limit of the briquette bed to be 200 mm and for the 1 m/s air flow rate a height of 100 mm (Mihaescu et al. 2011).

The average mass loading q_m of the briquette bed did not exceed 400 kg/(m²/h). For a briquette density ρ_c of 900 kg/m³ and a porosity, $\varepsilon = 0.45$–0.55, the relationship for computing the fuel bed height h, is:

$$h = \varepsilon \frac{q_m}{\rho_c} \left[\frac{m}{h} \right]$$

Linking the experimental data required a numerical application with reference to the combustion dynamics of maxi-briquettes of cereal straw on mobile grates.

Fig. 8: Grate assembly with fixed and mobile bars in the flame tube.

Fig. 9: Overview of the convective heat exchanger with the ash deposit cleaning installation.

The carbon dioxide formation speed is characterized by the equation:

$$C_{CO_2} = C_{O_2}^O e^{-a_4 z}$$

with $C_{O_2}^O$ - initial oxygen concentration and the kinetic coefficient constant for the burning reaction:

$$\alpha_4 = \frac{15}{ru}\left(3,9e^{-\frac{40000}{RT}}\right),$$

r is the briquette's equivalent geometrical dimension in m; $\frac{kJ}{m_N^3}$; T – temperature, in K; u – air flow rate through fuel bed.

If the general combustion equation expressed in the following form is used:

$$\frac{CO_2}{CO_2^{max}} + \frac{CO}{CO^{max}} + \frac{O_2}{O_2^{max}} = 1$$

where: $CO_2^{max} = \dfrac{21}{1+\beta}$; $CO^{max} = \dfrac{21}{0,605+\beta}$; $O_2^{max} = 21$,

the dynamics of formation of carbon monoxide CO can be defined.

Depending on the fuel's elemental analysis, the energy characteristic β can be calculated using the relation:

$$\beta = 2,37(H^i - 0,125O^i + 0,038N^i)/C^i$$

The following relationship result. It expresses the concentration of carbon monoxide at the end of the combustion bed (Mihaescu et al. 2011):

$$C_{CO} = \frac{C^O_{O_2} - C_{O_2} - (1+\beta)C_{CO_2}}{0,605 + \beta}$$

The obtained relation is very important because, in general, the fuel combustion using a bed is characterized by high carbon monoxide emissions, thus it enables us to highlight the amount of the pollutant emissions depending on the dynamics of the combustion, expressed by the amount of carbon dioxide resulted at the end of the fuel bed.

Figure 10 shows the variation of the O_2 concentration in the height of the fuel bed, based on the calculations carried out using the above equation. The calculation temperature was $T = 1100$ K, compulsory considering reduced temperatures for the combustion in order to avoid slagging.

Numerical applications were made for straw briquettes with elementally analysis:

$C^i = 42\%$, $H^i = 6.4\%$, $Ni = 0.64$, $O^i = 34\%$, $W^i_t = 11.99\%$, $A^i = 4.97\%$

The obtained results are:

$\beta = 0,12$

$CO^{2\,max} = 18,75\%$

$CO^{max} = 28,9\%$

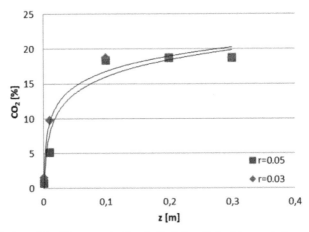

Fig. 10: Variation of the CO_2 concentration in the briquette bed for an air flow rate of 1 m/s.

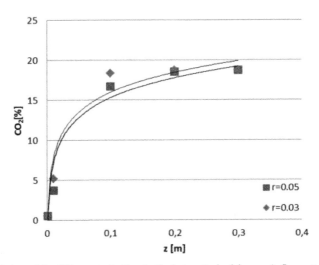

Fig. 11: Variation of the CO_2 concentration in the briquette bed for an air flow rate of 2 m/s.

Figures 10 and 11 show the concentration variation of the combustion's final product, CO_2 in the fuel bed (Mihaescu et al. 2011).

4. Economic and Technical Aspects of the Energy Use of Fast-Growing Sorghum and Willow Crops

As wood is less accessible for energy uses, the problem of finding alternative viable solutions is more frequent. From this point of view, efforts are being made today in Romania in order to implement plant crops with massive growth. Energy willow and sorghum are standing out among them.

In the future, for sorghum, it will have to be demonstrated if its use for energy production shall become more cost-effective than its use in the food industry (particularly in sugar manufacture). Until now, researches have shown that the direct burning of sorghum is not a solution for energy production.

Research related to energy willow cultivation has been carried out for over 50 years in Sweden and other European countries. The results have determined that several species (18 varieties) have a UPOV registration (Community Office for the Protection Plant Varieties) and to be grown on several thousand hectares in almost all European countries, providing a solution to ensure an environmentally friendly energy production.

Energy willow stands out due to several compelling properties, such as:

- rapid growth (3–3.5 cm/day);
- high heating value (similar to wood);
- remarkable adaptability to climate conditions;
- important harvest, constant over time; 25–30 years (25–30 t/ha/yr);
- except for the first year, it does not require special agricultural work;
- harvesting is done in winter (after the leaves fall off), when both manpower and machinery and transport means are in less demand;
- willow hash does not need covered storage, and can be used for a long time, without deteriorating.

The economic effects depend on the size of the cultivated areas.

In a hypothesis in which Romania would cultivate energy willow on high humidity land (500,000 ha), the effects over a period of 10 years would sum up to:

- costs for realization of plantations: 750,000,000 EUR
- biomass obtained: 50,000,000 t/in the first 10 years
- biomass produced during the plantation (25–30 years): 300,000,000 t
- the energy that can be produced from this biomass: 1,046,000,000 MWh
- the creation of plantations in question requires the contribution of about 1,500 employees, annually, for 10 years.

Energy willow is a plant for which, in the first year of cultivation needs the soil to be prepared properly. After planting the soil has to be cared of, so that the plantation grows cleanly.

Fig. 12: Harvesting operations in Poieni commune, Covasna County.

In the first year, the harvest is about 10% of the mature plantation harvest (a plantation is considered mature in year three).

After the first year, energy willow plantations require insignificant interventions (administration of fertilizers, harvesting) for 25–30 years.

In Romania there are currently over 80 companies or individuals who are creating energy willow plantations, but due to the lack of subsidies, parcels are limited today to about 16000 hectares, having rather an avant-garde character.

The harvesting device includes a pitchfork designed to lay down the willow and a capture system which comprises a horizontal and vertical knife. As shown in Fig. 12 (Mihaescu et al. 2012d, Mihaescu et al. 2013b), the harvesting device is fitted to a classical agricultural combine with an engine power in relation to that consumed by the harvesting device.

Harvesting yields a hash with a maximum size of 30 mm, which is loaded into a trailer that accompanies the combine.

The harvest period is, as a rule, towards the end of winter, when growth is slowed, and freezing temperatures are drying the wood.

Figure 13, shows open and closed storage in Poieni commune – Covasna.

There is already an extensive experience in the area of energy recovery in ERPEK boilers, built in the Covasna County.

The ERPEK company has made boilers in the range of 24–1000 thermal kW (Mihaescu et al. 2012d, Mihaescu et al. 2013b).

The 850 kW boilers, as shown in Fig. 14, have the mobile burning plant mounted to a set of greenhouses.

Fig. 13: Aspects of the open and closed storage work in the base in Poieni commune—Covasna.

Fig. 14: 850 kW boiler stoker.

The humidity that has yielded the best results was in the range of 14–22%.

The smallest boiler, intended to exploit biomass hash, with the output heat of 30 kW, includes a fluidized bed combustion plant (in air current). The boiler is fully automated, and is shown in Fig. 15.

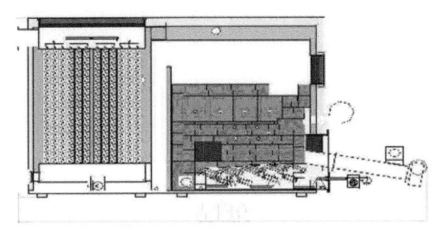

Fig. 15: 30 kW boiler construction.

Figure 16 shows the appearance of a flame obtained during energy willow hash test trials in this boiler.

The 400 and 500 kW boilers manufactured by the ERPEK Romania Company are intended for individual houses, apartment blocks, greenhouses, workshops or companies and they are operating with wood hash, including energy willow hash (Barta et al. 2011).

The stoker's resistance structure is built from sandwich steel sheet, with a high-volume combustion chamber.

The stoker where the mobile grate is located is clad inside in a refracting mantle that is resistant to high temperature. This is a solution enables to achieve a high temperature in the stoker in order to perform optimum combustion.

The convective heat exchanger of the ERPEK Boilers is vertical or horizontal, with two or three smoke passages for obtaining a maximum efficiency through an optimum heat exchange.

The grate assembly can be easily detached from the stoker, in order to maintain and check the actuating mechanism. The grate's stroke and motion time are adjustable depending on fuel quality in order to achieve an optimum fuel advance flow rate, thus admitting a relatively wide variation of the granulation and humidity limits. The stoker is of vaulted construction, made from firebrick, in order to increase the heat accumulation that is necessary for the fuel ignition phase.

Secondary, air is introduced into the combustion chamber in order to ensure the completion of combustion, and to control the NOx emissions.

Fig. 16: The appearance of the flame in the 35 kW boiler stoker.

Figure 17 (Mihaescu et al. 2012d, Mihaescu et al. 2013b), shows the wood storage system in transportable block containers. The boiler has two such containers. Replacing the filling container can be performed with the boiler in operation.

Pollutant emissions have been in accordance with European environmental regulation, reaching the representative values of:

- CO emission ranged between the limits of:
  CO = 10–98 mg/m^3, for O$_2$ = 11%
- Emission of NO$_x$ ranged within the limits of:
  NO$_x$ = 40–120 mg/m^3 for O$_2$ = 11%
- The emission of dust:
  dust < 47 mg/m^3

The fabrication of a 1 t/h boiler, in the version of those for energy willow hash intense burning is a challenge for the PIFATI SA company.

Fig. 17: View of fuel storage containers at a 500 kW boiler.

Currently, in the world, there are boilers destined for burning wood hash with grate combustion plants, with special construction, but for much higher flow rates than 1 t/h. In order to be able to reduce the boiler's flow, was considered (http:// www.energysavingcommunity.co.uk/coppiced-willow-for-energy.html, http://www.folkecenter.net/gb/rd/biogas/biomass-energy-crops/energy_willow/, http://www.heganbiomass.co.uk/).

This type of boiler can be manufactured in Romania, and will allow for the expansion of the use of energy willow in order to produce an electrical power of 200–250 kW.

In order to increase the thermal power of recovery plants, possibly with co-generation, it is necessary to use chopped solid biomass. The burning process can be in an air current that is similar to pulverized coal burning (Mihaescu et al. 2012d).

In terms of achievements we note the sawdust or wood hash burner, tested on the 2 MW pilot boiler at UPB, the plant is shown in Fig. 18.

This plant is operating in good conditions, for a heat output of 750–850 kW. Elementary calculation analysis for solid biomass was:

$C^i = 45.9\%$, $H^i = 3.5\%$, $O^i = 30.2\%$, $S_c^i = 0\%$, $N^i = 1.0\%$, $A^i = 4.4\%$, $W_t^i = 15\%$,

Low Heating Value: 15,500 kJ/kg.

For a better correlation between the ignition capacity and control of NO_x emission, the burner carries out combustion in stages. Swirling are used with two concentric swirling air channels. The biomass was centrally

Fig. 18: View of the front of the pilot boiler, with the location of the woody biomass burner.

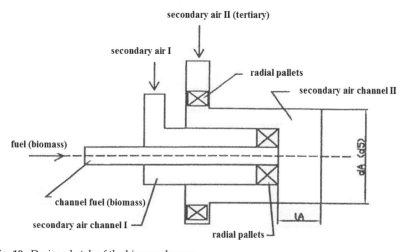

Fig. 19: Design sketch of the biomass burner.

introduced, along with the primary one, as can be seen in Fig. 19 (Mihaescu et al. 2012d).

A ratio of 0.5 to 0.8 is proposed for the primary and secondary air flows. Excess air at the burner it is necessary but within the limits of $\lambda = 1.2$–1.25.

Technical indicators achieved:

1. Transport capacity:

   Mass concentration of sawdust in the transport air

   $c = 0{,}24 \text{ kg}/\text{m}^3$ (0,19 kg/kg)

2. The carbon monoxide emission was: $CO = 15{,}9\text{--}34$ ppm.

   The resulting value was clearly lower than those of bed burning technologies.

3. Emission of NO_x ranged within the limits of: $NO_x = 150\text{--}155$ ppm.

Several burners coupling allows the fabrication of steam or hot water boilers with high thermal outputs, of up to tens of MW.

Conclusions

With a very high agricultural potential, around 10 million hectares of arable land, Romania stands out through the existence of an effort to use both a wide range of agricultural waste as well as rapid growing energy crops, for energy purposes.

Achievements in the field of the use of cereal straw for energy, in the form of briquettes and pellets, as well as that of energy willow, are noticed. To burn these unconventional fuels, specific plants have been developed.

The straw combustion plants have had to meet a tough challenge, represented by highly adhesive ash deposits at the end of the furnace. The problem was resolved through total combustion control, which has to occur at low temperatures.

Acknowledgement

This work was supported in part by a grant of the Romanian Ministry of Research and Innovation, CCCDI – UEFISCDI, project number PN-III-P1-1.2-PCCDI-2017-0404/31PCCD/2018, within PNCDI III.

References

Barta, S., I. Pîşă and L. Mihăescu. 2011. ERPEK company conception for boilers with wood biomass automatic feeding system. The 33-rd International Symposium of the Section IV of CIGR, Bucharest, June 2011.

Ciocea, Gh. 2010. Contributions to Optimization of Biomass Combustion Process Parameters. Univesty Dunarea de Jos, Galati, PhD thesis.

Commission of the European Communities, COM. 1997. 599 final: White Paper: Energy for the future—renewable sources of energy. European Communities, Brussels, 1997.

Commission of the European Communities, COM. 2006. 105 final: Green Paper—A European strategy for sustainable, competitive and secure energy. European Communities, Brussels, 2006.

Directive 2001/77/EC of the European Parliament and of the Council of 27 September 2001 on the promotion of electricity produced from renewable energy sources in the internal electricity market.

European Renewable Energy Council: Renewable Energy Target for Europe – 20% by 2020.

http://www.energysavingcommunity.co.uk/coppiced-willow-for-energy.html.

http://www.folkecenter.net/gb/rd/biogas/biomass-energy-crops/energy_willow/.

http://www.heganbiomass.co.uk/.

Leca, A. and V. Muşatescu. 2010. Energy-Environment Strategies and Policies in Romania. Academy of Technical Sciences of Romania, General Association of Engineers in Romania, AGIR Publishing House, Bucharest.

Leca, A. 2011. Considerations on the Energy Strategy of Romania. XXXVth National Conference on Thermo-Energy and Heating, Brasov.

Mihăescu, L., T. Prisecaru, M.E. Georgescu, G. Lazaroiu, I. Oprea, I. Pisa, G. Negreanu, E. Pop and V. Berbece. 2011. Construction and testing of a 600 kW burner for sawdust in suspension, revista Termotehnica, year XV, no. 2/2011, Ed. AGIR, ISSN 22471871, pp. 69–71.

Mihăescu, L., G. Negreanu and I. Oprea. 2012a. On social ethics in the use of renewable energy resources. 1st International Conference of Thermal Equipment, Renewable Energy and Rural Development-TE-RE-RD 2012, Bucharest, ISSN 1843-3359, pp. 31–37.

Mihăescu, L., G. Lazaroiu, T. Prisecaru, I. Pisa, E. Pop, V. Berbece and G.P. Negreanu. 2012b. Problems regarding the endurance of boilers when using straw from North-Eastern region of Romania. Proceedings of the 2nd International Conference on Environment, Economics, Energy, Devices, Systems, Communications, Computers, Mathematics (EDSCM'12), Saint Malo & Mont Saint-Michel, France, April 2–4, 2012, Published by WSEAS Press, ISBN: 978-1-61804-082-4, pp. 152–157.

Mihăescu, L., T. Prisecaru, M.E. Georgescu, G. Lăzăroiu, I. Oprea, I. Pişă, G. Negreanu, E. Pop and V. Berbece. 2012c. Construction and testing of a 600 kW burner for sawdust in suspension. Volume du Colloque Francophone sur l'Energie Environnement Economie et Thermodynamique COFRET'12, Université Technique de Sofia, pp. 120–125, ISBN 978-619-460-008-3, Sozopol, Bulgarie, 11–13 Juin 2012.

Mihăescu, L., T. Prisecaru, E. Enache, Gh. Lazaroiu, I. Pisa, G. Negreanu, V. Berbece and E. Pop. 2012d. Boilers made by E. Morarit—Romania for cereals straw briquettes. First International Conference of Thermal Equipment, Renewable Energy and Rural Development-TE-RE-RD 2012, Bucharest, 6 July 2012, ISSN 1843-3359, pp. 111–114.

Mihăescu, L., E. Enache, Gh. Lazaroiu, I. Pisa, V. Berbece, G. Negreanu and E. Pop. 2013a. Achievements in energy valorization of cereal straw in boiler manufactured in Romania. 5th International Conference TAE, 2013, 3–6 September, Prague, Czech Republic.

Mihăescu, L., A. Domokos, I. Pîşă, I. Oprea and S. Bartha. 2013b. Results of the first energetic willow crop in Romania, research people and actual tasks on multidisciplinary sciences, 12–16 June 2013, Lozenec, Bulgaria.

Negreanu, G., V. Berbece, L. Mihăescu, I. Oprea, I. Pîşă and D. Andreescu. 2014. Thermal power plant for energy willow use; design, performances. 3rd International Conference of Thermal Equipment, Renewable Energy and Rural Development TE-RE-RD 2014, Mamaia 12–14, June 2014, ISSN 2359–7941, pp. 109–113.

Nistorescu, M., A. Doba, V. Dragomir and M. Ventoniuc. 2011. Environmental Report. Romania's Energy Strategy for 2007–2020 Updated for 2011–2020, Ministry of Economy, Trade and Business Environment.

Forestry resources in EU (in Romanian: Resursele forestiere în UE), Mesagerul Energetic, year XIV, m-147, January 2014, ISSN 2066–4974.

Romania. Ministry of Energy. Romania's Energy Strategy 2016–2030, 2050 perspective. 2016.

Romanian Statistical Yearbook, 2007–2016, http://www.insse.ro/cms/ro/tags/anuarul-statistic-al-romaniei.

Stern, N.H. 2007. The Economics of Climate Change: The Stern Review. Cambridge, UK: Cambridge University Press, Print.

Toader, M. 2013. Research on cereal straw briquetting in Northeastern Moldavia. 2nd International Conference of Thermal Equipment, Renewable Energy and Rural Development TE-RE-RD 2013, Olănesti 20–22 June, 2013 Ed. Printech, pp. 119–122.

Toader, M. 2014. Analysis on the combustion dynamics of the straw briquette for boilers with mobile grate. 3rd International Conference of Thermal Equipment, Renewable Energy and Rural Development TE-RE-RD 2014, Mamaia 12–14, June 2014, ISSN 2359-7941, pp. 39–43.

Toader, M. 2015. Economic and technical aspects for the energetic exploitation of the solid biomass with high increasing speed, PhD thesis, UPB.

CHAPTER 12

New Efficient and Ecologic Energy Vectors (Solid Biomass-Hydrogen)

Gheorghe Lazaroiu,[1,*] *Lucian Mihaescu*[2]
and *Ionel Pisa*[2]

1. Introduction

Energy has a variety of definitions and according to the first principle of thermodynamics, energy is not lost and is not created but only transformed from one form to another.

Physics defines energy as the "property" that must be transferred to an object or to a system in order to perform a mechanical work or to heat an object or matter, moving from one system to another.

In the International Unit of Measurement System, the energy is measured in Joule: from a mechanical point of view, the energy is transferred to a system which is displaced by 1 meter in the direction of a force of 1 Newton.

As we know, between mass and energy, there is an equivalence relationship, so that the intrinsic energy is equal to the intrinsic mass multiplied by the square of the light velocity. To be noted that the intrinsic mass is that part of the mass of an object or system that remains invariant

[1] Department of Energy Production and Use, University Politehnica of Bucharest, Splaiul Independentei 313, 060042, Bucharest, Romania.
[2] Department of Thermotechnics, Engines, Thermal and Frigorific Equipments, University Politehnica of Bucharest, Splaiul Independentei 313, 060042, Bucharest, Romania.
* Corresponding author: glazaroiu@yahoo.com

to the global movement (Loretz transformations). Considering Plank-Enstein's relationship (Griffiths 1995) there is a relation of proportionality between the energy and the frequency of the photon.

Therefore, the energy is proportional to mass and frequency of radiation. From these considerations two great possibilities of obtaining energy can result: by consuming a certain mass of so-called classic fuel or by absorbing a certain amount of photons, either directly or indirectly.

The **energy obtained by** consuming a certain **mass** is considered to be **non-regenerative energy**, and the *energy obtained directly* or *indirectly* from the *energy of the photons* from the *sun* or the relative movement of the earth in relation to the sun is considered *renewable energy*.

Now, the primary energy sources are divided into non-renewable sources (NRESs) and renewable sources (RESs).

At present, fossil fuels (e.g., natural gas, crude oil/oil and coal) and mineral fuels (uranium or thorium) are considered to be NRESs.

RESs include solar energy, wind energy, hydropower and tidal energy, biomass energy and geothermal energy.

The concept of primary energy is used in energy statistics in order to determine the energy balance and it refers to that energy that has not undergone any transformation into another form of useful energy.

Total primary fuel consumed for fuel production experienced a steady increase between 1973 and 2014, as shown in the Fig. 1 (IEA 2016).

Total CO_2 emissions, from combustion, have risen continuously between 1973 and 2014, as shown in Fig. 2. These emissions are calculated by the IEA on the basis of global balances. Emissions from coal also include those due to the bituminous shale and peat, and others include CO_2 due to the burning of industrial and urban non-renewable waste. CO_2 is also emitted by non-energetic activities such as food production, water treatments, changes on the use of land, etc.: this contribution is about 1/3 of the overall emissions and should be taken into account in a global CO_2 scenario.

Because of this increase in CO_2 emissions, mankind is looking for solutions to reduce them. The new energy vector Hydrogen-Solid biomass can make an important contribution in reducing these CO_2 emissions.

A renewable resource is a natural resource that renews in order to avoid resource depletion due to its use and consumption. This renewal (regeneration) is done either by biological reproduction or by other naturally recurrent processes, in a finite time scale. Renewable resources are part of the natural environment of the earth and the largest components of its ecosystem.

The main renewable resources used by mankind for energy production are solar radiation, wind, river courses, water falls, waves, geothermal and

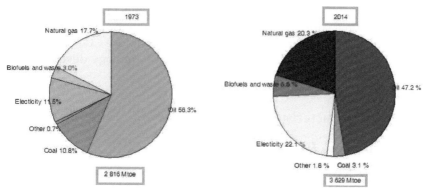

Fig. 1: Total primary energy.

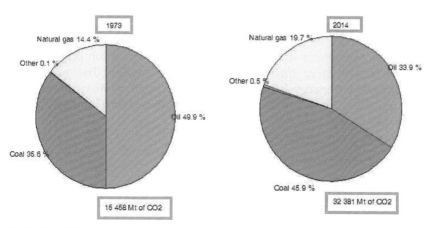

Fig. 2: Total CO$_2$ emissions.

biomass. They can be energy sources for electricity generation, heating or water/air cooling, transport and other rural energy needs.

To limit greenhouse gases, the EU has set a target, 20% out of total energy production to be renewable energy by 2020.

Biomass is a biological material derived from plants or animals. Therefore, we have a biomass of animal origin and one with vegetable origin. To be a renewable energy resource, this biomass should not be used as food.

Biomass can be used as a source of energy directly (combustion – heat production) or indirectly (after being transformed into sundry biofuel forms).

The most famous and the most important energy resource is vegetal biomass.

Currently, all European Union Member States have realized 27 various national support schemes. A necessary item for the development of a sustainable energy mix is the RES-E that can reduce CO_2 and diverse emissions and can contribute to the process of technology development, that are considered to be new.

2. Energy Characteristics of Solid Biomass

2.1 *The Biomass Energy Potential*

The biomass term, generally has more meanings. Most often, it is defined in antithesis with inorganic matter. Thus, biomass has biological origin, can be incorporated into a living organism and contains carbon atoms in its composition. In the broadest sense, biomass is organic matter, alive or dead.

Biomass has in its chemical bonds nitrogen, oxygen, carbon and hydrogen. Sometimes it also has sulfur in its composition but in minor proportions or large quantities of inorganic species. The process of capturing solar energy by plants is done through photosynthesis, and energy storage is made by carbohydrates that form biomass blocks (Demirbas 2001).

If we consider biomass as a general term, it is used for:

- Food or living beings in general;
- Production of chemicals necessary for life;
- Production of energy for heating and electricity;
- Fuel for transport.

An important datum is represented by the exponential population growth on Earth and the need to ensure vegetal food, animal food and furniture. After this use, to improve people's lives, it is also a priority to extract useful chemical compounds which otherwise cannot be synthesized.

The use of biomass as a source of primary energy in the energy or transport sector is therefore with the above restrictions. Even if there is a continuous productivity increase, it should be noted that agricultural areas are being degraded for various reasons, which determine a biomass reduction which is a burden in the global balance.

A significant source of primary energy is biomass and the prime source is wood. Beside wood, there are other resources (waste and by-products; methane from plats for wastewater treatment and landfills; energy crops; residue: wood, straw and stalks of grain, other residues from processing of food).

2.2 General Properties of Physical, Chemical and Energetically of Solid Biomass

An element that is similar to coal is biomass, but it is different in terms of organic and inorganic material that are in its content and also its physical properties and calorific value. An important fact is that biomass has more potassium, silica, oxygen and less aluminum, iron, carbon compared to coal. Biomass has also a lower calorific value, friability density and a higher water content.

The biomass combustion process involves changes to any plant due to the biomass composition (volatile content). The spay type biomass combustion, in the gaseous phase, is characterized by volatilization and combustion while coal combustion is characterized by gas-solid oxidation and coke forming.

Elemental analysis of biomass is determining its energy properties, its ecological influence and efficiency. For this analysis, a valuable equipment is necessary and also some trained personnel. A technical analysis does not require trained personnel and valuable equipment but it requires a standard laboratory. The combustion process is affected by the inorganic elements that are present in the biofuels and by the ash composition (heavy metals). These metals are decreasing the temperature and the size of the particles that are precipitated. This effect is independent for the use of biofuels. The reactions that are present in the boiler are influenced by the presence of K, Na, Cl and flying ash in biomass.

2.3 Energy Characteristics of Biomass

The fuel laboratory equipment has the capability to determine the energy characteristics for the biomass categories, such as: wood waste (Table 1), wheat and rye straw (Table 2), straw and composite briquette (Table 3), sorghum (Table 4).

Table 1: The characteristics of wood waste.

Characteristics		Wood waste	Variation with wood essence		
			Beech	Pine	Spruce
Carbon	C % (dry mass)	50	49.3	51	50.9
Hydrogen	H % (dry mass)	6.2	5.8	6.1	5.8
Oxygen	O % (dry mass)	4.	43.9	42.3	41.3
Nitrogen	N % (dry mass)	0.3	0.22	0.1	0.39
Sulf	S % (dry mass)	0.05	0.04	0.02	0.06
Clorine	Cl % (dry mass)	0.02	0.01	0.01	0.03
Cenuşă	A % (dry mass)	1	0.7	0.5	1.5
Volatile	V % (dry mass)	81	83.8	81.8	80

Table 2: Energy characteristics of wheat and rye straw.

Characteristics	Yellow straw	Grey straw
Moisture content, W_t^i	10–20	10–20
Volatile matter V^i,	≥ 70	> 70
Ash content, A^i	4	3
Carbon, C^i	42	43
Hydrogen H^i	5	5.2
Oxygen O^i	37	38
Chloride %	0.75	0.2
Nitrogen N^i	0.35	0.41
Sulph combustibile S_c^i	0.16	0.13

Table 3: The characteristics of straw and composite briquette.

No.	Fuel	Elemental analysis [%]						
		C	H	N	S	O	A	W
1	Acacia wood	49.6	6.0	0.9	0.1	33.8	4.20	5.4
2	Reed briquette	48.4	5.5	0.6	0.0	31.2	7.30	7.0
3	Sawdust briquette	50.0	5.9	1.8	0.0	33.6	2.60	6.0
4	Sawdust briquette 50% + stalks 50%	46.1	5.5	0.4	0.0	38.0	3.30	6.7
5	Sawdust briquette 50% + straw 50%	48.0	5.8	0.5	0.0	36.1	4.40	5.2
6	Sawdust briquette 25% + stalks 75%	48.2	5.9	0.6	0.0	34.5	3.40	7.5
7	Sawdust briquette 25% + straw 75%	48.5	5.7	0.7	0.0	36.4	1.40	7.3

Table 4: The characteristics of sorghum.

	Wt	C	H	N	S	O	A
SORGHUM 1 Hybrid Porumbeni 4	%	%	%	%	%	%	%
Leaf	9.08	44.7	5.8	2	-	28	10.42
Panicles	14.2	44.8	5.7	1.4	-	28.5	5.4
Stem	43.5	22.1	7.1	0.2	-	25.8	1.3
SORGHUM 2 Hybrid F135-ST Fundulea (sweet sorghum)	%	%	%	%	%	%	%
Leaf	11.66	45.4	5.8	1.1	-	27.5	8.54
Panicles	12.45	44	6.1	1.7	-	32.9	2.85
Stem	38.18	27.9	7.8	0.2	-	23.6	2.32

3. The Mathematical Model and Experimental Tests on the Combustion Characteristics of the New Hydrogen-Solid Biomass Vector

3.1 The Main Thermodynamic Properties of Hydrogen

The use of hydrogen in biomass burning (Considine 2005, Lazaroiu et al. 2014) has the role of improving the combustion conditions and to increase the reaction rate. In addition, this positively influences the concentrations of carbon oxides: hydrogen when it is used as fuel is characterized by the highest adiabatic flame temperature, if air is considered as oxidant, nitrogen oxides are increasing.

Porous wood biomass was used: cut wood and sawdust. Also, the pellets were tested but the amount of hydrogen absorbed was very low.

Etymologically, the word hydrogen is a combination of two Greek words that mean "to make water", and this means it is a completely non-polluting fuel.

Hydrogen is the first chemical element in the Periodic Table of Elements and has the symbol H, atomic number 1 and atomic mass equal to 1.00794 u.am. (Atomic mass units).

It is the lightest chemical element that is slightly flammable, colorless, insipid, odorless gas, and naturally occurring in the form of a diatomic molecule, H_2.

Elementary hydrogen is the main component of the Universe, it is predominantly present in the formation of stars in the form of plasma.

Hydrogen gas (in the diatomic H_2 state) is highly flammable, it ignites in air for volumetric concentrations between 4% and 75%, and in contact with pure oxygen between 4.65% and 93.9% (Carcassi 2005). The detonation range is between 18.2% and 58.9% and in the air and between 15% and 90% in oxygen. The variation of enthalpy after combustion is −286 kJ/mol (National Academy of Engineering 2004).

It can react spontaneously and violently at room temperature with chlorine and fluorine, forming HCl and HF (Clayton 2003).

Molar specific heat at constant pressure, c_{pm} $\left[\dfrac{kJ}{kmol\cdot K}\right]$, of hydrogen ($H_2$), as gas, as a function of temperature, T[K], valid on the domain −60 < t [°C] < 3500, deduced by interpolation from available data (Kuzman 1976, Zucrow and Hoffman 1977) is given by the relationship:

$$c_{pm}(T) = c_1 \cdot a^9 + c_2 \cdot a^8 + c_3 \cdot a^7 + c_4 \cdot a^6 + c_5 \cdot a^5$$
$$+ c_6 \cdot a^4 + c_7 \cdot a^3 + c_8 \cdot a^2 + c_9 \cdot a + c_{10} \tag{1}$$

where: $a = \dfrac{T[K]}{298.15}$, $c_1 = 1.083 \cdot 10^{-6}$, $c_2 = -6.8989 \cdot 10^{-5}$, $c_3 = 1.8892 \cdot 10^{-3}$, $c_4 = -0.029075$, $c_5 = 0.27565$, $c_6 = -1.6545$, $c_7 = 6.1785$, $c_8 = -13.373$, $c_9 = 15.27$, $c_{10} = 22.104$.

Based on the above relation we determine the specific heat mass at constant pressure, $c_p \left[\dfrac{kJ}{kg \cdot K}\right]$, depending on the temperature, $T[K]$, with the relationship:

$$c_p(T) = \frac{c_{pm}(T)}{M_{mol}} \tag{2}$$

where $M_{mol} = 2.016 \left[\dfrac{kg}{kmol}\right]$ is the molar mass of hydrogen.

The heat expresses the actual mass at constant volume, $c_v \left[\dfrac{kJ}{kg \cdot K}\right]$, respectively the molar specific heat at constant volume, $c_{vm} \left[\dfrac{kJ}{kmol \cdot K}\right]$ depending on the temperature, $T[K]$, is:

$$c_v(T) = c_p(T) - R_{H2} \qquad c_{vm}(T) = c_p(T) - R_{H2} \cdot M_{mol} \tag{3}$$

where $R_{H2} = 4.1242 \left[\dfrac{kJ}{kg \cdot K}\right]$ is the perfect hydrogen gas constant.

The adiabatic exponent, $\gamma_{H2}[-]$, depending on the temperature, is:

$$\gamma_{H2}(T) = \frac{c_p(T)}{c_v(T)} \tag{4}$$

Enthalpy-specific mass, $h_m \left[\dfrac{kJ}{kg}\right]$, of hydrogen ($H_2$) as temperature-dependent gas, $t[°C]$, valid for $-60 < t[°C] < 3500$, deduced by interpolation from available data (Kuzman 1976, Zucrow and Hoffman 1977), is given by the relationship:

$$h_m(t) = c_1 \cdot t^3 + c_2 \cdot t^2 + c_3 \cdot t + c_4 \tag{5}$$

where: $c_1 = 3.0431 \cdot 10^{-7}$, $c_2 = 9.6712 \cdot 10^{-5}$, $c_3 = 14.376$, $c_4 = 0.67534$.

Based on the above relationship the specific molar enthalpy is determined, $h_{mol} \left[\dfrac{kJ}{kmol}\right]$, depending on the temperature, $t[°C]$:

$$h_{mol}(t) = \frac{h_m(t)}{M_{mol}} \tag{6}$$

where $M_{mol} = 2.016 \left[\dfrac{kg}{kmol} \right]$ is the molar mass of hydrogen.

The specific volumetric enthalpy, $h_v \left[\dfrac{kJ}{m^3} \right]$, depending on the temperature, $t[°C]$, is:

$$h_v(T) = \frac{h_m(t)}{V_{mol}} \tag{7}$$

where $V_{mol} = 22.414 \left[\dfrac{m^3}{kmol} \right]$ is the perfect hydrogen gas constant.

The thermal conductivity of hydrogen (H_2) $k \left[\dfrac{W}{m \cdot K} \right]$, as a temperature-dependent gas, $T[K]$, valid for $115 < T[K] < 1470$, deduced by interpolation from the available data (Dickinson 1976), is:

$$k(T) = c_1 \cdot T^3 + c_2 \cdot T^2 + c_3 \cdot T + c_4 \tag{8}$$

where: $c_1 = 1.5620 \cdot 10^{-10}$, $c_2 = -4.1580 \cdot 10^{-7}$, $c_3 = 6.6890 \cdot 10^{-4}$, $c_4 = 8.0990 \cdot 10^{-3}$.

3.2 Diffusion of Hydrogen in Solid Fuel (Coal or Biomass)

Molecular diffusion is the molecule penetration of an object among the molecules of another, without flow. Therefore, diffusion is the process in which the concentration or temperature is equalized for both objects.

Molecular diffusion is usually mathematically described using Fick's law. Is the basis pf mass transfer, a transport phenomenon along with thermal conductivity and flow of fluids or impulse transfer to fluids.

The diffusion is of three types: molecular, brows and turbulence. Molecular diffusion takes place in the all of the three states of aggregation (solid, liquid and gas). It should be noted that this molecular diffusion occurs with any type of molecules, even for single gas and appears as an effect of the molecules thermal movement. It continues at utmost gas velocity, at a reduced speed for liquids and for solids, at a lower velocity. The differences are obtained depending on the nature of the thermal movement of these environments.

Solid fuels such as coal or biomass can be considered porous objects in relation to hydrogen diffusion.

There are three mechanisms: the majority molecular diffusion, Knudsen diffusion and surface diffusion. They can either act separately or concurrent in the same system.

In solids that have large pores (greater than the free path of the diffused gas molecules), molecular diffusion is predominant.

Knudsen diffusion occurs between gases and solids with reduced pores or reduced pressure when the free average path of the molecules is larger than the dimension of the pores and the molecules collide with the walls more often than they do between them. The molecular reflection between the gas molecules and the pore walls is normally diffuse, meaning that the molecules spread in all directions, and the diffusion of the molecules along the pores depends precisely on these collisions. This Knudsen diffusion can play an important role in a particular scale of pore dimensions and gas pressures.

Surface diffusion is analyzed during the adsorption of a diffused substance with a solid. A surface gradient of a diffused substance is in the layer of a pore surface due to the increasing of the gas concentration at equilibrium surface and of the partial pressure of adsorbed species. In some conditions this effect can improve the total flow of a broadcast component.

For molecular diffusion in porous solids, the effective diffusion coefficient is arbitrarily related to the substance concentration gradient that is normally diffusing at the outer surface of the object. The whole porous space must be considered.

Biomass and many other porous components are multiphase structures that include a network of pores or interconnected and irregular channels. It is difficult and time consuming to model these irregular structures in a three-dimensional approach.

For this reason, it is preferable to use simplified empirical relations that offer an easy and accurate model by introducing the term tortuosity. The porosity of a porous network is used to obtain efficient transport properties.

It is assumed that tortuousness affects transport properties, diffusivity and liquid phase conductivity, with similar functionality:

$$D_{ef}\left[\frac{m^2}{s}\right] = D \cdot \frac{\theta}{\tau^2} \tag{9}$$

The coefficient of molecular diffusion effective for transport through a pore structure, $D_{ef}\left[\frac{m^2}{s}\right]$, is estimated (Thorat et al. 2009), as follows:

$$D_{ef}\left[\frac{m^2}{s}\right] = D \cdot \frac{\theta}{\varsigma^2} \tag{10}$$

where $D\left[\dfrac{m^2}{s}\right]$ is the coefficient of diffusivity of the gas through individual pores; $\theta[-]$ is open porosity for transport; $\varsigma[-]$ is the generally determined, tortuosity factor in each specific case.

In plan (2-D), tortuousness is defined as the ratio of the pore length, L and the length between the pins of the right pillar, C,

$$\tau = \frac{L}{C} \tag{11}$$

Note that this ratio is equal to 1 for a straight line and is infinite for a circle.

The diffusion of hydrogen in biomass pores cannot be deduced from Fick's law, but only experimentally.

Typical diffusion coefficients $D_{ef}\left[\dfrac{m^2}{s}\right]$ are (Pisa et al. 2016): $10^{-5}\left[\dfrac{m^2}{s}\right]$ for gases, $10^{-9}\left[\dfrac{m^2}{s}\right]$ for liquids, and $10^{-12}\left[\dfrac{m^2}{s}\right]$ for solids.

Specifically, the diffusivity of hydrogen in a series of solids has values:

$$D_{H2\rightarrow steel} = 0.3 \cdot 10^{-12}\left[\frac{m^2}{s}\right], \; D_{H2\rightarrow nickel} = 1 \cdot 10^{-12}\left[\frac{m^2}{s}\right], \; D_{H2\rightarrow polyethylene} = 87000 \cdot 10^{-12}$$
$$\left[\frac{m^2}{s}\right], \; D_{H2\rightarrow biomass} = 1.5 \cdot 10^{-14}\left[\frac{m^2}{s}\right].$$

3.3 The Mathematical Model for Hydrogen Diffusion

For characterizing biomass combustion with diffused hydrogen in its pores, a mathematical model was developed during the researches which allows to develop a mathematical simulation of gas absorption (Pisa et al. 2016). A numerical simulation for the hydrogen uptake gas was performed.

The expression describing the diffusion of hydrogen (considering a one-dimensional diffusion) in the pores of a renewable fuel is:

$$D_i \cdot \frac{d^2C}{dx^2} + q = 0 \tag{12}$$

where: $D_i\left[\dfrac{m^2}{s}\right]$ is the diffusion coefficient, $C\left[\dfrac{kg}{m^3}\right]$ is the concentration, $x[m]$ the depth of the pore and $q\left[\dfrac{kg}{m^3 \cdot s}\right]$ the hydrogen flux.

Expression (12) can also be written as follows:

$$\frac{d^2C}{dx^2} = -\frac{q}{D_i} \tag{13}$$

By integrating expression (12), the solution is obtained:

$$C = -\frac{q}{2 \cdot D_i} \cdot x^2 + C_1 \cdot x + C_2 \tag{14}$$

The limit conditions for determination of constants C_1 and C_2 are:

$$x = 0;\ C = C_s \Rightarrow C_2 = C_s \tag{15}$$

where C_s is the oxygen concentration at the surface of the particle. For,

$$x = 1 \quad \Rightarrow \quad \frac{dC}{dx} = 0 \tag{16}$$

and thus, from equation (14) it results:

$$0 = -\frac{q}{D_i} \cdot l + C_1 \Rightarrow C_1 = -\frac{q}{D_i} \cdot l \tag{17}$$

where l[m] is the length of the pore.

   Finally, the variation in hydrogen concentration within the pores can be expressed as:

$$C = -\frac{q}{2 \cdot D_i} \cdot x^2 - \frac{q}{D_i} \cdot l \cdot x + C_s$$
$$= C_s - \frac{q}{D_i} \cdot l \cdot x - \frac{q}{2 \cdot D_i} \cdot x^2 \tag{18}$$

   The hydrogen concentration on the surface particle by time and by the hydrogen flux can be calculated with the relation, $C_s \left[\frac{kg}{m^3}\right]$:

$$C_s = q \cdot \tau \tag{19}$$

   From equations (16) and (17), the hydrogen concentration for one length of l is:

$$C_l = q \cdot \tau - \frac{3}{2} \cdot \frac{q}{D_i} \cdot l^2 = q \cdot \left(\tau - \frac{3}{2} \cdot \frac{l^2}{D_i} \right) \tag{20}$$

This concentration is zero for a time less than the absorption time and then proportional to it.

The time, τ[s], of the adsorption process is given by the relationship:

$$\tau = \frac{3}{2} \cdot \frac{q^2}{D_i} \tag{21}$$

The variation of concentration within the pore, $C_m \left[\dfrac{kg}{m^3}\right]$, is given by:

$$C_m = \frac{\int_0^l \left(-\dfrac{q}{2 \cdot D_i} \cdot x^2 - \dfrac{q}{D_i} \cdot l \cdot x + q \cdot \tau \right) dx}{l}$$

$$= \frac{\left(-\dfrac{q}{2 \cdot D_i} \cdot \dfrac{x^3}{3} - \dfrac{q}{D_i} \cdot l \cdot \dfrac{x^2}{2} + q \cdot \tau \cdot x \right)\Big|_0^l}{l} \tag{22}$$

$$= \frac{1}{l} \cdot \left(-\dfrac{q}{2 \cdot D_i} \cdot \dfrac{l^3}{3} - \dfrac{q}{D_i} \cdot l \cdot \dfrac{l^2}{2} + q \cdot \tau \cdot l \right)$$

$$= -\frac{2}{3} \cdot \dfrac{q}{D_i} \cdot l^2 + q \cdot \tau$$

Taking into account the surface concentration, $C_s \left[\dfrac{kg}{m^3}\right]$ results

$$C_s = C_m + \frac{3}{2} \cdot \frac{q}{D_i} \cdot l^2 \tag{23}$$

Based on this relationship it is possible to calculate the amount of hydrogen absorbed, Q[kg],

$$Q = V \cdot C_m \cdot S_i \cdot \varepsilon \tag{24}$$

where V[m^3] is the particle volume, S_i [m^2/m^3] the internal surface of the pores, and ε [m] is the Fourier criterion and represents the relative depth of the power during the reaction.

3.4 Results of Simulation of the Mathematical Model

Based on the previous model, adsorption of hydrogen has been simulated for biomass pores through a MATLAB® program.

Input data used to evaluate the model are reported in Table 5. Minimum time for the adsorption process, $\tau[s]$, calculated based on (21), function of pore's length l, is presented in Fig. 3.

There is an exponential growth along the length of the pore while reducing the increase time of the diffusion coefficient. For example, for

Table 5: Input data for evaluating mathematical model.

No. crit.	Parameter	Value
1	Hydrogen flux, $q \left[\dfrac{kg}{m^3 \cdot s} \right]$	$10^{-6} \cdot [1, 1.5, 2]$
2	The pore length, $l[m]$	$10^{-7} \cdot [1, 2, 3, 4, 5, 6, 7, 8, 9, 10]$
3	The particle volume, $V[m^3]$	$10^{-3} \cdot [1, 2, 5, 10, 20]$
4	The internal pore surface, $S_i \left[\dfrac{m^2}{m^3} \right]$	100, 150
5	Relative depth of the pore during the reaction, $\varepsilon[m]$	0.497
6	The diffusion coefficient $D_i \left[\dfrac{m^2}{s} \right]$	$10^{-14} \cdot [1, 1.5, 2]$

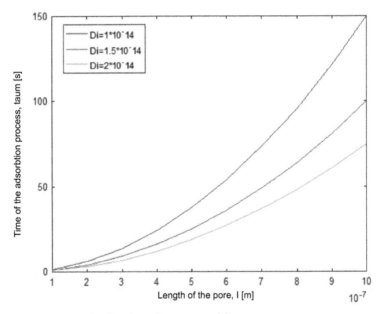

Fig. 3: Minimum time for the adsorption process, $\tau[s]$.

a diffusion coefficient, $D_i = 1 \cdot 10^{-14} \left[\frac{m^2}{s} \right]$ the duration of the absorption process is 150, for the diffusion coefficient, $D_i = 2 \cdot 10^{-14} \left[\frac{m^2}{s} \right]$ the duration of the absorption process is 75.

The variation of the hydrogen concentration on the pore surface for different hydrogen streams is shown in Fig. 4. A linear increase of this pore-length concentration and hydrogen flow is noted.

The hydrogen concentration for a pore with the length l is shown in Fig. 5. It is noted that this concentration, according to (9), is zero for any time less than the absorption time and after that it is linearly increasing along with the hydrogen flow and time.

Figure 6 illustrates the variation of the hydrogen concentration on the surface and the length of the pore for different hydrogen streams. This allows to observe the variation and the definition of the concentration on the surface and the linear one for the pore length. Both surface and pendulum concentrations increase with increasing of the hydrogen flow and pore length.

Figure 7 shows the concentration variation within the pore, C_m, depending on the pore length, l, for different hydrogen streams. Note

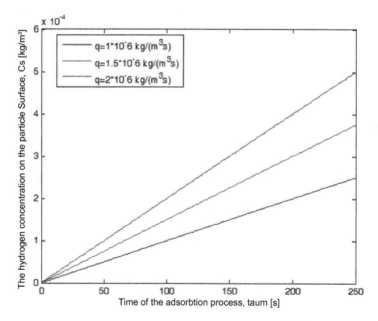

Fig. 4: Concentration of hydrogen on the surface of the particle, $C_s \left[\frac{kg}{m^3} \right]$.

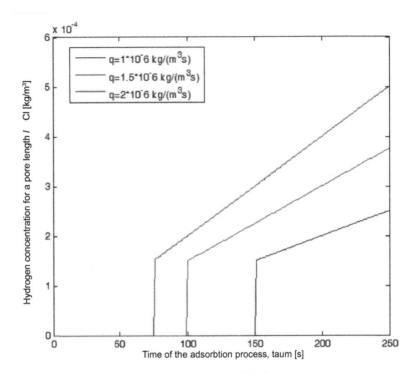

Fig. 5: Hydrogen concentration for a pore length l, $C \left[\dfrac{\text{kg}}{\text{m}^3} \right]$.

that hydrogen diffusion was considered constant. It is noted that this concentration is proportional to the hydrogen flow and it is increasing with the flow but decreases following a 2nd degree parabola along with the increasing of the pore length for the same hydrogen flux.

The variation of the amount of the absorbed hydrogen, depending on the pore length, for various volumes of hydrogen particles, is shown in Fig. 8. The flow of hydrogen, the diffusivity, the relative depth of the pore and the internal surface of the pores have been kept constant.

The amount of hydrogen absorbed decreases with increasing pore lengths. For a given pore length, Q increases with the increasing of the particle volume.

Figure 9 shows the variation in the amount of the absorbed hydrogen for an internal pore surface, increased by 50%.

The general shape maintains its tendencies but increases the absorbed quantity.

From the analysis of the relationships and the graphs, it is observed that the increase of the pore length leads to the decrease of the absorbed

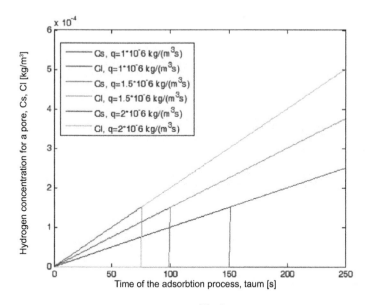

Fig. 6: Hydrogen concentration for a pore, C_s, $C_l \left[\dfrac{\text{kg}}{\text{m}^3}\right]$.

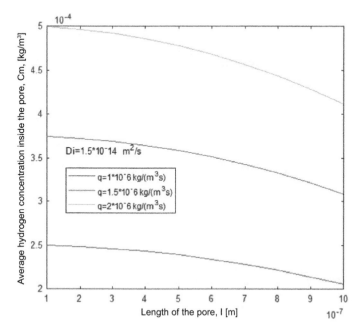

Fig. 7: Average hydrogen concentration inside the pore, $C_m \left[\dfrac{\text{kg}}{\text{m}^3}\right]$.

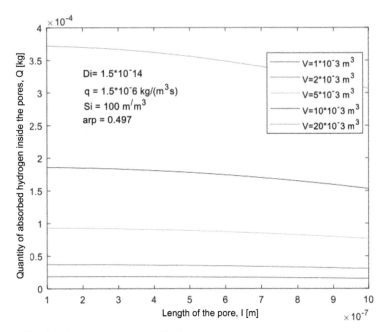

Fig. 8: Absorbed hydrogen quantity, Q[kg].

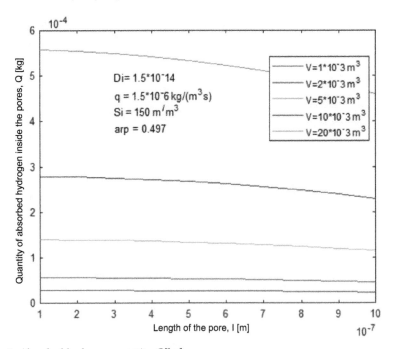

Fig. 9: Absorbed hydrogen quantity, Q[kg].

quantity of hydrogen. For a given value of the pore length, the total amount of the absorbed hydrogen increases along with the increasing of the hydrogen flow Q. Also, particles with larger volumes involve higher absorption quantities. If the internal surface of the pores grow, the amount absorbed will also increase.

The most important result is that the concentration inside the pores decreases as the pore length increases.

3.5 *Experimental Results*

We tested three types of biomass: cut wood, sawdust and energy willow chips, showed in Fig. 10. These samples were passed into a hydrogen stream and held for a sufficiently long time, higher than the minimum absorption time, $\tau[s]$, until saturation was reached. The duration of the considered period of time depends on the diffusion coefficient of each material.

The time ranged between 75 and 150 seconds, depending on the diffusion coefficient.

Fig. 10: Samples used in experimental tests.

Fig. 11: Experimental stand for hydrogen diffusion.

The hydrogen feed was composed from a reservoir and the samples were located in a tube (Fig. 11). Several tests have been carried out in order to obtaining sufficient quantities of cut wood, sawdust or energy willow.

Samples were weighed before and after hydrogen diffusion. In general, a mass loss of 4–5% was recorded. The explanation can be the removal and also the replacement of oxygen with hydrogen, in the particle pores (it has a lower density and obviously, a lower specific weight).

The samples thus prepared were burned in an experimental boiler with a thermal power of 20–40 kWt (Pisa and Mihaescu 2011).

The scheme and most important components of this boiler are shown in Fig. 12. The main components are:

The main performance of this type of boiler used for testing is shown in Table 6.

In order to burn small amounts of fuel (biomass-hydrogen), the feed is made with a screw feeder.

The combustion analysis was performed with MAXYLIZER analyzer (Fig. 13).

Tests were performed for all the biomass types showed in Fig. 10 with and without hydrogen injection. The evolution of the flame temperature was monitored with an infrared camera.

The results of the measurements are shown in Table 7.

The values are determined for a 7% O_2 content in the combustion gasses.

A higher moisture content of sawdust offers a high temperature of the flame and a lower CO concentration to the combustion of the minced wood. The data form Table 7 are showing that the biomass infusion

1- furnace; 2- heat exchanger; 3- double vault of refractory cement; 4- wool glass insulation; 5- door for grate cleaning; 6- door for vault cleaning; 7- door for heat exchanger cleaning; 8- mobile grate; 9- flue gasses exhaust; 10- worm-screw supplier; 11- first motor-gear transmission; 12- second motor-gear transmission; 13- worm-screw supplier from the storage; 14- reducing extractor; 15- pan extractor; 16- primary air; 17- secondary air; 18- flue gasses cleaning cyclone; 19- flue gasses fan; 20- chimney.

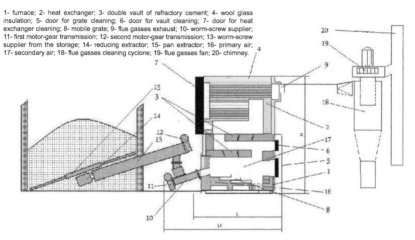

Fig. 12: Experimental stand for hydrogen diffusion.

Table 6: The main performance of the used boiler.

Performance	Value
Heat release rate/unit area	450–600 kW/m²
Net efficiency	83–87%
Flame temperature	680–850°C
Allowable heat release rate	300–400 kW/m³
Lower heating value	14000–18000 kJ/kg
Excess air ratio (end of furnace)	1.3–1.5
Heat loss with unburned carbon	0.5–1.5%
Automation level	95–100%

Fig. 13: Experimental boiler for biomass combustion.

Table 7: The measurement results (average values) with H_2 injections.

Biomass type	Parameter	Value
Chopped wood with H_2	Flame temperature Heat loss, q_{ch} CO concentration$_a$ SO_2 concentration$_a$ NO_x concentration$_a$	840°C 0.51% 1020 ppm 3–8 ppm 40–50 ppm
Chopped wood without H_2	Flame temperature Heat loss, q_{ch} CO concentration$_a$ SO_2 concentration$_a$ NO_x concentration$_a$	780°C 0.72% 1490 ppm 5–10 ppm 40–45 ppm
Sawdust with H_2	Flame temperature Heat loss, q_{ch} CO concentration$_a$ SO_2 concentration$_a$ NO_x concentration$_a$	810°C 0.63% 1242 ppm 3–5 ppm 40–50 ppm
Sawdust without H_2	Flame temperature Heat loss, q_{ch} CO concentration$_a$ SO_2 concentration$_a$ NO_x concentration$_a$	740°C 0.74% 1450 ppm 8–10 ppm 50–55 ppm
Energy willow with H_2	Flame temperature Heat loss, q_{ch} CO concentration$_a$ SO_2 concentration$_a$ NO_x concentration$_a$	823°C 0.57% 1174 ppm 6–8 ppm 40–55 ppm
Energy willow without H_2	Flame temperature Heat loss, q_{ch} CO concentration$_a$ SO_2 concentration$_a$ NO_x concentration$_a$	775°C 0.72% 1472 ppm 5–9 ppm 45–50 ppm

with hydrogen offers a flame temperature that is 10% higher. The CO concentration is 20%–25% lower and the SO_2 and NO_x concentrations are approximately the same. The heat loss with unburned carbon is reduced by 15%–30% with the decreasing of the CO concentration.

It can be noticed that with the use of hydrogen in the combustion of biomass, the reaction speed is increasing. The same process was applied to pallets but due to their small porosity the hydrogen absorption is insignificant.

4. Influence of Hydrogen Enriched Gas Injection on Polluting Emissions from Pulverized Coal Combustion

4.1 General Aspects

Decarbonization is an increasingly striking requirement for sustainable development (Zanganeh and Shafeen 2007). The main energy component of classical combustibles is carbon. The current efforts are in the direction of reducing the share of fuel energy with adverse effects. Hydrogen or hydrogen enriched gas can be used. Hence, hydrogen together with biomass are promising. Alternative fuels, as the hydrogen blend with fossil fuels that determines a growth of the calorific value, thus reducing emissions of greenhouse gases. Hydrogen has a very high calorific value, and biomass absorbs carbon dioxide.

Although hydrogen is omnipresent, a number of current hydrogen technologies use fossil fuels (like natural gas), as raw material. Besides that, the water can be dissociated in oxygen and hydrogen through thermolysis, electrolysis or photolysis (McKendry 2002, Hydrogendrich 2004).

After the first oil crisis, mankind began to seek alternative sources for the energy resulted from the combustion of fossil fuels. Moreover, because the level of pollution has reached alarming levels, partly due to the carbon dioxide produced by fossil fuels combustion, the research has been made to the use of hydrogen as a source of fuel, leading to the emergence of hydrogen economy (HE) (Tanksale and Beltramini 2010). Further researches on a more sustainable energy economy (sources, transporters and storage) is one of the most important challenges (Saxena et al. 2008).

Technically, hydrogen production from water is a feasible process. The crucial issue is the cost for producing hydrogen and the energy required during the production process. If the energy used to produce hydrogen originated from classical sources based on the fossil fuels combustion, then these technologies are certainly not sustainable and will not lead to the reduction of greenhouse gases (mainly carbon dioxide).

Since hydrogen fuel does not produce greenhouse gases and has a high calorific value, it is considered an important and a future energy vector. The development of production, storage, transport and use technologies have a high scientific and technical impact. Since biomass is a renewable fuel and is a natural source of carbon dioxide storage, the production of hydrogen from biomass is one of the most promising methods.

Hydrogen has the characteristics of a perfect fuel in the sense that it is inexhaustible and does not affect the environment through greenhouse gas emissions.

The main disadvantage is the high rate of reaction and but the use of hydrogen (pure or mixed) for the combustion of solid fuel (Zucrow and

Hoffman 1977) has the ability to improve combustion conditions, increase reaction speed and reduce CO, SO_2 and NO_x in the reaction products.

The following are the results of the research on the combustion of pulverized coal inside a boiler after treatment with a hydrogen-enriched gas (HRG). The pulverized coal was passed through a HRG stream before being injected in the burner. The immediate effect on coal flames was a significant decrease in sulfur oxides and a small increase in nitrogen oxides. As a consequence, some additional research has been carried out on the effect of reducing the SO_2 concentration in the coal flame after the pulverized coal is treated with HRG. An unknown reaction mechanism occurs in the flame.

To reduce costs, HRG is preferred. It is a cheaper synthesis gas with hydrogen-related characteristics. Combustion of HRG with coal dust contributes significantly to the reduction of nitrogen oxides emissions in by-products of combustion.

Experimental results on the use of enriched hydrogen gas that is injected into a pulverized coal stream are for the first time made by the authors. The lack of previous studies on the effect of HRG on coal combustion could be caused by the small number of industrial applications of HRG and the potential danger of injecting pure hydrogen into a boiler. The direct injection of only HRG in a boiler would not offer many significant advantages despite being less dangerous than pure hydrogen due to the very low density and it is not significantly increasing the flame temperature. However, by injecting HRG into the pulverized solid fuel stream, certain distinctive effects may occur. Combustion of HRG with solid fuel offers an advantage over burning each of these fuels separately and reduces greenhouse gas emissions (CO_2, NO_x and SO_2) in reaction products. Synergetic effects occur between coal and HRG before combustion.

The advantages of using HRG, which is only a quasi-stoichiometric mixture of hydrogen and oxygen resulting from electrolysis, lies in the simplicity of the electrolytic system used to produce it and its storage under safer conditions and at lower costs than hydrogen.

4.2 Theoretical Aspects

The enriched hydrogen gas (HRG) is produced by an electrolytic system using water. This electrolytic system is a dynamic one, keeping the fluid in a permanent flow and producing a quasi-stoichiometric gaseous mixture of hydrogen and oxygen. In fact, this gas is composed of a mixture of hydrogen and oxygen molecules, almost respecting the stoichiometric water ratio. HRG is a highly reactive gas which, by adsorption, diffuses into the coal dust particles. The power that is consumed to produce

1000 liters of HRG is between 3–3.5 kWh (Zucrow and Hoffman 1977). This means approximately 0.4 Euro/1000 liters.

The power consumed for the production of 1000 l HRG is 3.5–4 kWh (Zucrow and Hoffman 1977), representing approximately 0.4 Euro/1000 l. This cost is strongly influenced by the price of electricity and the gas production capacity. If the capacity increases, the cost will reach 0.35 Euro/1000 liters. The cost of producing hydrogen is about $2.87 per kilogram of hydrogen (Lomax et al. 2009, Pisa et al. 2014).

The future of hydrogen or HRG depends on their availability at low costs and with a minimal environmental impact during their production. The method used for producing hydrogen using fossil fuels (coal or natural gas) is not adequate and determines greenhouse gases (CO_2). Emphasis should be placed on the production of hydrogen from renewable sources (Muresan et al. 2013).

The main features of HRG are: colorless gas; density, $\rho = 0.503 \left[\frac{kg}{m^3}\right]$; molecular weight, $\gamma = 12.3 \left[\frac{kg}{kmol}\right]$; self-ignition temperature, $t_{out} = 591 \div 605[°C]$; flammability concentration between 7,3 and 100% (Hydrogendrich 2004).

In order to use hydrogen in admixture with other fuels, precautions are required due to the high explosion hazard or. The limit of flammability is used to assess the risk of explosion. There are two limits of flammability: the Lower Flammability Limit (LFL) and the Upper Flammability Limit (UFL), specific to fuel mixtures with combustion air, for an autonomous flame (Huang et al. 2016, Karim et al. 1985, Wierzba and Ale 2000).

The mix of a solid fuel (coal or biomass) with a gas is based on the free diffusion process and the adsorption process. A significant amount of gas (hydrogen or HRG) is attached to the surface of the solid particles, forming a generally adsorbed layer in the pores of the same particle. The amount of the adsorbed gas at the balance depends on temperature, gas pressure, and solid particle surface. Coal is a solid adsorbent that has a significant porosity and provides a large surface area on the coal bed. The forces involved in the physical adsorption of HRG are low intensity forces, known as the Van der Waals forces. Adsorption is carried out at low temperatures (100°C) and desorption takes place at 300°C. Dust burning is usually a heterogeneous combustion reaction. When the solid fuel is porous and the gas penetrates into the pores of the particles, the combustion reaction occurs both at the surface of the particles and inside the particle (the internal reaction). Due to its small molecular weight, hydrogen has the largest diffusion power. Low activation energy is required for the reaction to begin. Before and after the ignition point, an area of active reaction centers are being developed and that initiates a series of chain reactions.

Hydrogen reacts with oxygen at 180°C and sulfur at 250°C. At higher temperatures (500°C to 1200°C), hydrogen can react with other elements such as nitrogen or carbon. The affinity of hydrogen (electrophoresis) for strong electronegative elements, such as sulfur, determines a reduction activity that hydrogen exerts it on the combinations of these elements.

Sulfur and hydrogen form a series of compounds, generally called polyphosphates or hydrogen sulphides of which the most important is hydrogen sulfide (Prisecaru et al. 2010).

Sulfur hydrogen can reduce the amount of sulfur dioxide in the flue gases (up to 50%) based on the following reactions:

$$H_2S + \frac{3}{2}O_2 \Rightarrow H_2O + SO_2 \qquad 2H_2S + SO_2 \Rightarrow 2H_2O + 3S \qquad (25)$$

The hydrogen sulfide can be neutralized with a base to form sulfides through a two-step reaction,

$$H_2S + NaOH \Rightarrow NaHS + H_2O \qquad NaHS + NaOH \Rightarrow Na_2S + H_2O \qquad (26)$$

HRG injection in dust also contributes to the reduction of carbon monoxide as well as OH, in combustion gases,

$$CO + OH \Rightarrow CO_2 + H \qquad H + O_2 \Rightarrow OH + O \qquad O + H_2O \Rightarrow 2OH \qquad (27)$$

4.3 Experimental Results Obtained

The coal-gas combustion tests with HRG injection were performed on a pilot boiler with a thermal power of 2 MW, from the University POLITEHNICA of Bucharest, schematically shown in Fig. 14 (Pisa 2013). This plant is very complex and allows us to study several aspects of the combustion process of the pulverized coal. It is very similar to the industrial steam generators. The equipment allows high precision in setting the ignition period, combustion duration and post-combustion effects (mechanical losses, ash/sludge composition and pollutant emissions).

The pulverized coal is then fed to the burner (6), and prior to the burner HRG is injected into the coal mixture and primary air (A1 and A2). The flame is developing from inside the boiler to the tip of the boiler. Fly ash is captured by a cyclone (9) and cleaned combustion gases are discharged from the fan (12) to the chimney.

Measurements of CO_2, CO, O_2, SO_2 and NO concentrations were carried out with two continuous gas analyzers, HORIBA PG250, for two situations: (1) combustion gases from the combustion chamber (P8) and (2) at the end of the boiler (P3). The precision of the HORIBA PG250

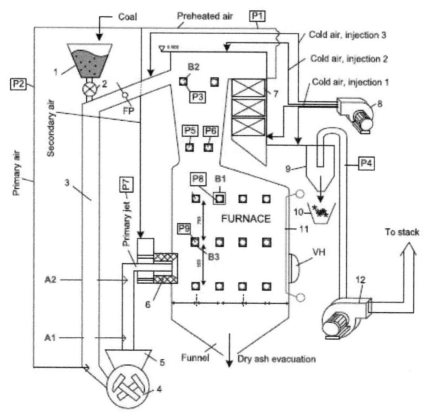

Fig. 14: Pilot boiler with a thermal power of 2 MW.

analyzer is ± 1% (full scale) for all measured species except for SO_2, for which the accuracy is ± 2%.

The HORIBA MEXA 7000 gas analyzer was used to correlate the amount of CO_2, CO and O_2. The H_2S content of the combustion gases (point P9 in the combustion chamber) was determined using the JEROME J605 gas analyzer with a precision of ± 0.3 ppm for every 5 ppm in the range of 1–10 ppm. The H_2 content of the combustion gases has not been measured.

The data was acquired on the LabView platform via the NI PXI-1000B 8-slot 3U PXI Chassiswith 10-32 VDC acquisition system).

HRG was injected through a special nozzle located inside the feed pipe of the primary mixer of the burner. Since HRG is a highly reactive gas, hydrogen should not be significantly heated before combustion (Birtas et al. 2009) and should therefore be injected as close as possible to the burner nozzle. Many preliminary tests have been done to find the optimal position. Brown coal was used as solid fuel, characterized by the initial elemental analysis (wet basis) reported in Table 8.

The elemental analysis was performed with the COSTECH ECS 4010 analyzer. 100 samples were analyzed from the pilot bunker and also their values deviations (taken from the technical specification of the apparatus) were ± 0.32% for carbon, ± 0.13% for Hydrogen, ± 0.24% for oxygen, ± 0.07% for sulfur, ± 0.06% for nitrogen, ± 0.74% for total moisture content and ± 0.34% for inorganic.

The values in the table are corrected for a concentration of 6% O_2 in the flue gases, and the sulfur and nitrogen oxides concentrations are mediated for the measuring points P3 and P8.

The experiments were carried out in the classic combustion process by spraying coal dust in a boiler with a thermal load of 1 MW. Numerous experiments were performed using HRG injections (as shown in Fig. 1) at a constant flow rate.

The temperature at the primary jet was approximately 70°C. The first injection (point A2) was located at 0.95 m distance from the burner nozzle, and the distance between A2 and second injection (A1) was 1.50 m. The distances were determined based on the time required for the absorption process.

All experiments on HRG injections were performed during several test sessions, during a hour.

The first one-hour test session was performed without HRG injection. This was followed by one hour with HRG injection, after which the system was operated for one hour without HRG injection while waiting for electrolytic system cooling. This process is repeated for 24 hours, which is the operating limit of the pilot bunkers.

The results of the measurements are shown in Table 9.

Elemental analysis of slag and ash was done by electron scanning microscopy (SEM) for all determinations to obtain the results in Table 10.

The main advantages of experimentally demonstrated HRG powder injection are:

- For the injection of HRG into the primary air, the concentration of sulfur dioxide in the combustion gases is significantly reduced as a result of the faster hydrogen diffusion, compared to the oxygen in the coal particles (Huang et al. 2006). This fall is between 35 and 40%. Oxygen molecules diffuse more slowly than the hydrogen molecules. It is found that hydrogen interacts with the sulfur atoms inside the coal particles before it is oxidized in the flame. The issue in this experiment is to establish the optimum distance from the burner nozzle in order to successfully inject HRG avoiding explosions. The

Table 8: Coal elementary analysis (wet basis).

C^i [%]	H^i [%]	O^i [%]	Sc^i [%]	N^i [%]	Wt^i [%]	A^i [%]	$LHV \left[\dfrac{MJ}{kg}\right]$
42.3	3.1	7.9	0.9	0.8	25.3	19.7	16.13

Table 9: Influence of HRG injection on the primary brown coal mixture.

Fuel consumption [kg/h]	HRG flow, 1000 [L/h]	Flue gas temp. in P8 [°C]	SO_2 conc. [ppm]	NO_x conc. [ppm]	H_2S conc. in P9 [ppm]
100	$HRG_{A1} = 0$ $HRG_{A2} = 0$	1032	795	190	1.05
	$HRG_{A1} = 0$ $HRG_{A2} = 1$	1051	704	198	2.46
	$HRG_{A1} = 3$ $HRG_{A2} = 0$	1079	610	205	4.03
	$HRG_{A1} = 3$ $HRG_{A2} = 1$	1098	475	224	6.22
150	$HRG_{A1} = 0$ $HRG_{A2} = 0$	1053	840	203	0.97
	$HRG_{A1} = 0$ $HRG_{A2} = 1$	1063	724	213	2.39
	$HRG_{A1} = 3$ $HRG_{A2} = 0$	1080	628	226	6.42
	$HRG_{A1} = 3$ $HRG_{A2} = 1$	1104	534	236	7.32
200	$HRG_{A1} = 0$ $HRG_{A2} = 0$	1073	881	228	1.21
	$HRG_{A1} = 0$ $HRG_{A2} = 1$	1081	738	237	3.14
	$HRG_{A1} = 3$ $HRG_{A2} = 0$	1118	647	249	6.98
	$HRG_{A1} = 3$ $HRG_{A2} = 1$	1136	552	258	8.53

Table 10: Average sludge chemical composition.

Fuel consumption [kg/h]	HRG flow, 1000 [L/h]	SiO_2 [%]	Al_2O_3 [%]	Fe_2O_3 [%]	FeO [%]	CaO [%]	MnO [%]	SO_3 [%]	Na_2O+K_2O [%]
100	$HRG_{A1} = 0$ $HRG_{A2} = 0$	49.87	28.78	9.04	0.88	3.34	1.86	2.88	3.32
	$HRG_{A1} = 3$ $HRG_{A2} = 1$	48.20	27.32	8.31	0.91	4.42	2.78	5.76	2.22
150	$HRG_{A1} = 0$ $HRG_{A2} = 0$	48.74	28.31	7.86	0.72	4.24	2.89	2.28	4.97
	$HRG_{A1} = 3$ $HRG_{A2} = 1$	47.99	29.18	7.74	0.78	4.19	2.65	5.24	2.25
200	$HRG_{A1} = 0$ $HRG_{A2} = 0$	49.19	29.37	8.16	0.65	4.04	2.11	1.92	4.54
	$HRG_{A1} = 3$ $HRG_{A2} = 1$	49.51	30.05	5.59	0.75	4.82	2.04	4.94	2.34

absorption of hydrogen inside the particles is strongly influenced by the temperature.

- The HRG injection modifies the average composition of the slag and ash as outlined in Table 10.
- When HRG is injected into the primary air, the concentration of sulfur dioxide in the flue gases is reduced by 35–40%, which is also monitored by reducing the amount of time required for desulphurization.

In the future, the HRG injection system will have to be improved in order to have a uniform blend of any solid fuel load. Table 2 shows that the efficiency of HRG injection is better at lower loads and when it is injected at both points (A1 and A2). This is explained by an increase in the ratio between HRG and solid fuel. Thus, for the 450 kW load, the ratio is 20 l/kg and for the 900 kW load the ratio is 40 l/kg.

- For using it as secondary desulfurization method, the costs of injection and desulphurization are analyzed. However, the advantages of the HRG injection are multiple and not just a reduction in SO_2 emissions.
- HRG injection costs are between 0.008 Euro/kg (if the ratio between HRG and solid fuel is 20 l/kg) and 0.016 Euro/kg (if the ratio between HRG and solid fuel is 40 l/kg).

Table 9 shows an increased concentration of NO_x since no primary NO_x reduction was considered in the experiment.

5. Influence of Hydrogen Enriched Gas Injection on Burning of Pulverized Biomass

5.1 General Aspects

Biomass was first used, especially for low-efficiency home combustion equipment.

Expanding the use of biomass in the power production requires high thermal power (over 1 MW). These thermal powers can be obtained by choosing a combustion solution that is similar to the combustion of the pulverized carbine dust and the injection of hydrogen or hydrogen enriched gas (HRG). The main research and conclusions of the influence of the HRG injection on the sawdust burning process are presented below.

The main energy component of classical combustibles, including biomass is carbon, and the current efforts are in the direction of reducing the share polluting energy and replace the polluting fuel with another one without adverse effects, namely hydrogen or hydrogen enriched gas.

To support the combustion process, the HRG injection solution was selected before the burner. HRG is obtained by electrolysis in water and represents a mixture of atoms and radicals (H, OH, O, HO_2) with a lower calorific value.

5.2 The Burning Plant of Pulverized Biomass

In principle, a pulverized biomass combustion plant, more specifically sawdust, has the following components: a fuel hopper for a 4 to 6 hours autonomy, a pneumatic system for feeding sawdust to the burner, a primary, secondary and tertiary fan, as well as a flue gas fan.

The research group at University POLITEHNICA of Bucharest proposed and tested a 620 kW burner with more than promising results (Mihaescu et al. 2012).

The burner provides a sawdust flow rate of 0.5 kg/s with a calorific power of 14115 kJ/kg.

The basic analysis of the sawdust was:

$$C^i = 42.7\% \quad H^i = 4.7\% \quad O^i = 36.3\%$$
$$Sc^i = 0\% \quad N^i = 1\% \quad A^i = 5.5\% \quad W^i = 9.8\%$$

(28)

A complete automatized boiler, with moving fire grate has the following auxiliary electrically acted: combustion feeder, acting moving grate, ash and slag discharge, air fan, gas fan.

For the suspension burning of sawdust, for feeding the combustion is a pneumatic system from the air circuit is used and the auxiliary acting installations are: air fan, gas fan, slag discharge (Barta et al. 2011, https://www.saacke.com/de/home/).

For the biomass (the sawdust) to be sprayed, the dimensions of the small combustible particles must be between 1–3.0 mm. There are a few burners that meet these requirements (Prisecaru et al. 2010), so a modular burner with 620 kW power has been proposed, of which 600 kW are produced by biomass combustion and 20 kW by the HRG contribution.

5.3 Burner Dimension

For design purposes, the essential elements are the displacement speed of the combustible particles and the rate of combustion.

The efficiency of the slurry burning (in the airflow) depends on the ratio of the velocity of the combustible particles, $W_p \left[\dfrac{m}{s} \right]$, the combustion rate, $K_s^c \left[\dfrac{m}{s} \right]$ and also depends on the residence time, $\tau_s [s]$ in the furnace.

The speed of the fuel particles is given by the relationship:

$$W_p = k_p \cdot d \cdot \left(\frac{\rho_p - \rho_a}{\rho_a}\right)^{0.5} \quad \left[\frac{m}{s}\right] \tag{29}$$

where: $k_p = 254$ is a constant, d [m]—the diameter of the particles, ρ_p and ρ_a, $\left[\frac{kg}{m^3}\right]$—the density of particles and air.

For sawdust, the values that were considered for the size calculation were: $d = 0.002$ [m], $\rho_p = 850 \left[\frac{kg}{m^3}\right]$, $\rho_a = 0.85 \left[\frac{kg}{m^3}\right]$, and the biomass particle velocity is:

$$W_p = 254 \cdot 0.002 \cdot \left(\frac{850 - 0.85}{0.85}\right)^{0.5} = 16.06 \quad \left[\frac{m}{s}\right] \tag{30}$$

Making the assumption that woody biomass in the form of sawdust has as fuel only carbon, then the reaction rate depends on the oxygen concentration in the reaction zone by a stoichiometric transform coefficient. As for any solid fuel, we have two components of combustion rate, a volatile burning rate and a carbon burn rate.

Therefore, the burning rate of carbon is,

$$K_s^c = k \cdot \frac{C_O}{\dfrac{1}{w_v} + \dfrac{1}{w_C}} \quad \left[\frac{m}{s}\right] \tag{31}$$

where: $k = \dfrac{12}{32}$—stoichiometric carbon-oxygen combustion coefficient with oxygen; $C_o \left[\frac{kg}{m^3}\right]$—oxygen concentration; $W_v \left[\frac{m}{s}\right]$—volatile burning rate; $w_c \left[\frac{m}{s}\right]$—carbon burning rate.

The burner made for experimentation has a thermal power of 620 kW and weight that is less than 50 kg. The construction is self-propelled, with all the welded subassemblies at the central channel of the primary agent, the channel that supports the entire constructive. Radial dimensions of the burner allow it to be integrated into the boiler, having a diameter of 168 mm.

The temperature of the pulverized biomass (rust) is within 60–90°C due to the humidity of the biomass at the entrance and depends on the temperature of the primary air.

The primary air temperature is between 150–200°C and depends on the heat load. In order to avoid accidental ignition of the biomass, mainly the ignition of the volatile matter emitted by the biomass in contact with the heated primary air, a dilution of the oxygen concentration (flue gas recirculation) can be used.

Secondary air temperature is between 150–220°C. Tertiary air temperature is between 150–220°C. The primary air represents 30–40% of the total air necessary for burning.

This air has the aim of pneumatic transportation of sawdust particles and also ensures the ignition and burning of volatile substances, which ensure the burning of fixed carbon of the sawdust wood matter.

Air ratio that is in excess at the end of the furnace is imposed, $\lambda f = 1.25$.

HRG injection is placed before the biomass burner. Thus, the flame also results as a biomass-hydrogen co-combustion process.

The axial blades at the end of the central channel tabulate the damp air.

The excess air ratio at the burner level is imposed within the limits: $\lambda = 1.2$–1.25.

The burner also has a rotating blade system for each of the air circuits.

The primary air velocity is recommended to be $w_1 = 25$–$35\left[\dfrac{m}{s}\right]$, for the secondary air $w_2 = 30$–$35\left[\dfrac{m}{s}\right]$, and for the tertiary air $w_1 = 20$–$30\left[\dfrac{m}{s}\right]$ $W_3 = 20$–$30\ m/s$.

The swirl ratio used for different jet categories: 1.2 for the secondary air and of 5.4 for the tertiary air.

The characteristics of burner working are:

- The ratio, $q_B \left[\dfrac{m_N^3}{kg}\right]$ of HRG flow and biomass flow (sawdust), gas volume/sawdust mass:

$$q_B = \frac{B_g}{B_b} \left[\frac{m_N^3}{kg}\right] \tag{32}$$

where: $B_g \left[\dfrac{m_N^3}{s}\right]$—the HRG gas flow capacity used as thermal support, $B_b \left[\dfrac{kg}{s}\right]$—the biomass (sawdust) flow capacity. The calculus of sawdust flow capacity was $B_b = 80 \left[\dfrac{kg}{h}\right]$.

- Heat quantity ratio developed by HRG and biomass:

$$q_B = \frac{B_g \cdot H_g}{B \cdot H_b} \tag{33}$$

where: $H_g \left[\dfrac{kJ}{m_N^3} \right]$—the lower calorific caloric value of the thermal support gas; $H_b \left[\dfrac{kJ}{kg} \right]$—lower calorific power of biomass (sawdust).

- The pneumatic transportation capacity, $c_B \left[\dfrac{kg}{m^3} \right]$ of sawdust from the stocking bunker to the burner, defined by volumetric concentration of the sawdust in the air transport,

$$c_B = \frac{B_g}{V_a} \quad \left[\frac{kg}{m^3} \right] \tag{34}$$

where: $V_a \left[\dfrac{m^3}{s} \right]$—the air flow capacity for pneumatic transportation.

- As mass ratio, $c_B^* \left[\dfrac{kg}{kg} \right]$ the sawdust concentration for transportation is:

$$c_B^* = \frac{B_g}{V_a \cdot \rho_a} \quad \left[\frac{kg}{kg} \right] \tag{35}$$

where: $\rho_a \left[\dfrac{kg}{m^3} \right]$—the density of the air for sawdust transportation (rectified for the real temperature).

- The thermal load, $q_a \left[\dfrac{kW}{m^3} \right]$ of the burner at embrasure (interface)

$$q_a = \frac{B_g \cdot H_g + B_b \cdot H_b}{S_a} \quad \left[\frac{kW}{m^2} \right] \tag{36}$$

where: $S_a [m^2]$—the aria of embrasure (for the embrasure diameter of 168 mm, the aria is $S_a = 0.022 \, [m^2]$).

- The thermal load of the furnace volume is:

$$q_v = \frac{B_g \cdot H_g + B_b \cdot H_b}{V_f} \quad \left[\frac{kW}{m^3} \right] \tag{37}$$

where: $V_f [m^3]$—the active furnace volume.

The designed thermal load of the furnace volume has the value $q_v = 41 \left[\dfrac{kW}{m^3}\right]$ and $V_f = 10 \, [m^3]$.

5.4 Experimental Results with the Proposed Burner

The experimental tests aimed to find the ignition capacity and burning efficiency (in pulverized state) in an medium power installation.

The tests were made in the pilot furnace that has 2 MWt form UPB, where it was implemented the new wood silvers burner and multiple swirling jets (620 kW thermal power).

Figure 15 show the saw blade placed at the front of the boiler, with assembly details (*a*) and front view (*b*).

Figure 16 present the flame in the furnace, visualized from the back door or the furnace, using a digital camera. It is noticed a strong intensity and radiation of the flame, similar to pit coal flame.

a)   b)

Fig. 15: The burner of boiler.

Fig. 16: The general aspect of the flame evolution.

The thermodynamics dimensions measured in the flame and the general aspect of the flame show an intensive burning, thus this burning technology is recommended.

The technology of sawdust suspension burning with swirling burners makes possible the increase of power from medium to high-energy production.

So, if the tested burner has a thermal power of 620 kWt, following the principles to realize it, the power can be increased to 3 MWt for the next burners.

The coupling of more burners permits the achievement of steam generators or hot water boilers with power over 10 MWt.

The main experimental elements tested are shown in Table 11.

Table 11: Characteristic data.

Measure	Minimum value	Maximum value
The ratio of HRG flow and biomass flow (sawdust), $q_B \left[\dfrac{m_N^3}{kg} \right]$	$q_B = 0 \left[\dfrac{m_N^3}{kg} \right]$	$q_B = 0.05 \left[\dfrac{m_N^3}{kg} \right]$
The pneumatic transportation capacity, $c_B \left[\dfrac{kg}{m^3} \right]$	$c_B = 0.24 \left[\dfrac{kg}{m^3} \right]$	$c_B = 0.24 \left[\dfrac{kg}{m^3} \right]$
Mass ratio, $c_B^* \left[\dfrac{kg}{kg} \right]$ the sawdust concentration for transportation	$c_B^* = 0.19 \left[\dfrac{kg}{kg} \right]$	$c_B^* = 0.19 \left[\dfrac{kg}{kg} \right]$
Excess air λ	1.06	1.45
Oxygen concentration in combustion gases, O_2 [%]	1.2	6.6
The concentration of carbon monoxide in the flue gases, $CO \left[\dfrac{mg}{m^3} \right]$	19.8	43
The concentration of nitrogen oxides in the combustion gases, $NO_x \left[\dfrac{mg}{m^3} \right]$	375	387

5.5 Experimental Conclusions

The designed burner successfully achieved the rated thermal power 620 kW, for a sawdust flow capacity of 80–85 kg/h and 4 Nm³ HRG. In all these stages, the flame had high temperatures, similar to pit coal flame.

The flame had high brilliancy aspect and filled the entire furnace volume. Under the influence of HRG, the burning was stable and the emissions were low.

The CO emission, which is high when using a layer burning biomass, was extremely low, under 43 ppm, when using suspension burning sawdust. This aspect offers an advantage to sawdust suspension burning as also the possibility to achieve a higher power.

6. Hydrogen—An Energy Vector in Efficient Combustion of Willow Wood

6.1 Introduction

An innovative and efficient technology for burning solid biomass, like willow, is its combustion in air jet. This technology allows the combustion of solid biomass in form of chips, as was harvested. We distinguish the following efficiency aspects:

- no need for mechanical processing for willow after harvest;
- increased thermal power that recommend the process for energy production;
- a swift and efficient combustion.

Regardless the technology, the CO emission is quite high for solid biomass combustion (Boyd et al. 2000). Combustion in a stream of hydrogen, with a thermal participation up to 8% allows controlling CO emissions. At the end of furnace the emissions reached the generally accepted values of max 100 ppm.

Achieving high power plants using solid biomass involves compliance with the environmental protection regulations. CO is the main pollutant that must be removed; consequently, the combustion of solid biomass with hydrogen is an effective way to solve this problem.

Combustion of solid biomass in airflow is similar to the pulverized coal combustion. Consequently, the combustion chamber (the furnace) will be equipped with a suitable burner. In our previous research, two burners of 600 and 620 kW have been built (Prisecaru et al. 2010, Mihaescu et al. 2012, 2013).

6.2 Willow Combustion Efficiency in Jet of Air and Hydrogen

Hydrogen was introduced with the willow chips in the burner, resulting a perfect blend between solid phase, air and hydrogen. The safety analysis concluded that the continuous hydrogen and solid fuel supply system,

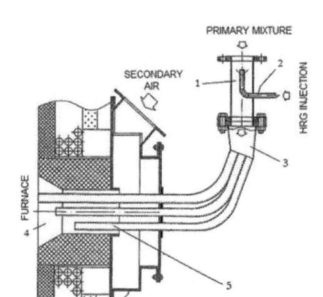

Fig. 17: HRG Injection system.
1-Injection channel; 2-HRG injector; 3-distribution channel; 4-embrasure; 5-fuel channel.

besides the furnace high-speed air jet, does not allow for the return of the hydrogen flame. The return of the flame may be possible due to the poor burn rate between hydrogen and biomass.

Figure 17 shows a scheme of the solid biomass burner and the hydrogen (HRG) injection.

Elemental analysis of the solid fuel is:

$$C^i = 40 - 44.8\% \quad H^i = 4.3 - 4.8\% \quad O^i = 33 - 36.3\%$$
$$Sc^i = 0\% \quad N^i = 0.7 - 1\% \quad A^i = 2.3 - 5.6\% \tag{38}$$
$$W^i = 9.8 - 13.7\%$$

Heating value varied in range 14115–19860 kJ/kg.

6.3 *Experimental Results on Burning the Energy Willow with Hydrogen*

Hydraulic resistance of pneumatic circuit connected to the burner was thus calculated to create compatibility with the boiler auxiliaries. Thus, for the primary agent (air and solid biomass particles), for the secondary and tertiary air circuit, hydraulic resistance is less than 100 mmH$_2$O and it is performed by the ventilation system.

The thermal regimes of fluid circuits are in the following range: the temperature of the primary agent: 60–90°C; The value is imposed by the solid biomass humidity at the entrance of the supply system and the primary air temperature.

The test data for willow and hydrogen is shown in Table 12.

Table 12: Test data for willow and hydrogen.

	Minimum value	Maximum value
Primary air temperature, t_1 [°C]	150	200
Secondary air temperature, t_2 [°C]	150	220
Tertiary air temperature, t_3 [°C]	150	220
Primary air accounted of total combustion air	30	40
The ratio of HRG flow and biomass flow (sawdust), $q_B \left[\dfrac{m_N^3}{kg} \right]$	0.04	0.0515
The pneumatic transportation capacity, $c_B \left[\dfrac{kg}{m^3} \right]$	0.22	0.24
Mass ratio, $c_B^* \left[\dfrac{kg}{kg} \right]$ the sawdust concentration for transportation	0.17	0.19
Excess air λ	1.05	1.40
Oxygen concentration in combustion gases, O_2 [%]	1.25	6.6
The concentration of carbon monoxide in the flue gases, $CO \left[\dfrac{mg}{m^3} \right]$	20	40
The concentration of nitrogen oxides in the combustion gases, $NO_x \left[\dfrac{mg}{m^3} \right]$	370	380

6.4 Theoretical Analysis of Impact of the Solid Biomass Combustion in Hydrogen Jet in Developing New Innovative Technologies

Experimental researches on the efficiency of solid biomass combustion using hydrogen intake has been conducted in the fuel laboratory and combustion installations from the University POLITEHNICA of Bucharest were used (Mihaescu et al. 2012, 2013). The research has been focused on two technologies for wood chips combustion.

The first technology is an application using fluidized bed combustion, the fluidization is taking place only in a room located in the furnace, and the second has similarities with the pulverized combustion conditions.

The research focused particularly on the combustion efficiency for the 0–30 mm wood biomass or willow chips (Hydrogendrich 2004), in order to reduce costs resulted from the additional machining after harvest (sort 0–30 mm resulting either once with harvesting or later after harvest).

The use of hydrogen was evaluated for reducing the flame length and carbon monoxide emissions performing experimental research.

The results for the use of hydrogen in order to support the combustion of solid biomass showed an excellent performance, particularly in reducing the emission of carbon monoxide by 30–100 times.

CO emissions were below 40 ppm for all cases that were taken into consideration.

Burning rate of wood biomass minced to 0–30 mm was 17 m/s, a very high value.

From the results of experimental research concerning solid biomass combustion in hydrogen jet, two innovative technologies have been proposed. The combustion of solid biomass with hydrogen in a tunnel or in a combustion chamber located before the furnace of the boiler.

This combustion technology has applications for low thermal power boilers, whose furnace have a small size. For this case, a complete combustion is not allowed. The technology refers to wood biomass combustion, including willow chips.

Such combustion is performed in two stages, the ignition and combustion of fine fraction in the tunnel and the completion of combustion in the burner combustion chamber.

Hydrogen will be admitted into the burner tunnel, where by its high speed combustion will catalyze the thermo-gas dynamic processes.

For overall efficiency, the combustion will be achieved in a proportion of not less than 70% in the burner tunnel. The thermal proportion of hydrogen determined from the previous program of experimentation will be maintained, 2.8 to 4.8%.

These data are for an averaged analysis of wood, close to those imposed by willow combustion.

With this technology, it is estimated that boilers with very low superheated steam flow (in range of 1.3 to 4.8 t/h) will be designed, enabling the development of micro electricity production plant of 100–600 kW.

Such a micro power plant, possibly in a cogeneration cycle will achieve our economic target "energy crop—local use in energy production".

This concept will shape the size of the renewable fuel culture and will guide the financial efficiency only on energy production.

Positive results from biomass combustion with hydrogen support can be extended to biomass combustion with gas producer support. Energy is thus obtained in a complex power plant comprising:

- Gas-producing or pyrolysis for solid biomass;
- Installation for producing mechanical work with gas producer or pyrolysis gas comprising an internal combustion engine or gas turbine;
- Steam boiler firing biomass with producer gas or pyrolysis gas support, coupled to a turbine.

This power plant guides gaseous fuel obtained from solid biomass to two users, one direct energy producer and the other one is a user for steam boiler thermal support. Figure 3 shows, schematically, the proposed power plant. The power plant allows greater heat generation in CHP scheme, with applications especially when using gas turbines.

Acknowledgement

This work was supported in part by a grant of the Romanian Ministry of Research and Innovation, CCCDI – UEFISCDI, project number PN-III-P1-1.2-PCCDI-2017-0404/31PCCD/2018, within PNCDI III.

References

Barta, S., I. Pisa and L. Mihaescu. 2011. ERPEK company conception for boilers with wood biomass automatic feeding system. The 33-rd International Symposium of the Section IV of CIGR, Bucharest, June 2011.

Birtas, A., I. Voicu, R. Chiriac, N. Apostolescu and C. Petcu. 2009. Constant volume burning characteristics of HHO gas. Theory Pract. Energetic Mater 8: 244–50.

Carcassi, M.N. 2005. Deflagrations of H2–air and CH4–air lean mixtures in a vented multi compartment environment. Energy 30(8): 1439–1451.

Clayton, D.D. 2003. Handbook of Isotopes in the Cosmos: Hydrogen to Gallium. Cambridge University Press. ISBN 0-521-82381-1.

Demirbas, A. 2001. Biomass resource facilities and biomass conversion processing for fuels and chemicals. Energy Convers. Manage 42: 1357–78.

Dickinson, E. 1976. Thermal conductivity of noble gas + hydrogen mixtures: the separability of translational and internal energy transport. Chemical Physics Letters 42(1): 64–68.

Griffiths, D.J. 1995. Introduction to Quantum Mechanics, Prentice Hall, Upper Saddle River NJ, ISBN 0-13-124405-1.

Huang, Z., Y. Zhang, K. Zeng, B. Liu, Q. Wang and D. Jiang. 2006. Measurements of laminar burning velocities for natural gashydrogen-air mixtures. Combust Flame 146: 302–11. 380 POWER ENGINEERING: Advances and Challenges, Part A Hydrogen. 2005.

Considine, G.D. 2005. Van Nostrand's Encyclopedia of Chemistry. Wiley-Interscience, Hoboken, NJ.

Hydrogendrich, D.A. 2004. Gas from fruit shale via supercritical water extraction. Int. J. Hydrogen Energy 29: 1237–43.

IEA 2016, KEY WORLD ENERGY STATISTICS, https://www.iea.org/publications/freepublications/publication/ KeyWorld2016.pdf.

Karim, G., I. Wierzba and S. Boon. 1985. Some considerations of the lean flammability limits of mixtures involving hydrogen. Int. J. Hydrogen Energy 10: 117–23.

Kuzman, R. 1976. Handbook of Thermodinamic Tables and Charts, Hemisphere Publishing Corporation, McGraww-Hill Book Company.

Lazaroiu, Gh., L. Mihaescu, I. Pisa, E. Pop, G.P. Negreanu and V. Berbece. 2014. Hydrogen—An Energy Vector In Efficient Combustion of Energy Willow, 2014 49th International Universities Power Engineering Conference (Upec), 49th International Universities Power Engineering Conference (UPEC), ClujNapoca, Romania, Sep. 02–05, 2014.

Lomax, F., M. Lyubovsky and Z. Wang. 2009. Low-cost hydrogen. Distributed production system. Development. Final Technical Report DE-FG36e05GO15026. Mckendry, P. 2002. Energy production from biomass (part 1): overview of biomass. Bioresour. Technol. 83: 37–46.

Mihaescu, L., T. Prisecaru, M. Georgescu, G. Lazaroiu, I. Oprea, I. Pisa, G. Negreanu, E. Pop and V. Berbece. 2012. Construction and testing of a 600 kW burner for sawdust in suspension, COFRET'12, 11–13 Juin Sozopol, Bulgary.

Mihaescu, L., E. Pop, M.E. Georgescu, Gh. Lazaroiu, I. Pisa, G.P. Negreanu and C. Ciobanu. 2013. Testing of a 620 KW burner for sawdust in suspension and gas enrichen in hydrogen (HRG). In Proc. 2nd Int. Conf. Thermal Equipment, Renewable Energy and Rural Development, June 20–22.

Muresan, M., C.C. Cormos and P.S. Agach. 2013. Techno-economical assessment of coal and biomass gasification-based hydrogen production supply chain system. Chem. Eng. Res. Des. 91: 1527–41.

National Academy of Engineering, National Academy of Sciences. 2004. The Hydrogen Economy: Opportunities, Costs. National Academies Press. ISBN 0-309-09163-2.

Pisa, I. 2013. Combined primary methods for NOx reduction to the pulverized coal-sawdust co-combustion. Fuel Process Technol 106: 429–438.

Pisa, I., Gh. Lazaroiu, L. Mihaescu, T. Prisecaru and G.P. Negreanu. 2016. Mathematical model and experimental tests of hydrogen diffusion in the porous system of biomass. International Journal of Green Energy 13: 774–780.

Pisa, I. and L. Mihaescu. 2011. The Romanian boilers endurance in the biomass combustion. Third International Conference on Applied Energy, Perugia, Italy, 978-889-058-430-5, 1735–1740.

Pisa, I., Gh. Lazaroiu and T. Prisecaru. 2014. Influence of hydrogen enriched gas injection upon polluting emissions from pulverized coal combustion. International Journal of Hydrogen Energy 39: 17702–17709.

Prisecaru, T., L. Mihaescu, Cr. Petcu and R. Popescu. 2010. Co-combustion of coal with hydrogen enriched gas. In: Proc. 9th TECN&URE, p. 145–8.

Prisecaru, T., L.C. Petcu, M. Prisecaru, E. Pop, C. Ciobanu and I. Pisa. 2010. Co-combustion of coal with hydrogen enriched gas.

Proc. IX National Conference of Nuclear and Classic Thermo-mechanical Equipment. 145–149.

Saxena, R.C., D. Seal, S. Kumar and H.B. Goyal. 2008. Thermo-chemical routes for hydrogen rich gas from biomass: a review. Renew Sustain Energy Rev. 12: 1909–27.

SSB-D Dust Burner, http://www.saacke.de/en/products/special-plants/products/index. phd. New Efficient and Ecologic Energy Vectors (Solid Biomass-Hydrogen) 381.

Tanksale, A. and J. Beltramini. 2010. A review of catalytic hydrogen production processes from biomass. Renew Sustain Energy Rev. 14: 166–82.

Thorat, I.V., D.E. Stephensona, N.A. Zachariasa, K. Zaghibb, J.N. Harba and D.R. Wheelera. 2009. Quantifying tortuosity in porous Li-ion battery materials. Journal of Power Sources 188: 592–600.

Tjaden, B., D.P. Finegan, J. Lane, D.J.L. Brett and P.R. Shearing. 2017. Contradictory concepts in tortuosity determination in porous media in electrochemical devices. Chemical Engineering Science 166: 235–245.

Wierzba, I. and B. Ale. 2000. Rich flammability limits of fuel mixtures involving hydrogen at elevated temperatures. Int. J. Hydrogen Energy 25: 75–80.

Zanganeh, K.E. and A. Shafeen. 2007. A novel process integration, optimization and design approach for large-scale implementation of oxy-fired coal power plants with CO2 capture. Int. J. Greenh Gas Control 1: 147–54.

Zucrow, M. and J. Hoffman. 1977. Gas Dynamics, ISBN-13: 978-0471984405, ISBN-10: 047198440X, SBN-13: 978-0471018063, ISBN-10: 0471018066,Wiley, 99th Edition.

III

Storage of Thermal Energy

CHAPTER 13

Thermal Energy Storage Technologies

Kostantin G. Aravossis and *Vasilis C. Kapsalis**

1. Introduction

Energy storage technology is an efficient and effective way to manage all types of energy, such as mechanical, biological, magnetic, chemical and thermal. Thermal Energy Storage (TES) systems are used particularly in buildings and industrial processes but there is also an increasing interest in micro and nano-scale applications due to their dominant role in the efficiency enhancement. In this chapter, we present energy storage from the thermal point of view. Indeed, energy storage leads to savings of premium fuels and makes the system more cost effective by reducing wastage of energy. It may lead to improvements of the performance of the energy systems due to the capability and the flexibility to smooth the supply and to increase the reliability. Historically, energy storage has become particularly attractive since 1973 because the rapidly increasing availability of power from nuclear power plants made the incremental cost of off-peak power low (Biyikoğlu 2002).

Although today's satellite data in conjunction with forecasting methods and simulation tools have been attaining at high accuracy levels, thermal energy storage is the most proper technology used to achieve the goals of the efficiency enhancement. The advantage of these systems is the

National Technical University of Athens 9, Iroon Polytechniou str, 15780 Zografou, Greece.
* Corresponding author: bkapsal@mail.ntua.gr

storing of thermal energy in a medium, namely heating or cooling, in order to use it at an appropriate time to match the demand and supply curves. TES systems may be used in a seasonal basis of any time scale duration and at the same time affect the peak demand, the consumption and the CO_2 emissions and the efficiency of the overall systems. In this context they also increase the renewables in the energy mix, becoming particularly important for electricity storage in combination with conventional and hybrid configurations where renewable energy resources may be stored for electricity production when they are not available. The thermal energy storage may be separated in sensible, latent and thermochemical technologies regarding the driving force mechanisms of activation.

2. Fundamentals

2.1 *Mechanisms of the Thermal Energy Storage*

Solar radiation exploitation is confined by geographical variations, seasonal cycles, intensity oscillations and intermittent availability. Following the photons conversion to heat or electricity temporary energy storage for later use is a common method to overcome these variations and to match the energy demand and the supply of that resource in a controllable way. Similar cyclic processes are also found in biological photosynthetic systems. A comprehensive analysis of the thermal storage characteristics and behavior may be essential to the understanding of the process and the optimization of its performance under different conditions and requirements. Typically, there are three driving mechanisms for the corresponding thermal energy storage types which are analyzed here. Figure 1 depicts the thermal energy storage implementation and the three main different driving mechanisms respectively.

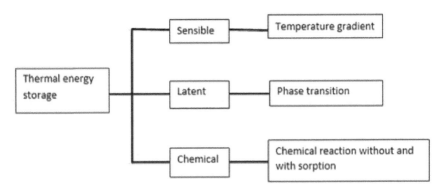

Fig. 1: Thermal energy storage and driving mechanisms.

2.2 The Thermal Heat Storage Cycle

TES technologies are based on the same procedure which is related to the charging, storing and discharging process, aiming to match the accumulated energy in a later time to better serve the demand. This seasonal variation of energy may apply to different scale capacities, depending on the specific applications. Since the thermal energy storage cycle always refers to the circular process, it can be defined by two instances: the decrement factor (DF) and the phase shift of the period or the time lag (τ). The former refers to the percentage reduction of the amplitude of the sinus curve and the later to the delay of the angle of the oscillations before and after the implementation of the thermal energy system. The significant role of thermal energy storage technologies in our everyday life can be realized by their implementation in a wide range of human activities as well as in natural and biological systems. TES systems are used to provide efficient energy supply which leads to sustainable and cost effective resources. The balance between supply and demand, in turn, provides a better performance of the demand and supply curves due to the effective use of the thermal equipment. The specific application of the thermal storage technology depends on many factors, such as temperature and environmental conditions, space requirements and availability, heat losses and economic evaluation, the construction and operation costs, the energy efficiency, the charging and discharging rates, and the storage density capacity (Rezaie et al. 2014).

3. Sensible Thermal Energy Storage (STES)

3.1 Basics in STES Process

When a change in a closed, adiabatic system's energy state condition is experienced, a thermal equilibrium of heat transfer rates is achieved, stating that, $\dot{E}_{in} + \dot{E}_{gen} = \dot{E}_{out} + \dot{E}_{st}$ where, $\dot{E}_{in}$, $\dot{E}_{gen}$, $\dot{E}_{out}$ and $\dot{E}_{st}$ is the incoming, internally generated, out coming and stored energy rates (Fig. 2).

The infinitesimal rate of energy stored in the incompressible control volume, $V = A\Delta x$ is proportional to the temperature gradient, $\dfrac{dE_{st}}{dt} = \rho c V \dfrac{dT}{dt} = \dot{E}_{in} - \dot{E}_{out} + \dot{E}_{gen}$, where, ρ denotes density and c the specific heat of solid. In case where there is no energy generated and following the integration of the first two terms, the thermal energy storage is described by the equation, $E_{st} = mc_p(T_f - T_i)$, where m (kg) is the mass of the material, T (K) states for the temperature in the final and initial state (f and i respectively), $c_p (^{kj}/_{kg.K})$ the specific heat capacity of the material used

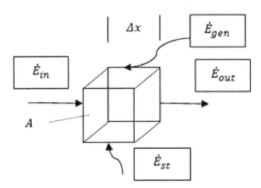

Fig. 2: Schematic energy stored in a solid control volume.

to store thermal energy and determines the amount of energy needed to change the temperature of 1 kg of substance by 1 K.

Nevertheless, the steady state consideration of the systems is useful for a rather bulk analysis. In the static and time independent state where no spontaneous change occurs the macroscopic quantities remain unchangeable, therefore the sum of entropy changes is equal to zero. According to the second law of thermodynamics the system gives the inequality such that the entropy is greater than zero. Therefore, the transient phenomena of the instantaneous charging and discharging cycles are to be taken into account to get a better insight of the sensible thermal storage processes. The behavior analysis during the charging and discharging cycles is a challenging task and reveals the influence between the key parameters during the process (Rezaie et al. 2017). Thus, the rate of heat released and extracted of a thermal storage system is another critical property to consider about and is expressed with the thermal conductivity of the material. As noted, thermal conductivity is dependent on the electron transport of energy (thermal diffusion) and the phonon interactions.

The thermal conductivity is a macroscopic or aggregate property of the materials which reveals what occurs in a molecular level. Moreover, the electron conduction theory and the quantum mechanics result in an analogy consideration between the thermal and the electrical conductivity, through the Wiedemann-Franz Law. Especially in metals, electrons do not obey Maxwellian statistics. They rather follow the Fermi-Dirac statistics and present a fraction of the absorbed energy.

3.2 The Impact of Thermophysical Properties in STES Technologies

Thermophysical properties of the materials, such as the heat transfer coefficients, density, diffusivity, viscosity and the specific heat capacity are involved with the potential of thermal energy storage (Fernandez et al. 2010, Navarro et al. 2012, Khare et al. 2013, Li 2016). The conductivity and the specific heat capacity are defined by the free electrons mobility and the thermal vibrations of the crystalline lattice, the energy quanta which are called phonons. The conduction electrons play a dominant role in metals and their movement is impeded by scattering which is the result of the interaction with phonons or impurities or other imperfections. In semi-conductors the thermal conduction is caused by the electrons where they are stimulated and go over the conduction band. In insulators the dominant role attributes to the phonons where they provide the mechanism for the thermal flux transportation through the material. Time-dependent conduction in materials obeys the first law of thermodynamics.

The transient process is induced by surface convection conditions, surface radiation conditions, a surface temperature or heat flux, and the internal energy generation. Heat transfer mechanisms act simultaneously in an infinitesimal control volume element and are time dependent. Thickness-based Ra critical value on thermal boundary layer triggers convection which draws the energy from the surface. The temperature distribution depends on velocity distribution, the fluid type and the flow regimes. The radiation can proceed even in the absence of continuous medium. Obviously, the sensible thermal energy storage of a given material depends on the value of specific heat capacity (c_p) or the volumetric energy density (ρc_p). The former characterize, for example, the high capability of water and the later the excellent properties of the iron which is also found to present reversible latent heat transition and controlled conductivity (Grosu et al. 2017), as we will see later.

3.3 Sensible Thermal Energy Storage (STES) Technologies

Beside the heat transfer rates and the high specific capacity, the compatibility of the containment and the cost effectives of the cycling process is desirable for long term stability. The sensible TES technologies may be classified on the basis of the heat storage media. Therefore, they are liquid (e.g., water, oil based fluids, molten salts), solid (e.g., bricks, rocks, metals and others) and gas technologies. The thermal storage with liquid media is achieved by heating accumulation of the bulk material (pressurized water, molten salt, etc.) without state changing during the accumulation and later energy recovery which is used as the heat source to drive the demand.

3.3.1 Liquid and Gaseous Fluids

High temperature thermal storage for solar power plants or industrial process heat technologies may use liquid media or two phase heat transfer fluids. Development of high temperature heat storage and ceramic heat exchangers for gaseous heat transfer fluids increase the energy efficiency of power plant and process technologies. For example, the later technology in power plants increase the flexibility of the combined cycle gas turbine plants through deploying high-temperature heat storage tanks.

3.3.2 Low Temperature Fluids

The low temperature fluids are used for thermal energy storage, instead of water, around the temperature of 4°C or below and refer to aqueous solutions containing chemical additives or no aqueous chemicals. They support low temperature air conditioning and some food process applications while they have performed good behavior against corrosion and microbiological control properties.

3.3.3 Aquifers

Aquifers Thermal Energy Storage (ATES) is a proven sustainable technology to provide space heating and cooling when it is coupled with heat pumps. The climatic conditions, the availability of the aquifer and the feasibility of the specific application are the most critical factors for the implementation of such a technology. Aquifers are used as a sink heat pump or sources to store energy ambient air, waste heat or renewable sources. The ATES systems use natural water in a saturated and permeable underground layer as the storage medium. The extraction of the water from a well and the reinjection of it in an appropriate temperature in a nearby well is the main principle of this technology. They may be divided to open and closed or borehole systems. The former is cheaper and provides a greater transfer capacity than the latter and is preferred for a longer period.

The volume of storage is dependent of the thickness and the porosity of the aquifer. The cost effectiveness of the technology is based on the avoided equipment and the lower operation costs while the specific application provides flexibility to the designer incorporating augmentation facilities and combination with dehumidification or desiccant systems. In deep sedimentary basins the temperature gradient from the earth crust or the confined hot water or vapor provides thermal energy storage opportunities and exploitation with several technologies. The heat flow of the hot fluids or the magma structures which are surrounded by low thermal conductivity sediments, results in hydrothermal conduction and convection systems of temperature and pressure gradient, looking for passage towards the surface through the rock permeability and pores. The thermal energy storage potential in geothermal resources primarily exists in rocks and secondarily in fluids that fill the pores and the structure. The three regions of temperatures (low at below 90°C, medium between 90°C and 150°C, and high above 150°C) classify the storage technologies to low, medium and high temperature.

The coupling of these storage technologies with heat exchangers and heat pump systems is a well-known renewable technology and may be configured in vertical, horizontal or hybrid scheme. A review in concepts and applications of ATES systems may be found in the literature (Lee 2010).

3.3.4 Solid Media Storage

The solid medium storage is preferred in several cases of low and high temperatures range because these systems provided the right design and have advantages against the liquid medium storage, for example the container does not leak, freeze or boil. Moreover, the vapor pressure of water and other liquids limited properties may be avoided, while at the same time low cost, high capacity, conductivity and density may be achieved in rocks and peddle beds. Cast iron is also used but the payback period is longer.

Rock beds

Packed rock beds using air or water as heat transfer fluid (HTF) is a very attractive technology where high temperatures may are required. It performs promising applications in CSP (concentrating solar plants) but is also used in hybrid configurations in conjunction with other thermal energy storage technologies. Beside the high storage capacity, there are the chemical and thermo physical properties which play a role in the cycling stability and performance. The quartz and the calcite are the principle minerals controlling rock physical properties while Rhyolite and quarzitic sandstone have performed excellent mechanical ability in thermal cycling. On the other hand the limestone, marble and granite show decreased hardness after each cycle. A number of characteristics are used to describe the thermal energy storage of these technologies, such as the porosity, the cross-sectional of bed and the length, the superficial air velocity and the Reynolds of the heat transfer fluid. The grain size, the void fraction, the particle shape and the grain distribution are the most significant micro structural parameters which influence the porosity. The latter governs the mechanical properties and the density of the rocks (Tiskatine et al. 2016).

Fluidized bed

Fluidized bed thermal storage technology performs faster heat rates than rock beds. This technology is also applied to the waste heat recovery systems (Tiskatine et al. 2016). It is based on predictions of a rigorous method of kinetic theory to derive particular phase viscosities and granular

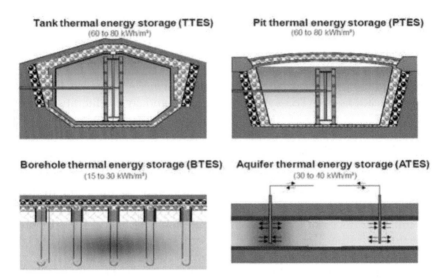

Fig. 3: Different configurations and the specific storage range of STES (Source: Solites).

conductivities. The gas fluidization is observed when gas continuously flows upward through a bed of particles in an appropriate flow rate.

Metals

Metals, as well as molten salts, are used in high temperatures where a high conductivity is needed and the cost is of secondary importance.

3.3.5 Solar Ponds

Solar ponds are shallow bodies of water in which an artificially maintained salt concentration gradient prevents convention. The combination of heat collection, through the radiation adsorption passing the water layers, with long term storage can provide sufficient heat for the entire year (Rabl and Nielsen 1975). The absorbed radiation may be further increased with the dark colored bottom while the cover with an impermeable light transmitting heat insulating layer like the similar technique in swimming pools to avoid evaporation of water. Typically, there are three distinct convection layers, namely the lower zone (LCZ), the non-convective zone (NCZ) and the upper zone (UCZ), Fig. 4.

It is interesting to observe that the insulating layer is the water itself. The vertical salinity which is created in the pond making the deeper layers to contain more salt and become correspondingly denser. In this way it is possible to impede convection and to achieve high bottom temperatures. Instead of polymer covers some others use gels, sufficiently viscous to impede convection, too. This technology is used as a heat sink in large

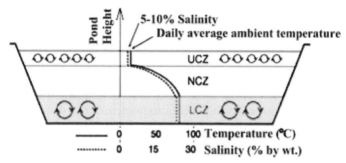

Fig. 4: Typical solar pond technology, salinity and temperature profile. Reproduced with permission from Elsevier (Leblanc et al. 2011).

areas or may be coupled with solar energy applications in roof ponds, in agriculture, as well as in thermally driven separation processes with sustainable desalination membranes (Rahaoui et al. 2017).

3.3.6 Stratification TES Technologies

The variations of temperature distribution cause what is called stratification in the direction of the implemented gradients. For example, in a quiescent fluid reservoir which exhibit a temperature gradient in any direction, such as a core fluid in an enclosure heated from any side or the bulk air in a sealed room. Stratification of TES has a substantial effect on the efficiency of TES systems (Abdoly and Rapp 1982). Significant parameters for a TES, or a set of TESs, include discharging temperature and recovered energy. Series, parallel and general grid (simultaneous series and parallel) TES configurations are considered.

In the parallel configuration, the TES behaves independently. This suggests that the TES consists of different storage media types and sizes, and that there is no restriction on initial temperature of the TES.

In the series configuration, the situation is different because the TESs are connected directly or indirectly through a heat exchanger. If there is no heat exchanger between the TESs, the TES storage media should be the same, because the outlet of one TES is the inlet to the next one in the series. The initial temperature of the second TES must be smaller than the discharge temperature of the first (Rezaie et al. 2013).

There are many factors which affect the efficiency in a stratified storage system. Due to the significance of these technologies regarding the efficiency improvement, there are plenty of investigations using dimensionless numbers which are designed to include the basic parameters of the heat transfer and the thermo physical properties of the process and the materials used.

The Richardson number best defines the stratification in a water tank and the MIX number characterize some problems and bad behavior (Castell et al. 2010). Since the analysis and the investigation of the stratification is very important technology, many developments have been performing advanced analytical procedures to provide a better understanding of the performance parameters which influence the charging and the discharging process. Recent advances in this technology include the use of techniques to improve the stratification in the tank by delaying the mixing process. For example, an equalizer in the dynamic inlet of the water tank may improve the stratification and the fill efficiency while at the same time reduce the mixing process.

Other dimensionless numbers have been investigated in the boundaries of the buildings with air as the heat transfer fluid and the dependence on the temperature gradients are the Rayleigh (Ra) and the Prandtl (Pr) number showing the dependence of the stratification process to the convection and the viscosity of the interactive media (Kapsalis and Karamanis 2015).

3.3.7 Thermocline TES Technology

This technology combines stratification and rock bed thermal storage. Separation of stratified temperature gradients due to a smart exploitation of buoyancy forces led to the thermocline thermal energy storage technologies. The investigation of this method is in an infant stage and a lot of research has to be done in a lab and industrial scale in order for it to be better understood. The key point in this technology is the combination of water storage and solid quartzite rocks at the same tank to cut costs. The storage efficiency of this technology is depended of the mass flow, the inlet velocity and the size of the particles. In general, the smaller the size of the particle the better the storage efficiency due to the better heat transfer between the fluid and the solid (Hoffmann et al. 2017).

4. Latent Heat Thermal Energy Storage (LHTES)

4.1 Basics in LHTES

The stored energy during a latent storage process can be evaluated as:

$$Q = mL \text{ (kJ)} \tag{1}$$

where m denotes the mass (kg) and L (kJ.kg^{-1}) is the specific latent heat of the phase change material (PCM). Examples of PCMs are water/ ice, paraffin and eutectic salts. An example of an industrial PCM is the hand warmer (sodium acetate trihydrate). PCMs are usually packed in tubes, plastic capsules, and wall board and ceilings and they are supplied mainly in three shapes: powder, granulate and board. Specific objectives include embedding thermosiphons and/or heat pipes (TS/HPs) within appropriate PCMs to significantly reduce thermal resistances within the TES system of a large-scale CSP plant and, in turn, improve the performance of the plant. Manufacturing of the proposed heat transfer devices presents a novel opportunity for economic development.

4.1.1 Solidification and Melting Process

Heat is absorbed and released in materials by melting and crystallization in solids or vaporization and condensation in liquids, respectively. In most cases variations of enthalpy with temperature depends on the direction considered and is different for melting and solidification. Solidification is the most critical process, as several problems are addressed and associated with super cooling, nucleation and conduction of heat through the frozen crystalline mass. Three main stages are involved in the crystallization (solidification) process during the phase transition of the material: Induction or nucleation, crystal growth and recrystallization or crystal re growth (Lane 1992). Induction or nucleation includes the formation of nuclei which are grown in appropriate size to be stable. Nucleation could be primary when it refers to all cases which do not contain crystalline matter or secondary where nuclei are often generated in the vicinity of crystals present in a supersaturated system. Furthermore, corresponding to the liquid–solid phase transformation, primary nucleation is separated in to homogeneous and heterogeneous one. The former refers to nuclei formed by the PCM itself and the latter by the dust or impurities in the liquid phase container wall or the additive crystals incorporated in the PCM to initiate the crystallization or to maintain the frozen state in contact with the liquid. Once the nucleation centers have been formed crystal growth occurs. Thus, the material diffuses to a nucleus and is absorbed to its surface, migrates along its surface and incorporates at preferred locations. As the process continues, the size of small crystals formation increases and becomes large enough to sustain a rapid rate of crystal growth, until the solidification procedure reaches to completion and the rate of crystallization slows.

The process continues with the recrystallization which modifies the particle shape and size distribution after the material has totally solidified. Recently, solidification process has been further examined as a heat transfer process for internal convection and external conduction under a statistical thermodynamic kinetic theory in the interfacial boundaries. The distinct regimes of convection, nucleation, transient, and film solidification are explored (Roh 2014). Three heat fluxes and four corresponding activation temperatures are implemented to model the control mechanisms of solidification which, in turn, play a major role in the dendrites formation as well as in other phase transitions, like growth or melting, as well as to a wide range of demanded applications such as laser process, welding, foundry casting, in an analogy of boiling mechanisms (Roh 2014). In the melting point with no heat gain or loss no growth or dissolution occurs. Added crystals above the melting point, where the solution is saturated, tend to dissolve. Convective heat transfer dominates the process here and superheating phenomena are associated with. In encapsulated PCMs contact melting occurs when the solid is free moving within the capsule due to the density difference between the solid and the liquid phase. The shrinking solid affect the geometrical shape and size and, therefore, the melting process. This phenomenon affects the time of the process (Dincer 2002). Recently Ho et al. (2015) reported the complexity of the transient transport which includes density gradients at early stages and free moving boundaries.

4.1.2 Phase Segregation

The importance that the energy storage had been receiving in the latter years led to a great number of studies on the solid–liquid interface where more parameters involved in the phenomenon had been examined. Colloidal suspension may appear after incongruent melting around nano particles used for thermal conductivity enhancement and the precipitation of impurities after several cycles of the charging/discharging process. Origin theories to explain the mechanisms of phase change materials segregation were based on the critical velocity concept (Uhlmann 1964). Later Peppin (2006) analyzed the nonlinear functional dependence of the diffusion coefficient on the volume fraction and found that for small particles, typically below micrometer scale, Van der Waals forces and Brownian diffusion dominates and constitutional super cooling leads to the instability of the interface. For larger particles, the above interactions are weak and the particles form a porous layer above the interface. In this case constitutional super cooling reaches a maximum near the surface of the layer and the porous medium itself becomes potentially unstable. In case of stable systems there exists the possibility of secondary nucleation. Arias (Arias and Wang 2015) in recent research also showed that in PCM storage systems the non-gravitational segregation due to motion of the solidification front plays an important role when the particle diameter is small while the gravitational segregation plays more an important role when the particle diameter increases to a certain size. It seems that beside the dynamics, a thermodynamic condition must be accomplished during the engulfing process namely that the net free energy change of the system is negative. In any case additives are needed for the decomposition to be avoided.

4.1.3 Super Cooling

The high degree of super cooling is due to the fact that either the rate of nucleation or the rate of growth of these nuclei (or both) is very slow. The thermodynamic equilibrium in a given temperature becomes unstable, leading to supersaturated or super cooled situations, both of them are considered two sides of the same phenomenon. When isothermal processes occur, the former name is used, while the latter name is preferred in poly-thermal ones. Lane (1992) presents some empirical evidence which connects the tendency to super cool with the viscosity of the melt in the melting point. Materials with high viscosity in the liquid state, have low diffusion coefficients for their constituent atoms (or ions) and these are unable to rearrange themselves to form a solid and, instead, the liquid super cools. The advantage of the material to store energy is reduced because the melt does not solidify at the thermodynamic melting point. Proper nucleating agents return the solution in equilibrium. Some other concepts permit PCM to come in direct contact with heat carriers providing good heat exchange without heat exchanger or capsules. An immiscible fluid is bubbled to the bottom of a fused PCM and heat is transferred as droplet rise. The immiscible fluid agitates the PCM and the disadvantage of super cooling is minimized (Sokolov and Keizman 1991). Tian (2013) proposes the employment of metallic surfaces to promote heterogeneous nucleation and this reduces super cooling as well.

4.1.4 Compatibility

Packaging of encapsulated PCMs and containers with PCMs are crucial tasks because of the effect between materials. Compatibility tests are required to match the appropriate encapsulated material with the packaging one.

4.1.5 Stability

The usefulness of PCMs is associated with the ability to retain and keep their properties stable for a long time. The stability, according to the classical theory (Gibbs 1948), occurs above a maximum overall excess free energy which corresponds to a certain critical size of crystal nuclei. The constituent molecules are coagulated, resisted to tendency to dissolve and orientated into fixed lattice. Moreover, the thermal stability is associated with the vapor pressure of the material surface. In recent research (Behzadi and Farid 2014) has been reported that the exposure of PCMs in temperatures well above the transition point may affect the long term stability. They pointed out the need to take care of the melting point of PCMs or to encapsulate them. The usage of nucleating and thickening agents as well as supporting materials for shape and form stability of the PCM system can be found in bibliography (Efimova et al. 2014, Memon 2014).

4.2 The Impact of Thermophysical Properties in LHTES

Practically, we use their capability for thermal energy storage where we need it. An extended research for a wide range of temperatures can be found, but not limited to, industry (Peiró et al. 2015), agriculture (Nishina 1984, Reyes 2014, Takakura 1981), cold storage and transportation (Oró et al. 2012, 2014), textiles (X. 2001, Sarier 2012, Gao 2014), vehicles (Ramandi 2011, Jankowski 2014), electronics (Kandasamy et al. 2008), batteries (Babapoor et al. 2015), biomedical (Mondieig 2003, Wang 2010, Pielichowska 2010, Lv 2011) and thermoelectric (Omer 2001, Riffat 2001). Plentiful applications for thermal energy storage exists such as their utilization within the industry of the water usage and heating, ventilation and air conditioning, in building applications. They focus in the range of the operating temperatures of the weather, the systems and the comfort conditions of the built environment. Table 1 depicts the design properties of PCMs and the expected properties of phase change materials.

4.2.1 Heat Transfer Enhancement

Conductivity enhancement is a significant factor for thermal storage technologies in order to reduce the charging and discharging response time of the process and consequently the response time of the system. Many attempts have been made to improve PCM properties such as heat transfer enhancement, the increasing of temperature (e.g., in salt hydrates) or the use of a heat pump which raises the temperature of the heat extracted high enough to satisfy the thermal needs. Another common method that plays a key role in the electricity production as well as the industrial heat management is the specific design of finned heat exchanger tubes which improved the heat transfer due to the used phase change material. In this technology, of great significant is the exact investigation of the role of the conduction and the convection dominated mechanisms to define in a better way the impact of the fins design. Moreover, the use of bimetallic materials in the design of the tubes and the fins performed advantages regarding to the operation conditions in medium temperature range and the stresses.

4.2.2 Design Properties

Some of the most significant properties to design the LHTES technologies are depicted in the table below.

Table 1: Design properties of PCM in latent thermal energy storage technologies and expected benefits.

PROPERTIES	EXPECTED BENEFITS
Thermal ✓ Phase change transition temperature ✓ High transition latent heat ✓ High (Low) thermal conductivity, depending on the application ✓ Exergy optimized	✓ Operation and transition temperature matching ✓ Minimizes the physical size of heat store ✓ Enhanced (Decreased) heat transfer at charging/discharging ✓ Improves energy and environmental performance
Physical ✓ Sufficient phase equilibrium ✓ High density ✓ Small volume change and low vapor pressure	✓ Phase stability and heat transfer settlement ✓ Smaller size of storage container ✓ Reduces the containment problem
Kinetics ✓ Elimination of super saturation or super cooling and capable crystallization rate	✓ Affects the proper heat recovery from the store, especially at salt hydrates
Chemical ✓ Long term stability and compatibility with material of the operational system ✓ Non toxics, nonflammable and non-explosive	✓ Deduces degradation due to hydration loss and chemical decomposition ✓ Health and Safety
Economics ✓ Abundant resources ✓ Large scale availability ✓ Cost effective ✓ Financial return	✓ Environmental acceptance ✓ Demand management ✓ Market penetration ✓ Bankability

4.3 Materials in LHTES

The energy which is stored by nanocomposites, based on Fe_3O_4-functionalized graphene nano sheet (Fe_3O_4–GNS) embedded form-stable polymer phase change materials may be combined with novel magnetic-

and sunlight-driven energy conversion. The excellent magneto caloric performance of Fe_3O_4 and the universal photo absorption and photo thermal conversion of graphene, may effectively convert magnetic or light energy into thermal energy under an alternating magnetic field or solar illumination. Furthermore, they exhibit excellent thermal stability with high melting–freezing enthalpy and excellent reversibility and the novel nanocomposites show the characteristics of form-stable phase transformation. The Fe_3O_4–GNS embedded phase change material composites for energy conversion and storage are expected to open up a rich field of energy materials (Wang et al. 2017).

Advances in the field of LHTES regarding the efficient integration within the corresponding technologies deal with the enhancement of the above properties in order to better implement the cycling process in an efficient matter using composite materials to combine phase stabilizers and high conductivity content or ultrathin materials such as graphene oxide (Wang et al. 2012, Qi et al. 2014, Xu et al. 2017).

The confinement of organic shape stabilized PCM within the matrix of mesoporous composites has been widely investigating due to the behavior change in crystallization and the effects on stability and thermal storage capability, for example in PEG alkyl ether (Brij)/porous silica (MCM-41) where, additional to the bulk, a new peak at $18.8°$ performed in XRD patterns (Zhang et al. 2013).

The supporting and the protective role of graphene oxide nano platelets (GNPs) may be further enhanced when combined with comp-like polymers. The hydrogen bonding and the physical adsorption between the former and the latter results in advanced thermal storage efficiency during the melting and solidification process. The thermal cycling and shape stability also enhanced between the pristine and the composite while a linear dependence of the thermal stability with the GO content has been revealed showing a good barrier effect of the GO nano fillers (Liu et al. 2016). The prevention of leakage in encapsulated PEG above the melting point also is benefited by the combination of the organic PCM with a highly porous and strong three dimension network of lightweight cellulose/GNP aerogel (Xu et al. 2017). The correlation of the stability, the storage cycling and the thermal repeatability with the degree of oxidation of the GO has also been established and provide tunable structures of GAs, and also affect the transformation of hydroxyl groups into carboxyl and epoxy groups. Moreover, the employment of a novel eutectic of Li_2CO_3-Na_2CO_3-K_2CO_3 improved by LiF, as a heat transfer fluid, may eliminate the disadvantage of limited operating temperature range, the low specific heat and the thermal conductivity and is applied as the admirable heat transfer fluid based on molten carbonate salt for thermal storage and heat transfer in high-temperature CSP plants (Zhang et al. 2017).

4.4 Latent Thermal Energy Storage (LHTES) Technologies

The LTES technologies are based on the utilization of the phase change states in order to derive the desired output. Figure 5 depicts an analysis of phase states of materials towards a potential for LTES technologies and emphasizes to phase change materials. Technologies of latent thermal storage systems with PCMs cover a wide range of modern industry and innovative applications. Figure 5 also presents the classification of the different states of LHTES technologies (Kapsalis and Karamanis 2016).

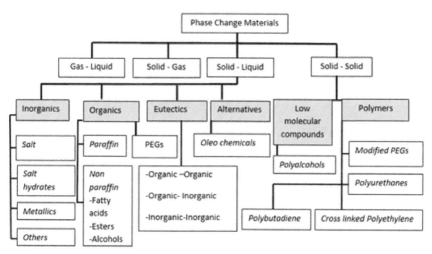

Fig. 5: Phase change materials and states for LHTES technologies (Kapsalis and Karamanis 2016).

4.4.1 Heat Pipe Embedded PCMs

Turnpenny (Turnpenny 2000) used salt hydrate (e.g., $Na_2SO_4.10H_2O$, 21°C m.p.), embedded in heat pipes, maintained sufficient DT between discharging air and melting temperature of PCM and the system provided heat storage capacity of 0.240 kWh for 8 h. Higher temperature difference between melting point and charging air or higher air flow rates are beneficial to the solidification process and can adjust the given time span. The environmental impact of PCM utilization is presented in another experiment. He proposed that if the air conditioning units were replaced in 2000 London offices by free cooling units of 1000 Wh latent heat storage within 2–3 hours and heat transfer rate of 200 W, this would have had the potential of saving 430 t of CO_2. In Fig. 6 the examples depict the utilization of LHTES surrounding the HTF tubes, embedded in heat pipes and in industrial process efficiency improvement (right).

Also, LTES are embedded with heat pipes and PCM's stored within the framework of porous metal foam. A transient, computational analysis of the metal foam enhanced LTES system during charging and discharging process can be found in the literature (Nithyanandam and Pitchumani 2014). In Fig. 7, the combined PCM and Heat Pipe (HP) heat sink is investigated as anti-thermal shock technology in cycling LED operation,

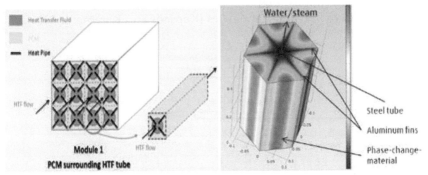

Fig. 6: Left. Latent heat thermal energy storage system with embedded heat pipes to reduce thermal resistance (Source: iea.org) **Right.** Latent heat thermal energy storage system to facilitate enhancement efficiency to industrial processes (Source: Linde Group).

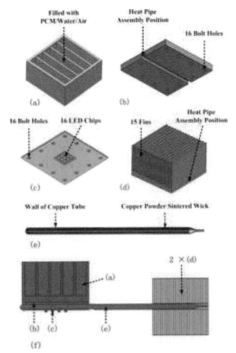

Fig. 7: PCM–Heat Pipe heat sink as anti-thermal shock technology in LEDs (Wu et al. 2016).

is depicted (Wu et al. 2016). The results showed an enhanced thermal performance owing to the reduction of heat rate and peak temperature.

4.4.2 Building Embedded

Passive

First experiments with PCMs tested passive solar heating strategy (Telkes 1980) and resulted in better performance than ordinary walls (Chandra 1985, Ghoneim et al. 1991), increased conductivity and overall efficiency beside the efforts to optimize the thickness (Knowles 1983, Stritih and Novak 1996). Later Ismail (Ismail and Henriquez 2001) succeeded in achieving a highly responsive behavior and reduced heat loss during the cold seasons. The solar wall has a lot of advantages over other systems such as:

a) The high efficiency of conversion solar energy into latent heat and absorption of solar radiation directly into paraffin through transparent plastic glass which acts as an insulation material and, at the same time, prevents convective and radiation losses into the surrounding

b) Heat storage in the form of low energy also decreases lost heat and causes less failure on the wall construction, in comparison with existing systems and

c) Conductive heat losses from the room are also decreased on the surface where the wall is situated.

Since 80s, Lane (1988) researched the latent storage in buildings envelope using PCMs in the range of 15–35°C. Neeper (2000) found that the maximum result occurs when the melt temperature equals the wallboard temperature and noticed that may be there is a limit in the diurnal storage succeeded in practice. Feldman (1989a,b, Feldman 1991) and Kissock et al. (1998) reported early beneficiary results for passive strategies. Athienitis et al. (1997) increased the thermal capacity up to 100–130%, in the range of 16–20.8°C. Scalat et al. (1996) with the use of wallboard PCM showed that the room comfort temperature was maintained for sufficient time after the heating or cooling off mode in Quebec climate conditions.

Hawes et al. (1993) enhanced the storage capacity by 200–230% from the conventional within a 6°C change. Lee (2000) examined two types of concrete blocks impregnated in butyl stearate and paraffin PCM. One regular, being of Portland type cement, steam cured at atmospheric pressure while the other autoclaved consisting of Portland cement and silica, were steam cured under high pressure. The latter absorbed more PCM than the former, while both of them performed doubled thermal storage capacity under the presence of PCM compared with the conventional types. Hadjieva (2000) impregnated autoclaved porous concrete in sodium thiosulphate pentahydrate ($Na_2S_2O_3.5H_2O$) and the porous and capillary spaces found to absorb 60 wt% of the PCM while in repeated thermal cycling only 10% reduction of thermal capacity occurred. It also eliminated the problems of super cooling and phase seperation. Applications in porous aggregates clay and tiles can be found in bibliography (Zhang 2005) and (Cerón et al. 2011) respectively. Schossig (2005) reduced the cooling demand when used different micro encapsulated PCM products in construction materials (dispersion based plaster with 40% weight PCM-6 mm thickness and gypsum plaster with 20% weight PCM-15 mm thickness).

Active

Ice-slurries are suspensions of water and ice particles, in which additives prevent the aggregation of ice in the slurry. They have been used in building air-conditioning and industry cooling process applications with phase transition from ice to water at 0°C or vice versa and store or release large quantities of heat. As the same concept, Gschwander et al. (2005), use PCM—slurries with dispersed micro encapsulated paraffin in water which have higher temperatures melting point, depending on the mixture fraction of the different paraffin.

Yanbing and Yinping (2003) developed and placed fatty acid between the floor and the hung ceiling and the floor and coupled with night ventilation. The maximum amount of the cold discharged from PCM to room was about 300 W during the hot day times and during night time it was about 1 kW. Marin et al. (2004) tested commercially available paraffin based PCMs, encapsulated RT25, using methacrylate in a flat plate type heat exchanger, to allow phase change visualization. The thermal performance improved by using aluminum fins attached to the PCM rectangular container. Recently, the thermal energy storage with PCMs has regained considerable attention in the area of heat pump operation. Real et al. (2014) uses two storage tanks with different PCM melting temperatures, at the operation temperatures of the system. The high temperature tank works as a heat sink and takes advantage of the latent heat to store energy in constant temperature. The low temperature tank works as a cold accumulation and provides more steady operation of the heat pump. COP operation is independent from external conditions and electricity savings in the warm mode are obtained. From the available data a more efficient operation is expected.

4.4.3 Ionic PCMs

Latest advances refer to the composites PCM nanofluids behavior with ionic liquids (ILs), which are organic salts with some interesting properties, such as low vapor pressure and high thermal conductivity. They have been used in the synthesis of nanoparticles as supramolecular solvents due to their advantage to form H-bonds in the liquid phase with a distinct structure. The related applications refer to the sorption cycle, CO_2 capture or as an electrolyte in transparent dye-sensitized solar cells and batteries. Recently, they have been performing a promising potential in energy storage in conjunction with anions as phase change materials in solar thermal applications. The combination of nanoparticles with PCMs in order to develop more efficient heat transfer fluids seems to be an innovative idea because it tends to enhance the thermodynamic efficiency of solidification by decreasing the specific entropy generation rate. The fabrication method and the ionic/solid nanoparticles additives affect the nanofluids thermophysical properties and the critical characteristics of the thermal management PCMs systems. The investigation of the properties of the mixed PCM/ionic surfactant composites in terms of morphology, storage density, mechanical strength and thermal stability resulted in preference of the mixed against the nonionic and ionic surfactants.

5. Thermochemical Energy Storage (TCTES)

5.1 Basics in TCTES

The thermochemical energy stored in the material is expressed by eq. (2):

$$Q = n\Delta H \text{ (kJ)} \tag{2}$$

where n is the moles number of the reactant A (mol^{-1}) and ΔH the reaction enthalpy kJ^{-1}.

The TCTES technologies store more energy density than sensible and latent TES technologies. The volumetric comparison of the energy density of the three technologies, namely sensible, latent and thermochemical, is approximately 50, 100 and 500 $Kwh.m^{-3}$, respectively. Storage in chemical processes occurs at ambient temperature with minimum heat losses while the other two take place at the charging temperature with limited cycling performance due to the losses. However, the applicability of the former is mainly at the laboratory scale (limited industrial applications) compared with the extended industrial use of the others. The TCTES technologies present increased complexity that are rather simple in comparison with the others. When a chemical reaction takes place, there is a difference between the enthalpies of the substances presented at the end and the start of the reaction. This enthalpy difference is known as the heat of reaction. If the reaction is endothermic, it will absorb this heat while it takes place; if the reaction is exothermic, it will release this heat. Any reverse chemical reaction can be used for thermal energy storage, given that the products of the reaction can be stored during the reaction and released when the reverse reaction takes place or vice versa. The amount of heat stored when using chemical reactions can be calculated, using the appropriate enthalpy change, as we already mentioned. As the binding energy in a chemical reaction is usually large, the temperature necessary to destroy the bonds is usually high. Advanced materials are used for the enhancement of properties in thermochemical thermal energy storage.

Chemical reactions accompanied with energy changes. They absorb or release energy as heat with sorption or without sorption process. These processes may be considered and analyzed by thermochemistry which is the study of the energy transformations and transfers accompanying chemical and physical changes. The sorption phenomenon is applied for the same purpose as the sensible and latent heat in thermal storage systems. One way of improving the sorption capacity and kinetics of a sorbent is to increase the surface area by reducing the sorbent size. In the sorption desorption cycle the heat is stored in materials using water vapor taken up by a sorption material. If the material is solid it is called adsorption and when is a liquid it is called absorption. Adsorption thermal energy storage is considered to be a promising technology that can provide an excellent solution for long-term thermal energy storage in a more compact and efficient way. Suitable materials can be organic or inorganic, as long as their reversible chemical reactions involve absorbing and releasing a large amount of heat. Three basic criteria are required to be considered when designing such a thermochemical thermal storage system: excellent chemical reversibility, large chemical enthalpy change and simple reaction conditions to be realized (Tian and Zhao 2013). Absorption is the penetration of the adsorbate through the surface layer of an absorbent with a change of composition.

The adsorption desorption cycle based technologies are well established in several processes. The hysteresis between the reversible cycles provides thermal heat storage potential and is widely used in the industry.

Generally, the hysteresis loop in the desorption process is associated with the emptying and filling of the pores and past research connect it with multilayer effects or a variation of pore diameter along single channels. Delay condensation in the adsorption branch may also be caused by

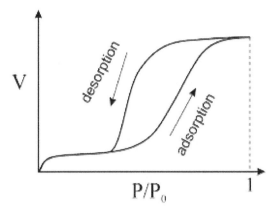

Fig. 8: The sorption desorption hysteresis loop.

metastable multilayers which persist especially in slit formed pores. The classical approach uses a corrected form of the Kelvin equation to evaluate the pore width from the pore filling pressure whereas the liquid–form of adsorptive fills the mesoporous. The solid–fluid and fluid–fluid interactions control the pore filling which takes the form of micro pores filling or the capillary condensation in meso and macro porous materials (Sing 2001). For example, capillary condensation is the mechanism of pore filling of the mesoporous materials with widths pore size in the range of 2–50 nm and occurs in high relative partial vapor pressure ratios $\frac{P}{P_0}$ (> 0.4) where the temperature control is significant.

5.2 Materials in TCTES

The most representative and promising materials used in TCTES technology are $CaCl_2.H_2O$, zeolite and silica gel. They give the high heat for adsorption and ability to hydrate and hydrate while maintaining the structural stability. The hydroscopic property coupled with rapid exothermic reaction which occurs in the transformation from dehydrated to hydrate form makes them effective in thermal storage. The structure of the honeycomb and the high internal surface area are positively related with their ability to store thermal energy or to convert under-utilized resources into useful energy. A breakthrough property during the zeolite and others materials during the hydration/dehydration process refers to the ability to tune the liberated heat with the controllable environment and the amount of water vapor injection. Beside these materials, new sorbents like aluminophosphates (AIPOS), silico-aluminophosphates (SAPOs) and metal organic frameworks (MOFs) have been proposed for heat storage due to their high energy density performance.

The GO/rGO is also a promising material which has been explored in many TCTES technologies. The importance of this material is derived from the unique physical and chemical properties both to the oxygen reduction reaction (ORR) and the oxygen evolution reaction (OER) which is a significant pair in chemical reactions, due to the high surface area, access to large quantities, tunable electronic/ionic conductivity, unique graphitic basal plane structure, and the easiness of modification or functionalization (Wu and Gao 2015).

Moreover, the discovery of Ti3C2 in 2011 boosted the research in the family of 2D transition metal carbides, carbonitrides and nitrides (collectively referred to as MXenes). The availability of solid solutions, the control of surface terminations and a recent discovery of multi-transition-metal layered MXenes offer the potential for the synthesis of many new structures. The versatile chemistry of MXenes allows the tuning of properties for applications including thermochemical energy storage due to surface terminations, such as hydroxyl, oxygen or fluorine, which impart hydrophilicity to their surfaces (Anasori et al. 2017).

5.3 Thermochemical Energy Storage (TCTES) Technologies

The main principle of TCTES is based on a reaction (eq. (3)) that can be reversed:

$$C + heat\ (D) \leftrightarrow A + B \tag{3}$$

In this reaction, a thermochemical material (C) absorbs energy and is converted chemically into two components (A and B), which can be stored separately. The reverse reaction occurs when materials A and B are combined together and C is formed. Energy is released during this reaction and constitutes the recovered thermal energy from the TES. The storage capacity of this system is the heat of reaction when material C is formed. The three stages of the storage cycle are also repeated here, namely the charging, storing and discharging process (Fig. 9).

5.3.1 Sorption Technology

The sorption cycle is classified in open and closed technology depending on the thermodynamic boundaries with the ambient air or in some cases the humidifier from which the sorbate (water vapor) is obtained. The open systems have an easier design and operation. Recent years have seen the sorption cycle technology efficiently integrating in power generation turbines/expanders (Aydin et al. 2015).

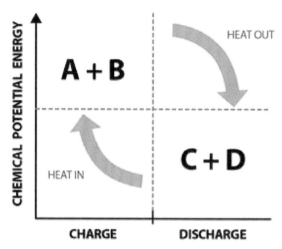

Fig. 9: Typical cycling process of TCTES technologies (Stekli et al. 2013).

5.3.2 *Thermochemical Heat Pumps*

Chemical heat pump

A chemical heat pump that uses a reversible reaction system and is based on the equilibria of a hydration/dehydration process is called a chemical heat pump. Thermal drivability, which does not require mechanical work, is one of advantages of this heat pump. The environmentally friendly and economical nature of the reactants is also advantageous. The functionalization treatment and a proper $Mg(OH)_2$ load were fundamental to better the dispersibility of $Mg(OH)_2$ into the carbon nanotubes bundles which in turn enhanced the thermochemical performance of the active material, fully exploiting for the first time its maximum potential heat storage capacity, that is ~ 1300 kJ/kg$_{Mg(OH)_2}$, thus bringing the development of this technology to a level closer to its industrial application (Mastronardo et al. 2016).

Absorption heat pump (AHP)

The absorption heat pump (AHP) cycles are more appropriate for low temperature heat transfer. Some of the most promising sorbent/sorbate couples for absorption cycles are water–lithium bromide (H_2O–LiBr) and ammonia–water (NH_3–H_2O). Absorption use a two-component working fluid and the principles of boiling-point elevation and heat of absorption to achieve temperature lift and to deliver heat at higher temperatures. The operating principle is the same as that used in steam-heated absorption chillers that use a coupled mixture as their working fluid. Key features of absorption systems are that they can deliver a much higher temperature lift than the other systems, their energy performance does not decline steeply at higher temperature lift, and they can be customized for combined heating and cooling applications. An absorption heat pump uses high temperature prime energy into the desorber, which produces high pressure vapor. The high-pressure vapor is condensed in the condenser where the heat is recovered into a process stream. Subsequently, the high-pressure condensate from the condenser is throttled to a lower pressure in the evaporator, where the waste heat is recovered to vaporize the low-pressure condensate. Concentrated working fluid from the desorber contacts the low-pressure vapor from the evaporator in the absorber. This creates heat that is recovered into a process stream. The working fluid is then returned to the desorber to complete the cycle.

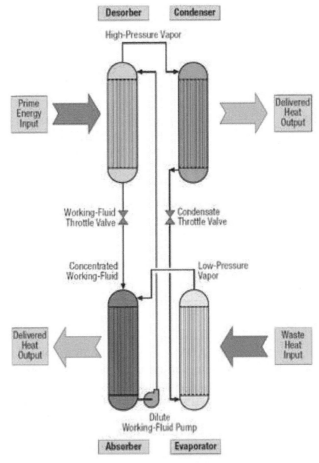

Fig. 10: Absorption cycle technology in industry waste energy recovery application (Source: ITP).

Thermal vapor recompression

This method works in a similar fashion as mechanical vapor recompression. The difference is that thermal vapor compression uses thermal energy as the external energy source to compress vapor instead of mechanical energy. Common applications of this method are evaporation, e.g., in industry drying processes.

Adsorption heat pumps

Adsorption heat pumps is a well-known technology and has been extensively analyzed with several working pair and different configurations (single stage, double stage, cascade, etc.). It is a low grade heat transportation technology and, respectively, it is based to the single effect adsorption cycle, the double effect or the multi effect providing flexibility to the heat source temperatures and the range of applications. A typical, one stage adsorption heat pump with a regenerative cycle of the thermal wave process is analyzed in the literature (Sun et al. 1997).

6. Hybrid Thermal Energy Storage Technologies (HTES)

The utilization of excess energy in a storable manner that could efficiently and effectively balance the demand and the supply curve, has been extensively used in engineering processes, such as mechanical (Lumentut and Howard 2016, Wei and Jing 2017), biological (Ebenhard et al. 2017, Yan et al. 2017), magnetic (Ung et al. 2015, Chen et al. 2016) and chemical (Khademi et al. 2015). Moreover, the energy harvest, which efficiently store energy for a later usage, have been identified as a breakthrough technology in industrial recovery systems (Gutierrez et al. 2016) as well as in renewable processes (Yu et al. 2016).

The technology developments of combined thermal systems boost the efficiency and result in increased renewable fraction coefficient within the system operation while at the same time reduced running life cycle costs. The prerequisite for this configuration is to integrate the waste energy from one resource within the heat pump operation. The configuration of different kind of TES technologies in a compact scheme obtaining a specific purpose, may be characterized as HTES. Therefore, we can recognize plenty of hybrid configurations in the literature. In fact, the innovative utilization of different TES technologies is limited only from the creativity of the designer and the cost effectiveness of the specific applications. So far, the building integration of renewables may provide multifunctional prefabricated elements in façade, windows (Skandalos and Karamanis 2016), wallboards and roofing combined with heat pumps

(Jradi et al. 2017), PV/T (Al-Waeli et al. 2017) and thermoelectricity. The wide range of thermal storage implementation and exploitation within different configurations is a challenging procedure and worthy of further investigation (Fig. 11).

The integration of solar energy in building hybrid applications is becoming a necessary technology to be further investigated (Kalogirou et al. 2016, Dong et al. 2017).

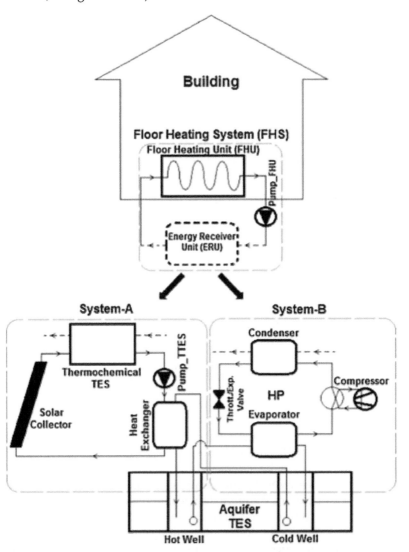

Fig. 11: Combined technologies of thermal energy storage in STES and TCTES in buildings studied by (Caliskan et al. 2012).

Recently, a combined PV-ASHP system employing a seasonal underground thermal energy storage for building purpose has been presented (Jradi et al. 2017). The proposed system consists of three main components: solar photovoltaic units, air-source heat pump and the soil storage medium. In the charging phase, PV panels harness solar irradiation allowing renewable-based electric power generation. Part of this electric power is utilized to fulfil the residential project electric power and lighting demands and the other part is used to run the heat pump. Part of the heat produced at the condenser of the heat pump is utilized to provide the hot water needs where the remaining portion is transferred using a charging loop heat exchanger to heat an underground soil storage bed. Water is used as the heat transfer working fluid allowing heating the soil storage bed. The charging phase occurs in the summer months when there is plenty of solar irradiation and surplus of electric power production. In the winter months, heat stored in the soil bed is discharged through the discharging loop (Jradi et al. 2017). Another hybrid technology refers to the design and usage of a special accumulation device, which is composed of thermal panels based on phase change materials (PCMs) combined with thermoelectricity and heat pumps. The thermal panels have an integrated tube heat exchanger and heating foils. The technology can be used as a passive or active system for heating and cooling (Fig. 12). It is designed as a "green technology", so it is able to use renewable energy sources, e.g., photovoltaic (PV) panels, solar thermal collectors and heat pumps. Moreover, an interesting possibility is the ability to use thermoelectric coolers. The technology for the accumulation of the thermal energy can work in many different modes, which depend on the requirements. The technology can be used as a standard passive system. In this mode,

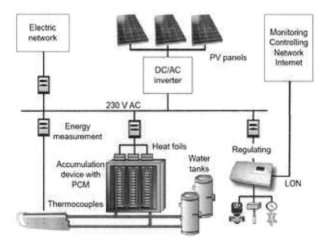

Fig. 12: Configuration of the whole combined technology (Skovajsa et al. 2017).

the technology accumulates the heat or cold. This mode can reduce the temperature peaks and keep a stable temperature in the monitored room during the day and night. The technology can also work in active mode. The measurement indicates that the technology improves the thermal capacity of the building, and it is possible to use it for active heating and cooling (Skovajsa et al. 2017).

7. Assessment of TES Technologies

So far, we presented the most challenging and promising TES technologies which are used in our everyday life. Our concerns in energy, environmental economic and social issues make them appropriate for the integration within the production and consumption systems in a matter which promote efficiency and sustainability. In this context many assessment tools have been useful for proposing the design and the selection process. The majority of TES technologies include combined forms and the decision making process have to take into account advanced methodology and multi criteria analysis in order to approach an optimized solution.

Therefore, the appropriate procedure in the evaluation of TES technologies is becoming a challenging task. A review of the decision making process in the evaluation of such systems in specific applications can be found in the literature (Strantzali and Aravossis 2016). A methodology to build a decision making procedure for this purpose is described in to the recent literature for typical industrial refrigeration coupled with LHTES technology such as the Cristopia Energy System (Xu et al. 2017).

8. Conclusions

Several of the most common TES technologies were presented in this chapter, namely the sensible, the latent and the thermochemical ones. A brief description on the fundamentals was depicted while the properties related with the thermal energy storage were analyzed.

In general, latent and thermochemical storage systems are more expensive than sensible heat systems and are economically viable only for applications with a high number of cycles.

The most efficient technology to store energy is by using thermochemical materials (TCM). The high energy density which is observed in the thermochemical processes in conjunction with the heat losses provide a promising thermal energy store technology.

Special chemicals can absorb/release a large amount of thermal energy when they break/form certain chemical bonds during endothermic and exothermic reactions. The market thus needs such a "thermal battery", which should have a variety of kWhs capacities.

Several key challenges remain in the way of the development of an efficient sorption thermal battery, which are as follows: sorption materials with high storage density and low cost, sorption bed with good heat and mass transfer to ensure charging and discharging power, and being stable after repeated cycles with minimum heat capacity ratio between the inert materials to the sorption thermal energy.

In mature economies (e.g., OECD countries), a major constraint for TES deployment is the low construction rate of new buildings, while in emerging economies TES systems have a larger deployment potential. Nowadays, the developments in thermal energy storage are fully concentrated to enhance the heat storage capacity and bridge the gap between the laboratory and the industrial readiness level.

References

Abdoly, M.A. and D. Rapp. 1982. Theoretical and experimental studies of stratified thermocline storage of hot water. Energy Conversion and Management 22(3): 275–285.

Al-Waeli, A.H.A., K. Sopian, H.A. Kazem and M.T. Chaichan. 2017. Photovoltaic/Thermal (PV/T) systems: Status and future prospects. Renewable and Sustainable Energy Reviews 77: 109–130.

Anasori, B., M.R. Lukatskaya and Y. Gogotsi. 2017. 2D metal carbides and nitrides (MXenes) for energy storage. Nature Reviews Materials 2: 16098.

Arias, F.J. and X. Wang. 2015. Segregation due to motion of front of solidification in phase change materials systems and dependence with shape and dimension factors. Applied Thermal Engineering 75: 366–370.

Athienitis, A.K., C. Liu, D. Hawes, D. Banu and D. Feldman. 1997. Investigation of the thermal performance of a passive solar test-room with wall latent heat storage. Building and Environment 32(5): 405–410.

Aydin, D., S.P. Casey and S. Riffat. 2015. The latest advancements on thermochemical heat storage systems. Renewable and Sustainable Energy Reviews 41: 356–367.

Babapoor, A., M. Azizi and G. Karimi. 2015. Thermal management of a Li-ion battery using carbon fiber-PCM composites. Applied Thermal Engineering 82: 281–290.

Behzadi, S. and M.M. Farid. 2014. Long term thermal stability of organic PCMs. Applied Energy 122(0): 11–16.

Biyikoğlu, A. 2002. Optimization of a sensible heat cascade energy storage by lumped model. Energy Conversion and Management 43(5): 617–637.

Caliskan, H., I. Dincer and A. Hepbasli. 2012. Energy and exergy analyses of combined thermochemical and sensible thermal energy storage systems for building heating applications. Energy and Buildings 48: 103–111.

Castell, A., C. Sole and L.F. Cabeza. 2010. Dimensionless numbers used to characterize stratification in water tanks for discharging at low flow rates. Renewable Energy 35(10): 2192–2199.

Cerón, I., J. Neila and M. Khayet. 2011. Experimental tile with phase change materials (PCM) for building use. Energy and Buildings 43(8): 1869–1874.

Chandra, S., R. Kumar, S. Kaushik and S. Kaul. 1985. Thermal performance of a non-air-conditioned building with PCCM thermal storage wall. Energy Conversion and Management 25(1): 15–20.

Chen, D.C., S.-H. Kao and K.C. Huang. 2016. Study of piezoelectric materials combined with electromagnetic design for bicycle harvesting system. Advances in Mechanical Engineering 8(4): 1–11.

Dincer, I. and M.A. Rosen. 2002. Thermal Energy Storage: Systems and Applications. England, J. Willey & Sons Ltd.

Dong, X., Q. Tian and Z Li. 2017. Experimental investigation on heating performance of solar integrated air source heat pump. Applied Thermal Engineering 123: 1013–1020.

Ebenhard, T., M. Forsberg, T. Lind, D. Nilsson, R. Andersson, U. Emanuelsson, L. Eriksson, O. Hultåker, M. Iwarsson Wide and Göran Ståhl. 2017. Environmental effects of brushwood harvesting for bioenergy. Forest Ecology and Management 383: 85–98.

Efimova, E., P. Pinnau, M. Mischke, C. Breitkopf, M. Ruck and P. Schmidt. 2014. Development of salt hydrate eutectics as latent heat storage for air conditioning and cooling. Thermochimica Acta 575: 276–278.

Feldman, D., M. Shapiro, D. Banu and C.J. Fuks. 1989a. Fatty acids and their mixtures as phase change materials for thermal energy storage. Solar Energy Materials 18: 201–221.

Feldman, D., M.A. Khan and D. Banu. 1989b. Energy storage composite with an organic phase change material. Solar Energy Materials 18: 333–341.

Feldman, D., D. Banu, D. Hawes and E. Ghanbari. 1991. Obtaining an energy storing building material by direct incorporation of an organic phase change material in gypsum wallboard. Solar Energy Materials 22: 231–242.

Fernandez, A.I., M. Martínez, M. Segarra, I. Martorell and L.F. Cabeza. 2010. Selection of materials with potential in sensible thermal energy storage. Solar Energy Materials and Solar Cells 94(10): 1723–1729.

Gao, C. 2014. 9—Phase-change materials (PCMs) for warming or cooling in protective clothing. Protective Clothing. F. Wang and C. Gao, Woodhead Publishing: 227–249.

Ghoneim, A.A., S.A. Klein and J.A. Duffie. 1991. Analysis of collector-storage building walls using phase-change materials. Solar Energy 47(3): 237–242.

Gibbs, J.W. 1948. New Heaven, Yale University Press.

Grosu, Y., A. Faik, I. Ortega-Fernández and B. D'Aguanno. 2017. Natural Magnetite for thermal energy storage: Excellent thermophysical properties, reversible latent heat transition and controlled thermal conductivity. Solar Energy Materials and Solar Cells 161: 170–176.

Gschwander, S., P. Schossig and H.M. Henning. 2005. Micro-encapsulated paraffin in phase-change slurries. Solar Energy Materials and Solar Cells 89(2-3): 307–315.

Gutierrez, A., L. Miró, A. Gil, J. Rodríguez-Aseguinolaza, C. Barreneche, N. Calvete, X. Pyf. I. Fernández, M. Grágeda, S. Ushak and L.F. Cabez. 2016. Advances in the valorization of waste and by-product materials as thermal energy storage (TES) materials. Renewable and Sustainable Energy Reviews 59: 763–783.

Hadjieva, M., R. Stoykov and T. Filipova. 2000. Composite salt-hydrate concrete system for building energy storage. Renew Energy 19: 111–115.

Hawes, D., D. Feldman and D. Banu. 1993. Latent heat storage in building materials. Energy Build 20: 77–86.

Ho, C.J., K.C. Liu and W.-M. Yan. 2015. Melting processes of phase change materials in an enclosure with a free-moving ceiling: An experimental and numerical study. International Journal of Heat and Mass Transfer 86(0): 780–786.

Hoffmann, J.F., T. Fasquelle, V. Goetz and X. Py. 2017. Experimental and numerical investigation of a thermocline thermal energy storage tank. Applied Thermal Engineering 114: 896–904.

Ismail, K.A.R. and J.R. Henríquez. 2001. Thermally effective windows with moving phase change material curtains. Applied Thermal Engineering 21(18): 1909–1923.

Jankowski, N.R. and F.P. Mc Cluskey. 2014. A review of phase change materials for vehicle component thermal buffering. Applied Energy 113.

Jradi, M., C. Veje and B.N. Jørgensen. 2017. Performance analysis of a soil-based thermal energy storage system using solar-driven air-source heat pump for Danish buildings sector. Applied Thermal Engineering 114: 360–373.

Kalogirou, S.A., S. Karellas, K. Braimakis, C. Stanciu and V. Badescu. 2016. Exergy analysis of solar thermal collectors and processes. Progress in Energy and Combustion Science 56: 106–137.

Kandasamy, R., X.-Q. Wang and A.S. Mujumdar. 2008. Transient cooling of electronics using phase change material (PCM)-based heat sinks. Applied Thermal Engineering 28(8-9): 1047–1057.

Kapsalis, V. and D. Karamanis. 2015. On the effect of roof added photovoltaics on building's energy demand. Energy and Buildings 108: 195–204.

Kapsalis, V. and D. Karamanis. 2016. Solar thermal energy storage and heat pumps with phase change materials. Applied Thermal Engineering 99: 1212–1224.

Khademi, F., İ. Yıldız, A.C. Yıldız and S. Abachi. 2015. Advances in algae harvesting and extracting technologies for biodiesel production. Progress in Clean Energy. Novel Systems and Applications 2: 65–82.

Khare, S., M. Dell'Amico, C. Knight and S. Mcgarry. 2013. Selection of materials for high temperature sensible energy storage. Solar Energy Materials and Solar Cells 115: 114–122.

Kissock, J.K., J.M. Hannig, T.I. Whitney and M.L. Drake. 1998. Early results from testing phase change wallboard. Phase Change Materials and Chemical Reactions for Thermal Energy Storage First Workshop, IEA Annex 10.

Knowles, T.R. 1983. Proportioning composites for efficient thermal storage walls. Solar Energy 31(3): 319–326.

Lane, G.A. 1988. Solar heat storage: Latent heat materials. Florida, CRC Press.

Lane, G.A. 1992. Phase change materials for energy storage nucleation to prevent supercooling. Solar Energy Materials and Solar Cells 27: 135–160.

Leblanc, J., A. Akbarzadeh, J. Andrews, H. Lub and P. Golding. 2011. Heat extraction methods from salinity-gradient solar ponds and introduction of a novel system of heat extraction for improved efficiency. Solar Energy 85(12): 3103–3142.

Lee, K.S. 2010. A review on concepts, applications, and models of aquifer thermal energy storage systems. Energies 3(6): 1320.

Lee, T., D.W. Hawes, D. Banu and D. Feldman. 2000. Control aspects of latent heat storage and recovery in concrete. Sol. Energy Mater. Sol. Cells 62: 217–237.

Li, G. 2016. Sensible heat thermal storage energy and exergy performance evaluations. Renewable and Sustainable Energy Reviews 53: 897–923.

Liu, L., L. Kong, H. Wang, R. Niu and H. Shi. 2016. Effect of graphene oxide nanoplatelets on the thermal characteristics and shape-stabilized performance of poly(styrene-co-maleic anhydride)-g-octadecanol comb-like polymeric phase change materials. Solar Energy Materials and Solar Cells 149: 40–48.

Lumentut, M.F. and I.M. Howard. 2016. Electromechanical analysis of an adaptive piezoelectric energy harvester controlled by two segmented electrodes with shunt circuit networks. Acta Mechanica 1–21.

Lv, Y., Y. Zou and L. Yang. 2011. Feasibility study for thermal protection by microencapsulated phase change micro/nanoparticles during cryosurgery. Chem. Eng. Sci. 66: 3941–3953.

Marin, J., B. Zalba, F. Cabeza and H. Mehling. 2004. Free-cooling of buildings with phase change materials. International Journal of Refrigeration 27: 839–849.

Mastronardo, E., L. Bonaccorsi, Y. Kato, E. Piperopoulos, M. Lanzad and C. Milone. 2016. Thermochemical performance of carbon nanotubes based hybrid materials for MgO/H_2O/Mg(OH)$_2$ chemical heat pumps. Applied Energy 181: 232–243.

Memon, S.A. 2014. Phase change materials integrated in building walls: A state of the art review. Renewable and Sustainable Energy Reviews 31(0): 870–906.

Mondieig, D., F. Rajabalee, A. Laprie, H.A. J. Oonk, T. Calvet and M.A. Cuevas-Diarte. 2003. Protection of temperature sensitive biomedical products using molecular alloys as phase change material. Transfus Apher Sci. 28: 143–148.

Navarro, M.E., M. Martíneza, A. Gilb, A.I. Fernández, L.F. Cabeza, R. Olives and X. Pyc. 2012. Selection and characterization of recycled materials for sensible thermal energy storage. Solar Energy Materials and Solar Cells 107: 131–135.

Neeper, D.A. 2000. Thermal dynamics of wallboard with latent heat storage. Sol. Energy 68: 393–403.

Nishina, H. and Takakura, T. 1984. Greenhouse heating by means of latent heat storage units. Acta Hort (Energy in Protected Cultivation Ill) 148: 751–754.

Nithyanandam, K. and R. Pitchumani. 2014. Computational studies on metal foam and heat pipe enhanced latent thermal energy storage. Journal of Heat Transfer 136(5): 051503-051503-051510.

Omer, S.A., S.B. Riffat and X. Ma 2001. Experimental investigation of a thermoelectric refrigeration system employing a phase change material integrated with thermal diode (thermosyphons). Appl. Thermal Eng. 21(12): 1265–1271.

Oró, E., A. de Gracia, A. Castell, M.M. Farid and L.F. Cabeza. 2012. Review on phase change materials (PCMs) for cold thermal energy storage applications. Applied Energy 99: 513–533.

Oró, E., L. Miró, M.M. Farid, V. Martin and L.F. Cabeza. 2014. Energy management and CO_2 mitigation using phase change materials (PCM) for thermal energy storage (TES) in cold storage and transport. International Journal of Refrigeration 42: 26–35.

Peiró, G., J. Gasia, L. Miró and L.F. Cabeza. 2015. Experimental evaluation at pilot plant scale of multiple PCMs (cascaded) vs. single PCM configuration for thermal energy storage. Renewable Energy 83: 729–736.

Pielichowska, K.B.S. 2010b. Bioactive polymer/hydroxyapatite (nano)composites for bone tissue regeneration.

Peppin, S.S.L., J.A.E. Elliott and M.G. Worster. 2006. Solidification of colloidal suspensions. J. Fluid Mech. 554: 147–166.

Qi, G.-Q., C.L. Liang, R.-Y. Bao, Z.-Y. Liu, W.Y. Bang-Hu and X.M.-B. Yang. 2014. Polyethylene glycol based shape-stabilized phase change material for thermal energy storage with ultra-low content of graphene oxide. Solar Energy Materials and Solar Cells 123: 171–177.

Rabl, A. and C.E. Nielsen. 1975. Solar ponds for space heating. Solar Energy 17(1): 1–12.

Rahaoui, K., L.C. Ding, L.P. Tan, W. Mediouri, F. Mahmoudi, K. Nakoa and A. Akbarzadeh. 2017. Sustainable membrane distillation coupled with solar pond. Energy Procedia 110: 414–419.

Ramandi, M.Y., I. Dincer and G.F. Naterer. 2011. Heat transfer and thermal management of electric vehicle batteries with phase change materials. Heat Mass Transfer 47(777-788).

Real, A., V. García, L. Domenech, J.Renau, N. Montés and F. Sánchez. 2014. Improvement of a heat pump based HVAC system with PCM thermal storage for cold accumulation and heat dissipation. Energy and Buildings 83(0): 108–116.

Reyes, A., A. Mahn and F. Vásquez. 2014. Mushrooms dehydration in a hybrid-solar dryer, using a phase change material. Energy Conversion and Management 83: 241–248.

Rezaie, B., B.V. Reddy and M.A. Rosen. 2013. Configurations for multiple thermal energy storages in thermal networks. IEEE International Conference on Smart Energy Grid Engineering, SEGE 2013.

Rezaie, B., B.V. Reddy and M.A. Rosen. 2014. Energy analysis of thermal energy storages with grid configurations. Applied Energy 117: 54–61.

Rezaie, B., B.V. Reddy and M.A. Rosen. 2017. Thermodynamic analysis and the design of sensible thermal energy storages. International Journal of Energy Research 41(1): 39–48.

Riffat, S.B., S.A. Omer and X. Ma. 2001. A novel thermoelectric refrigeration system employing heat pipes and a phase change material: an experimental investigation. Renew Energy 23: 313–323.

Roh, H.-S. 2014. Heat transfer mechanisms in solidification. International Journal of Heat and Mass Transfer 68(0): 391–400.

Sarier, N. and E. Onder. 2012. Organic phase change materials and their textile applications: A review. Thermochimica Acta 540(20): 7–60.

Scalat, S., D. Banu, D. Hawes, J. Paris, F. Haghighata and D. Feldman. 1996. Full scale thermal testing of latent heat storage in wallboard. Solar Energy Materials and Solar Cells 44: 49–61.

Schossig, P., H.M. Henning, S. Gschwander and T. Haussmann. 2005. Micro-encapsulated phase-change materials integrated into construction materials. Sol. Energy Mater Sol. Cells 89(2-3): 297–306.

Sing, K. 2001. The use of nitrogen adsorption for the characterisation of porous materials. Colloids and Surfaces A: Physicochemical and Engineering Aspects 187-188: 3–9.

Skandalos, N. and D. Karamanis. 2016. Investigation of thermal performance of semi-transparent PV technologies. Energy and Buildings 124: 19–34.

Skovajsa, J., M. Koláček and M. Zálešák. 2017. Phase change material based accumulation panels in combination with renewable energy sources and thermoelectric cooling. Energies 10(2).

Sokolov, M. and Y. Keizman. 1991. Performance indicators for solar pipes with phase change storage. Solar Energy 47(5): 339–346.

Stekli, J. 2013. Technical challenges and opportunities for concentrating solar power with thermal energy storage. Journal of Thermal Science and Engineering Applications 5(2): 021011-021011-021012.

Strantzali, E. and K. Aravossis. 2016. Decision making in renewable energy investments: A review. Renewable and Sustainable Energy Reviews 55: 885–898.

Stritih, U. and P. Novak. 1996. Solar heat storage wall for building ventilation. Renewable Energy 8(1–4): 268–271.

Sun, L.M., Y. Feng and M. Pons. 1997. Numerical investigation of adsorptive heat pump systems with thermal wave heat regeneration under uniform-pressure conditions. International Journal of Heat and Mass Transfer 40(2): 281–293.

Takakura, T.N.H. 1981. A solar greenhouse with phase change energy storage and a microcomputer control system. Acta Hort (Energy in Protected Cultivation) 115: 583–590.

Telkes, M. 1980. Thermal energy storage in salt hydrates. Solar Energy Materials 2(4): 381–393.

Tian, Y. and C.Y. Zhao. 2013. A review of solar collectors and thermal energy storage in solar thermal applications. Applied Energy 104: 538–553.

Tiskatine, R., A. Eddemani, L. Gourdo, B. Abnay, A. Ihlal, A. Aharoune and L. Bouirden. 2016. Experimental evaluation of thermo-mechanical performances of candidate rocks for use in high temperature thermal storage. Applied Energy 171: 243–255.

Turnpenny, J.E.D. and D. Reay. 2000. Novel ventilation cooling system for reducing air conditioning in buildings. Part I: testing and theoretical modeling. Applied Thermal Engineering 20: 1019–1037.

Uhlmann, D.R., B. Chalmers and K.A. Jackson. 1964. Interaction between particle and solid-liquid interface. J. Appl. Phys. 35: 2986–2993.

Ung, C., S.D. Moss and W.K. Chiu. 2015. Electromagnetic energy harvester using coupled oscillating system with 2-degree of freedom. Proceedings of SPIE—The International Society for Optical Engineering.

Wang, C., M. Hossain, L. Ma, Z. Ma, J.J. Hickman and M. Su. 2010. Highly sensitive thermal detection of thrombin using aptamer-functionalized phase change nanoparticles. Biosensors and Bioelectronics 26: 437–443.

Wang, C., L. Feng, H. Yang, G. Xin, W. Li, J. Zheng, W. Tian and X. Li. 2012. Graphene oxide stabilized polyethylene glycol for heat storage. Physical Chemistry Chemical Physics 14(38): 13233–13238.

Wang, W., B. Tang, B. Ju, Z. Gao, J. Xiu and S. Zhang. 2017. Fe_3O_4-functionalized graphene nanosheet embedded phase change material composites: efficient magnetic- and sunlight-driven energy conversion and storage. Journal of Materials Chemistry A 5(3): 958–968.

Wei, C. and X. Jing. 2017. Vibrational energy harvesting by exploring structural benefits and nonlinear characteristics. Communications in Nonlinear Science and Numerical Simulation 48: 288–306.

Wu, G. and W. Gao. 2015. GO/rGO as Advanced Materials for Energy Storage and Conversion. Graphene Oxide: Reduction Recipes, Spectroscopy, and Applications. W. Gao. Cham, Springer International Publishing: 97–127.

Wu, Y., Y. Tang, Z. Li, X. Ding, W. Yuan, X . Zhao and B. Yu. 2016. Experimental investigation of a PCM-HP heat sink on its thermal performance and anti-thermal-shock capacity for high-power LEDs. Applied Thermal Engineering 108: 192–203.

X., Z. 2001. Heat-storage and thermo-regulated textiles and clothing. Smart fibres, fabrics and clothing. Cambridge, Woodhead Publishing Ltd and CRC Press LLC.

Xu, H., A. Romagnoli, J.Y. Sze and Xavier Py. 2017. Application of material assessment methodology in latent heat thermal energy storage for waste heat recovery. Applied Energy 187: 281–290.

Xu, Y., A.S. Fleischer and G. Feng. 2017. Reinforcement and shape stabilization of phase-change material via graphene oxide aerogel. Carbon 114: 334–346.

Yan, S.H., W. Song and J.Y. Guo. 2017. Advances in management and utilization of invasive water hyacinth (Eichhornia crassipes) in aquatic ecosystems—a review. Critical Reviews in Biotechnology 37(2): 218–228.

Yanbing, K., J. Yi and Z. Yinping. 2003. Modeling and experimental study on an innovative passive cooling system—NVP system. Energy and Buildings 35: 417–425.

Yu, Z., Y. Zhang, S. Hao, J. Zhang, X. Li, B. Cai and T. Xu. 2016. Numerical study based on one-year monitoring data of groundwater-source heat pumps primarily for heating: a case in Tangshan, China. Environmental Earth Sciences 75(14).

Zhang, D., J. Zhou, K. Wu and Z. Li. 2005. Granular phase changing composites for thermal energy storage. Sol. Energy 78: 471–480.

Zhang, L., H. Shi, W. Li, X. Han and X. Zhang. 2013. Structure and thermal performance of poly(ethylene glycol) alkyl ether (Brij)/porous silica (MCM-41) composites as shape-stabilized phase change materials. Thermochimica Acta 570: 1–7.

Zhang, Z., Y. Yuana, N. Zhang, Q. Suna, X. Cao and L. Suna. 2017. Thermal properties enforcement of carbonate ternary via Lithium fluoride: a heat transfer fluid for concentrating solar power systems. Renewable Energy.

CHAPTER 14

Impact of Energy Storage Systems Value-Added Options

*David Bullejos Martín** and
Jorge M. Llamas Aragonés

1. Introduction

International laws, such as the European Directive 2001/77/EC (Promotion of electricity generated from renewable energy in inner electricity market) (Directive 2001/77/CE Council), and the U.S. Public Law 109–58 (US Congress Energy Policy act of 2005) focus their efforts on promoting the development and use of renewable energy, and more specifically on the implementation of solar thermal energy for electric energy generation. Such is the Spanish case, where the implementation of solar thermal technologies is prioritized for optimizing the energy mix choice from renewable sources (Bullejos et al. 2015).

Thermal storage systems used in thermal power plants give the possibility of developing electrical power generation by improving the intermittence and increasing the profitability of the plant (Trieb 2000). This is an important advantage which offers the opportunity to extend the electricity production to periods without solar radiation by adapting the operation procedures. Although the number of hours of direct electric power generation increases notably, the high set up expenses, the maintenance costs, and the long investment payback period raise interest in studying different technologies for thermal storage systems.

University of Cordoba, Campus de Rabanales, 14071 Córdoba, Spain.
* Corresponding author: bullejos@uco.es

The development of thermal storage systems lets the projection of thermal plants increase in order to generate electricity in partial cover moments even after the sunset, so that they are able to meet the requirements of electricity consumption.

For the coordination of the generators, as well as for the integration of generation from renewable resources, it is necessary to ensure the administration of the commitment of evacuation in addition to the combination between sources of energy and control of energy demand.

The storage system allows to optimize the benefit of the plant by adapting their production to load surplus and market prices, saving power restrictions and avoiding overproduction penalties. The investment evolution of Thermal Storage Systems in power thermal plants is shown in Fig. 1.

Figure 2 shows the possibility to adapt the production of electricity by thermal storage in solar thermal plants.

The selection of parameters and specific variables, such as storage capacity or generation systems, can obtain high efficiency proposals for optimal operation according to different technologies of generation.

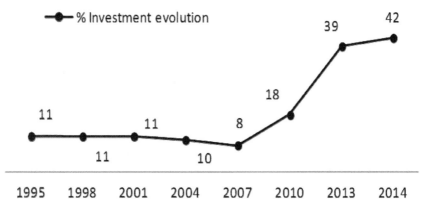

Fig. 1: Double tank molten salt TES investment evolution in power plants (own elaboration based on Compa Oró 2012).

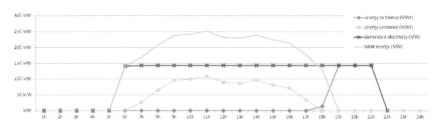

Fig. 2: Solar thermal plant adaptation by thermal storage (Trieb 2000, Teske 2005).

2. Thermoelectric Conversion Processes

For the generation of electricity by using thermal storage, there are several systems based on their technological complexity, durability and acceptable costs. Regarding the above parameters, such main systems are:

- Direct system of double tank (thermal exchange between two tanks HT/LT with the same transmitting fluid).
- Indirect double tank system (heat exchange between different fluids, transfer fluid and storage fluid, by forced coil systems).
- Thermocline system with single tank (heat exchange HT/LT by high-speed thermocline surface, which reduces the cost of installation and increases its performance up to 65%—still not available for commercial use).

The following figures show schematically the composition of different thermal storage for different sources of generation (Teske 2005).

For electricity generation in thermal plants, the power block is responsible for the conversion of thermal energy in electrical energy through the use of steam. Currently, it is responsible for 30% of the unplanned shutdowns, so high attention is required in its design and operation. The power block consists of a steam generation system, a steam

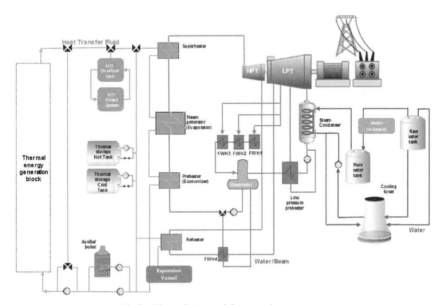

Fig. 3: Direct system with double tank (own elaboration).

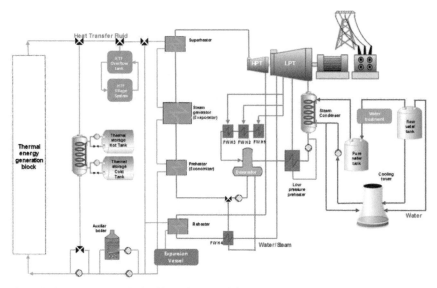

Fig. 4: Indirect system with double tank (own elaboration).

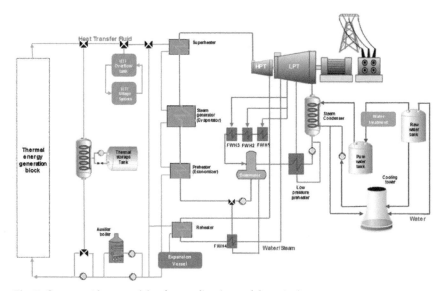

Fig. 5: System with one tank by thermocline (own elaboration).

a steam turbine, some power exchange systems, and some auxiliary systems which allow its operation. Each of these components is described below.

2.1 Steam Generation System

The standard Rankine cycle diagram of the Fig. 6 shows that the expansion of steam in the turbine reaching conditions of saturation. The condensation of water and its subsequent collision with the blades of the turbine would lead to a subsequent deterioration of the turbine. To avoid this, the steam is extracted, reheated and directed to the lower pressure block of the turbine.

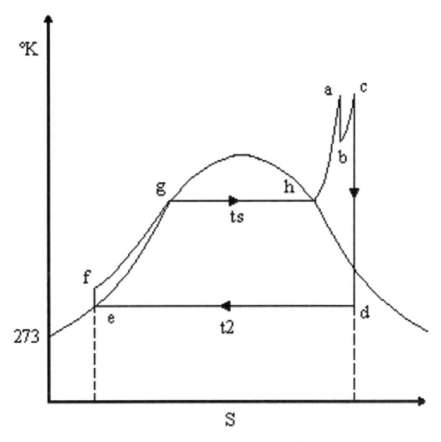

Fig. 6: Temperature-entropy diagram in Rankine cycle with reheating.

With the intermediate overheating the efficiency increase of the cycle of Rankine is achieved by reduction of the flow of steam. However, some disadvantages are that on the one hand the greater length of the turbine and, on the other hand, its higher cost with additional cost for the intermediate overheater.

The Rankine cycle usually applied to thermal power plants is based on the previous described cycle. For these kinds of power plants the cycle includes overheat, reheating and regeneration, working at two levels of pressure of steam, 103 bar and 18 bar approximately. The disposition of different devices engaged in the Rankine cycle (preheater, steam generator superheater, reheater, two stages steam turbine and condenser), is shown in the Figs. 3, 4 and 5.

The greater part of the steam flows throughout the turbine and reaches the condenser, while the remaining fraction of steam is extracted from a buffer and used to preheat the water supply in the heat exchanger before sending it to the element for heat generation.

2.2 Steam Turbine

The steam turbine converts the potential energy from the steam pressure on kinetic energy of rotation. The output steam of the turbine has lower pressure and temperature. Part of the energy lost by the steam is used to move the rotor of the turbine.

Direct Action turbines are currently used for applications of power generation from steam. In this type of turbines, the working fluid has an important change of pressure and speed through the turbine with thermo-mechanic transformation in the fixed edges. In reaction turbines, the thermo-mechanic transformation occurs in fixed and mobiles edges at once.

The turbines for thermal power plants (turbines of direct action or reaction, with overheating) are divided into two bodies, high and low pressure, with reduction devices.

The high-pressure turbine is a direct action multi-stage turbine that works between 103 bar for steam inlet and 18.5 bar for steam output in full load operation. Its nominal speed usually exceeds 3000 rpm, so it is connected with synchronous generators through a reducer.

The low-pressure turbine works after a process of overheating (380°C and 18.5 bar) that prevents condensation that would erode the turbine. This reaction turbine works at low pressure (18.5 bar) at a speed of 3000 rpm, so it can be directly connected to the generator.

2.3 Steam Generator

The thermal power plant steam circuit has the following elements: the economizer (preheater), where water temperature working at more than 100 bars of pressure rises from 240°C to 310°C; and the evaporator, where it produces the change of state of the water from of the sparing to steam of water to 314°C and 104 bars of pressure. The steam goes to the superheater, raising its temperature until 380°C. Finally, the reheater collects gases from the turbine discharge of high pressure at about 200°C, by raising its temperature up to 380°C, and making its dump to the low-pressure turbine possible.

The configuration is a steam generation train with two bodies in parallel producing each one the half of the total steam generated. This configuration allows to work in partial load 0–50–100%, and its main advantage is that in case of breakdown of one of the trains, the other one is available to continue producing electricity (although its production capacity can be reduced).

The construction of two steam trains entails a higher cost than the one-steam train option. Both the preventive and corrective maintenance is duplicated, which does not entail that the probability of failure or fatigue decreases.

2.4 Heat Exchanger

Currently, the multitube semi-hermetic exchanger is the most accepted option for thermal power plants instead of others, such as plate heat exchanger. This type of exchangers is difficultly obstructed, which increases the efficiency of the process by decreasing the loss of load. Therefore, the multitube heat exchanger is set as the most appropriate exchanger for this type of installation.

2.5 Power Net Coupling

The heat energy from the block of generation moves the electric generator, usually three-phase alternator through its rotor, coupled with a continuous current source in independent excitation mode. This source of continuous current of the inductor is located inside of the rotor and supplied with alternating current by a generator energized from a rectifier by permanent magnets in its own rotor (AC pilot excitation). The nominal speed of rotation for electric generators in thermal power plants is 3000 rpm for 50Hz electricity generation. The voltage of generation often reaches 11 KV, being thus elevated by the main transformer.

The generator can also be used at the starting boot of the plant to move the turbine and compressor until it reaches the speed needed to begin introducing steam to the turbine to operate the generator and compressor.

3. Dynamics of Electricity Generation

The control system which is currently implemented in thermal plants is based on a system of control for blocks in which each element of the plant incorporates a system of precise control to that process, but where the interrelationship with the rest of the blocks is low or virtually non-existent.

The control system of the power block regulates the fluids through the different devices to the exchange of heat between the HTF and the water-steam cycle, controlling variables such as temperature, pressure and flow levels in each device. This control system regulates the main pumps of the HTF and the outlet valves of steam, by-pass systems, the feed pumps of the water-steam cycle, pumps of condensates, as well as the levels of all tanks.

The use of different tools and models allows to evaluate the potential of optimal control techniques applied to the operation of thermal storage plants. These models allow to obtain a strategy of optimal operation through the analysis of all the predictable combinations of energy resource, storage and system capacity. This optimal operation strategy will improve the dynamic availability of the plant depending on the price of energy in the market. It increases the annual capacity factor and improves the plant performance by maximizing the work cycles during full load.

For a generation plant with thermal storage, there are four configurations, depending on the quantity and availability of resource, needs of load and production, or specific location of the plant. The Intermediate Load Configuration (ILC) is designed to produce electricity when the available thermal resource is sufficient to provide thermal energy in periods of specific needs. It requires a small amount of thermal storage and lower investment cost.

The Delayed Intermediate Production System (DIPS) produces thermal power during the entire day, producing electric energy in periods of higher demand of electricity. It requires a large amount of heat storage. The Based Load Configuration (BLC) works for twenty-four hours a day during the greater part of the year. As it needs a greater amount of thermal storage it is an appropriate production system when the limits of energy generation are predefined, which implies a lower use of the capacity of production of the thermal plant. The Peak Load Plant (PLP) is designed to generate electrical energy just for few hours, taking into account the daily periods of high demand of electricity. This type of plant will require

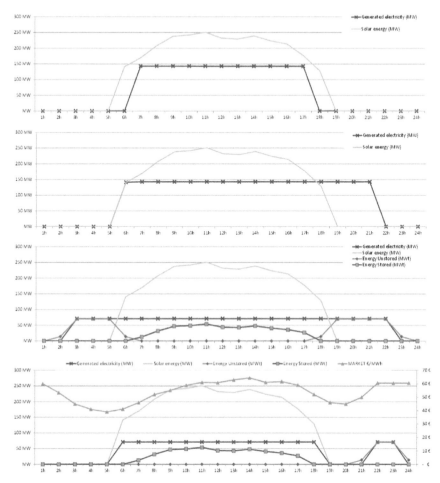

Fig. 7: Production of electricity according to the dynamics of the solar power plant. ILC (a). DIPS (b). BLC (c). PLP (d) (own elaboration).

a turbine of greater size and a high thermal storage, the production system is more expensive but produces electric energy during higher purchase prices (International Energy Agency Technology Roadmap 2010). The Fig. 7 shows the power electricity production for each configuration.

For these four configurations of the plant there are the three modes of operation which manage the production of energy and heat storage optimizing the operation of the plant. In the mode *generation and storage*, solar radiation is sufficient to operate the steam turbine and the power block at full load. Due to the availability of resource, the overproduced thermal energy may be stored. In the mode of *generation and recovery*, when the resource is not enough, the thermal storage system adds enough energy to the system to allow the operation at full load of the steam turbine. In the *recovery* mode (periods of very low thermal resource), the thermal energy stored goes to the block of power to produce electricity, also ensuring the latent heat of maintenance of the recirculation system and circuits of the thermal plant.

The limits of operation of the plant are conditioned by the minimum and maximum values of power and energy in each block of the plant. For the thermal storage system, there is a threshold of energy available. The maximum heat energy available in this storage system must be equivalent to the product of the maximum thermal power that can reach the power block, with the number of hours that, according to the design of the plant, the system is able to work in full load without receiving thermal energy from the block of generation. Finally, for the power block, the thermal energy as input of the turbine consists of the heat flow from the block of generation and the heat flow from the storage system which must supply a limited amount of thermal energy to the power block according to the design of the turbine.

4. Techno-Economic Assessment of Thermal Energy Storage for Generation of Electricity

Thermal storage systems invested in thermal power plants offer the possibility of developing the electrical power generation, improving the intermittence and increasing the profitability of the plant (Brandon 2003). This is an important advantage which gives the opportunity to extend the electricity production to periods of low availability or even high prices of resources by adapting the operation procedures. The demand for electric energy should match efficient generation, as peaks of consumption demand higher production of electricity. The utilization of thermal storage systems avoid the need to install total power several times greater than the electrical power required.

The selection of parameters and specific variables such as storage capacity or generation systems let us obtain high efficiency proposals for optimal operation according to the load curves and availability of resource.

To analyze the summary cost for a thermal power plant with thermal storage it is usual to consider a power block with 37% of average conversion efficiency, 391°C inlet temperature, 293°C outlet temperature, 100 bar boiler operating pressure, and 20% of thermal power fraction for standby or startup.

The thermal storage by double-tank direct system is performed with 20000 m³ storage volume, 36 m diameter, 20 m tank height, and 391°C fluid temperature for seven hours of equivalent full load thermal energy for a 50 MW power plant (Brandon 2003, Blake 2003).

Table 1: Numerical summary for three storage operations in thermal power plant with double-tank direct system.

Cost Concept	0h TES	2h TES	7h TES
Site	6,004,789€	6,004,789€	6,004,789€
Solar Field	105,084.000€	105,084,000€	105,084,000€
Power Plant	35,420,000€	35,420,000€	35,420,000€
HTF System	15,011,990€	15,011,990€	15,011,990€
Thermal Storage	0€	14,281,960€	49,986,720€
Fossil Backup	25,700€	19,250€	16,250€
Contingency	16,152,220€	17,580,430€	21,150,920€
Indirect Cost	43,885,660€	47,766,040€	57,467,060€
O & M	32,970,840€	33,465,530€	33,573,540€
Ins. and Prop.	12,167,190€	13,243,020€	15,932,630€
Whole	**266,722,389€**	**287,887,009€**	**339,647,899€**

With these conditions the evaluation of the implementation costs has the numerical results shown in Table 1.

The economic analysis is focused on the most profitable operation situation. The average lifetime leveled generation cost of electricity (LCOE) is determined for different values of storage capacity. The following eq. (1) is used to get the value of LCOE for the different thermal power plant configurations considered:

$$LCOE = \frac{\sum_{t=1}^{n}(I_t + O\&M_t + F_t)}{\sum_{t=1}^{n} E_t} \tag{1}$$

The capital cost in the year (t) is calculated in eq. (2):

$$I_t = crf \cdot I_c \tag{2}$$

The capital recovery factor is calculated according to eq. (3):

$$crf = \frac{i \cdot (1 + i)^n}{(1 + i)^n - 1} - k \tag{3}$$

where i is the debt interest rate; n is the depreciation period; k is the annual insurance rate; $\sum_{t=1}^{n}()$ is the values buzzer along the depreciation period; I_c is plant investment cost; $O\&M_t$ is the combined fix and variable operation and maintenance cost in the year t which can be calculated as $O\&M = O\&M_{fix.t} + O\&M_{var.t}$; $O\&M_{fix.t}$ is the operation and maintenance cost referenced to the plant capacity; $O\&M_{var.t}$ is the operation and maintenance

Table 2: Main data for thermal power plant LCOE calculation.

Concept	Value
Site cost ($€/m^2$)	226.33
HTF system ($€/kWh_{th}$)	210.95
Power plant Investment ($€/kW_e$)	643.20
Investment Indirect cost and contingencies surcharge (%)	16.00
Fixed O&M cost ($€/kW_e/year$)	45
Variable O&M cost. ($€/MWh_e$)	3.50
Debt interest rate (%)	8.00
Annual insurance rate (%/year)	0.50
Capital recovery factor (%)	8.38
Plant lifetime (n)	25

Table 3: Thermal energy storage.

Thermal storage size (Eq. hours)	Net diary Thermal storage size (MWh_{th})
1.14	164.38
1.49	213.69
1.72	246.57
1.94	279.45
2.17	312.32
2.40	345.20

cost referenced to the electric energy production; F_t is the fuel consumption cost in the year t; and E_t is the net electric energy production in the year t.

Main data assumptions used for economic analysis are shown in Table 2 (U.S. Department of Energy 2017). Cost due to investment, fixed and variable operation and maintenance and fuel consumption are different according to the thermal storage dimension as this value directly affects the amount of electricity generated. Table 3 represents the amount of diary thermal energy stored as a function of the double tank size. The data in Table 2 have been estimated considering a depreciation period of 25 years and a debt interest rate of 8.0%.

5. Electric Generation

After programming these models and analyzing their structuring and economic parameters we can observe (Table 4) the energy production and

Table 4: Numerical results for three options of thermal management: Double Direct Tank, direct production of electricity without storage and Increased Overflow Tank.

Thermal power plant with 7h TES	
Energy (kWh)	3,668,325,059.10
Energy Value	1,014,611,132.93€
After Tax Cash-flow	610,828,986.80€

Thermal power plant without thermal Storage	
Energy (kWh)	2,884,336,332.30
Energy Value	797,770,019.50€
After Tax Cash-flow	462,665,940.56€

Thermal power plant with greater overflow tanks	
Energy (kWh)	3,527,827,138.20
Energy Value	975,751,230.21€
After Tax Cash-flow	573,611,437.39€

cash-flow due to the use of different storage systems. We consider 30 years of operation as useful life for the plant (Teske 2005).

To conclude, the models described in this work are not mutually exclusive; rather they can be applied in different economic situations. For plants with free market operation the use of double-tank storage systems is necessary. This allows adaptation of production to higher price periods. In double direct storage systems, the energy production is not enough to justify the solutions in which higher storage than 7 equivalent hours is needed.

The capacity factor of a power plant is the ratio of its output over a period of time to its potential output if continuous operations over the same period of time were possible. The capacity factor should not be confused with the availability factor of the power plant, as it is the amount of time that it is able to produce electricity over a certain period, divided by the amount of the time in the period.

Table 5 shows the comparative results of production for different options for the construction of a thermal power plant, direct production without storage, storage direct thermal and thermal storage tanks by fluid (U.S. Department of Energy 2017).

This analysis shows that the technology of thermal storage can raise the capacity factor of a plant and thus increase the production of electricity. As peaks of consumption demand higher production of electricity there

Table 5: Numerical results for three options of thermal management: double direct tank, direct production of electricity without storage and increased overflow tank.

Thermal power plant with 7h TES	
Energy (GWh)	3,668.32
Energy Value (Mio.€)	1,014.61
After Tax Cash-flow (Mio.€)	610.82

Thermal power plant without thermal Storage	
Energy (GWh)	2,884.33
Energy Value (Mio.€)	797.77
After Tax Cash-flow (Mio.€)	462.66

Thermal power plant with greater overflow tanks	
Energy (kWh)	3,527.82
Energy Value (Mio.€)	975.75
After Tax Cash-flow (Mio.€)	573.61

should be more efficient generation of electric energy to match its demand and no need to generate excess power than required.

References

Blake, D. 2003. Overview on use of a Molten Salt HTF in a Trough Solar Field. NREL: Parabolic Trough Thermal Energy Storage Workshop, Golden, CO, Feb. 03.

Brandon, O. 2003. The value of thermal storage. NREL: Parabolic Trough Thermal Energy Storage Workshop, Golden, CO, Feb. 03.

Bullejos, D., J. Llamas and M. Ruiz de Adana. 2015. Spanish regulated scenarios for renewable energy and csp plants. ARPN J. of Engineering and Applied Sciences 10: 7217–7223.

Compa Oró, E., A. Gila, A. Gracia, D. Boer and L.F. Cabeza et al. 2012. Comparative life cycle assessment of thermal energy storage systems for solar power plants. Renewable Energy Commun. 44: 166–173.

Directive 2001/77/CE Council of 27th September 2001. On the promotion of electricity produced from renewable energy sources in the internal electricity market. European Parliament Official J. L283/33, 2001.

International Energy Agency Technology Roadmap. Concentrating Solar Power. Paris, 2010. http://www.iea.org.

Teske, S. 2005. Concentrated solar thermal power. Greenpeace-European Solar Thermal Power Industry Association (ESTIA), Sept. 2005, pp. 7–24.

Trieb, F. 2000. Competitive solar thermal power stations until 2010 the challenge of market introduction. Renewable Energy Commun. 19: 163–71.

US Congress. 2005. Energy Policy Act of 2005. US Public Law 109: 54.

U.S. Department of Energy. Energy efficiency and renewable energy (www.energy.gov), (Last access February, 2017).

Index

ABOUT THE BOOK

The future energy systems must cope with the new changes and advanced developments in technology like improvements of natural gas combined cycles and clean coal technologies, carbon dioxide capture and storage, advancements in nuclear reactors and hydropower, renewable energy engineering, power-to-gas conversion and fuel cells, energy crops, new energy vectors biomass-hydrogen, thermal energy storage, new storage systems diffusion, modern substations, high voltage engineering equipments and compatibility, HVDC transmission with FACTS, active grids and smart grids, power system resilience, power quality and cost of supply, plug-in electric vehicles, smart metering, control and communication technologies, new key actors as prosumers, smart cities.

Printed and bound by CPI Group (UK) Ltd, Croydon, CR0 4YY

01/11/2024

01782623-0013